AF594999

Selected Papers from Coastlab18 Conference

Selected Papers from Coastlab18 Conference

Special Issue Editors

Javier López Lara
Maria Maza

MDPI • Basel • Beijing • Wuhan • Barcelona • Belgrade • Manchester • Tokyo • Cluj • Tianjin

Special Issue Editors
Javier López Lara
Universidad de Cantabria
Spain

Maria Maza
Universidad de Cantabria
Spain

Editorial Office
MDPI
St. Alban-Anlage 66
4052 Basel, Switzerland

This is a reprint of articles from the Special Issue published online in the open access journal *Journal of Marine Science and Engineering* (ISSN 2077-1312) (available at: https://www.mdpi.com/journal/jmse/special_issues/Coastlab18).

For citation purposes, cite each article independently as indicated on the article page online and as indicated below:

LastName, A.A.; LastName, B.B.; LastName, C.C. Article Title. *Journal Name* **Year**, *Article Number*, Page Range.

ISBN 978-3-03936-096-3 (Pbk)
ISBN 978-3-03936-097-0 (PDF)

Cover image courtesy of IHCantabria.

Contents

About the Special Issue Editors

Javier López Lara is an associate professor at the University of Cantabria (UC). He is also responsible for the Hydrodynamics and Coastal Infrastructures Group of IHCantabria. Javier graduated as a Civil Engineer (1998) from the University of Cantabria and obtained his Ph.D. in 2002 from the same university. After holding various positions as a researcher at the University of Cantabria and at Cornell University (USA), he obtained a position as Ramón y Cajal (2008–2012) in the Department of Water Science and Technology at UC, where he is currently a sub-director. Javier has developed his research in the field of wave hydrodynamics, specializing also in surf zone hydrodynamics and wave–structure interactions, using both experimental and numerical methods, the latter focusing mainly on CFD codes. In recent years, he has extended his research to environmental flows, hydrodynamics of aquatic ecosystems, and hydraulics. Through his research he was awarded the Modesto Vigueras Prize 2004, granted by the Technical Association of Ports and Coastal Areas, Spanish branch of the International Ports Association (PIANC) and the International Research Prize Paepe-Willems (PIANC) in 2005, granted to researchers under 40 years old. Javier is currently the Principal Investigator for several projects funded by the Spanish Science Ministry, competitive projects financed by the Spanish Innnovation Ministry and different EU Proyects. He has published 52 SCI research papers and 8 book chapters. He has advised 45 M.S. and B.S. theses and 12 PhD dissertations. He has coauthored more than 100 national and international congress communications.

Maria Maza is a Juan de la Cierva researcher in the Hydrodynamics and Coastal Infrastructures group at IHCantabria. She graduated as a Civil Engineer in 2010, received her M.S. in Coastal and Ports Engineering in 2012, and her Ph.D. in 2015 by the Universidad de Cantabria (Universidad de Cantabria) receiving the award Premio extraordinario de doctorado from the Universidad de Cantabria. Then, she completed her postdoc at the Massachusetts Institute of Technology (MIT) in Boston, USA, in the Civil and Environmental Engineering Department (CEE). Maria has developed new numerical tools to analyze flow-ecosystem interaction and she is part of IHFOAM and IH2VOF development team. She has led different experimental campaigns run at the Cantabria Coastal and Ocean Basin (CCOB) and at the MIT Parsons Laboratory. She teaches at the Universidad de Cantabria and advises undergraduates and graduated students. She has worked in more than 20 competitive projects, she has presented her work in more than 30 international conferences, including 8 as an invited speaker. She has published 16 SCI papers (all Q1) and 2 book chapters. She is a specialist in physical and numerical modelling of flow–ecosystem interactions.

Journal of
Marine Science and Engineering

Editorial

Selected Papers from Coastlab18 Conference

Javier L. Lara and Maria Maza *

IHCantabria—Instituto de Hidráulica Ambiental de la Universidad de Cantabria, C/Isabel Torres nº 15, 39011 Santander, Spain; jav.lopez@unican.es

* Correspondence: mariaemilia.maza@unican.es

Received: 30 March 2020; Accepted: 2 April 2020; Published: 3 April 2020

The 7th International Conference on The Application of Physical Modelling in Coastal and Port Engineering and Science, Coastlab18, was organized in Santander, Spain, from 22 to 26 May 2018, by the Instituto de Hidráulica Ambiental de la Universidad de Cantabria, IHCantabria. The conference was organized under the auspices of the International Association of Hydro-Environment Engineering and Research (IAHR).

Coastlab18 continued a successful tradition, after Porto (2006), Bari (2008), Barcelona (2010), Ghent (2012), Varna (2014) and Ottawa (2016), to provide a forum to discuss the latest developments in physical modelling in the field of coastal and port engineering. In 2018, the Coastlab conference was held by IHCantabria in Santander (Spain), which has a long tradition in experimental hydraulics associated with the field of coastal engineering.

The main objectives of the Coastlab18 conference were:

- To provide a stimulating and enriching forum to discuss the latest developments in physical modelling applied to coastal and port engineering and in new trends coastal science.
- To increase the link between industry and the academia in the field of experimental modelling.
- To engage industry professionals from Coastal and Port Engineering and Marine Energy sector to be part of the Coastlab community.
- To attract new people from Latin America for the Coastlab network, especially young researchers and institutions from emerging coastal and port laboratories.

Coastlab18 welcomed 175 attendees from 18 different countries. The conference was preceded by a short course covering the most relevant aspects in wave generation, new techniques in laboratory measurements and best practices in innovative laboratory tests. Attendees had also the opportunity to participate in a field trip to visit Santander Harbor and Somo beach.

The Technical program included three renowned keynote lectures and 120 presentations focused on theoretical and practical aspects related to physical modelling in the field of coastal and ocean engineering. Coastal and ocean structures, breakwaters, revetments, laboratory technologies, measurement systems, coastal field measurement and monitoring, combined physical and numerical modelling, physical modelling case studies, tsunamis and coastal hydrodynamics were the main topics covered in the conference. The presentations and discussions, together with the social events held during the conference, helped to create a fruitful and friendly atmosphere, promoting close working relationships between coastal researchers and engineers from different countries and with different expertise. Thus, Coastlab18 continued its tradition of acting as a recognized platform for the direct exchange of knowledge and experience between early-stage researchers and senior scientists in the field of coastal and port engineering.

This book gathers 16 selected papers [1–16] from the conference, attempting to cover, as completely as possible, all the topics presented during Coastalb18. The papers have been accepted after a peer-review process based on their full text.

Special thanks are due to the Editorial and Reviewer Board, which had most of the responsibility for reviewing the documents and providing constructive comments. The quality of the presented papers gives good reasons to believe that this meeting contributes to promote a greater collaboration between researchers and engineers in the field of coastal and port engineering.

The organizers of Coastlab18 express their sincere acknowledgments to the Sponsor, the International Association for Hydro-Environment Engineering and Research (IAHR). We would also like to acknowledge to all members of the Local Organizing Committee and the Scientific Committee, as well as to all technical and administrative staff members for supporting the organization of Coastlab18.

Author Contributions: Writing—original draft preparation: M.M.; Writing—review and editing: J.L.L. All authors have read and agreed to the published version of the manuscript.

Funding: This research received no external funding.

Acknowledgments: This work has been funded under the Retos Investigación 2018 (grant RTI2018-097014-B-I00) program of the Spanish Ministry of Science, Innovation and Universities. M. Maza is indebted to the Spanish Ministry of Science, Innovation and Universities for the funding provided in the grant Juan de la Cierva Incorporación (BOE de 27/10/2017). The authors also express their acknowledgments to the Coastlab18 Sponsor, the International Association for Hydro-Environment Engineering and Research (IAHR).

Conflicts of Interest: The authors declare no conflict of interest.

References

1. Corvaro, S.; Crivellini, A.; Marini, F.; Cimarelli, A.; Capitanelli, L.; Mancinelli, A. Experimental and Numerical Analysis of the Hydrodynamics around a Vertical Cylinder in Waves. *J. Mar. Sci. Eng.* **2019**, *7*, 453. [CrossRef]
2. Mulligan, R.P.; Take, W.A.; Bullard, G.K. Non-Hydrostatic Modeling of Waves Generated by Landslides with Different Mobility. *J. Mar. Sci. Eng.* **2019**, *7*, 266. [CrossRef]
3. Lashley, C.H.; Bertin, X.; Roelvink, D.; Arnaud, G. Contribution of Infragravity Waves to Run-up and Overwash in the Pertuis Breton Embayment (France). *J. Mar. Sci. Eng.* **2019**, *7*, 205. [CrossRef]
4. Salauddin, M.; Pearson, J.M. Experimental Study on Toe Scouring at Sloping Walls with Gravel Foreshores. *J. Mar. Sci. Eng.* **2019**, *7*, 198. [CrossRef]
5. Landmann, J.; Ongsiek, T.; Goseberg, N.; Heasman, K.; Buck, B.H.; Paffenholz, J.-A.; Hildebrandt, A. Physical Modelling of Blue Mussel Dropper Lines for the Development of Surrogates and Hydrodynamic Coefficients. *J. Mar. Sci. Eng.* **2019**, *7*, 65. [CrossRef]
6. Van Gent, M.R.; de Almeida, E.; Hofland, B. Statistical Analysis of the Stability of Rock Slopes. *J. Mar. Sci. Eng.* **2019**, *7*, 60. [CrossRef]
7. Lykke Andersen, T.; Eldrup, M.R.; Clavero, M. Separation of Long-Crested Nonlinear Bichromatic Waves into Incident and Reflected Components. *J. Mar. Sci. Eng.* **2019**, *7*, 39. [CrossRef]
8. Ruju, A.; Passarella, M.; Trogu, D.; Buosi, C.; Ibba, A.; De Muro, S. An Operational Wave System within the Monitoring Program of a Mediterranean Beach. *J. Mar. Sci. Eng.* **2019**, *7*, 32. [CrossRef]
9. Lira-Loarca, A.; Baquerizo, A.; Longo, S. Interaction of Swell and Sea Waves with Partially Reflective Structures for Possible Engineering Applications. *J. Mar. Sci. Eng.* **2019**, *7*, 31. [CrossRef]
10. Michalzik, J.; Liebisch, S.; Schlurmann, T. Development of an Outdoor Wave Basin to Conduct Long-Term Model Tests with Real Vegetation for Green Coastal Infrastructures. *J. Mar. Sci. Eng.* **2019**, *7*, 18. [CrossRef]
11. Eldrup, M.R.; Lykke Andersen, T. Applicability of Nonlinear Wavemaker Theory. *J. Mar. Sci. Eng.* **2019**, *7*, 14. [CrossRef]
12. De Almeida, E.; van Gent, M.R.A.; Hofland, B. Damage Characterization of Rock Slopes. *J. Mar. Sci. Eng.* **2019**, *7*, 10. [CrossRef]
13. Kerpen, N.B.; Schoonees, T.; Schlurmann, T. Wave Impact Pressures on Stepped Revetments. *J. Mar. Sci. Eng.* **2018**, *6*, 156. [CrossRef]
14. Gómez-Martín, M.E.; Herrera, M.P.; Gonzalez-Escriva, J.A.; Medina, J.R. Cubipod® Armor Design in Depth-Limited Regular Wave-Breaking Conditions. *J. Mar. Sci. Eng.* **2018**, *6*, 150. [CrossRef]

15. Van der Werf, I.M.; Van Gent, M.R. Wave Overtopping over Coastal Structures with Oblique Wind and Swell Waves. *J. Mar. Sci. Eng.* **2018**, *6*, 149. [CrossRef]
16. Argente, G.; Gómez-Martín, M.E.; Medina, J.R. Hydraulic Stability of the Armor Layer of Overtopped Breakwaters. *J. Mar. Sci. Eng.* **2018**, *6*, 143. [CrossRef]

Article

Separation of Long-Crested Nonlinear Bichromatic Waves into Incident and Reflected Components

Thomas Lykke Andersen [1,*], Mads R. Eldrup [1] and Maria Clavero [2]

[1] Department of Civil Engineering, Aalborg University, 9220 Aalborg Ø, Denmark; mre@civil.aau.dk
[2] Instituto Interuniversitario de Investigación del Sistema Tierra (IISTA), Universidad de Granada, Avda. del Mediterráneo s/n, 18006 Granada, Spain; mclavero@ugr.es
* Correspondence: tla@civil.aau.dk; Tel.: +45-9940-8486

Received: 11 December 2018; Accepted: 31 January 2019; Published: 7 February 2019

Abstract: Methods for the separation of long-crested linear waves into incident and reflected waves have existed for more than 40 years. The present paper presents a new method for the separation of nonlinear bichromatic long-crested waves into incident and reflected components, as well as into free and bound components. The new method is an extension of a recently proposed method for the separation of nonlinear regular waves. The new methods include both bound and free higher harmonics, which is important for nonlinear waves. The applied separation method covers interactions to the third order, but can easily be extended to a higher orders. Synthetic tests, as well as physical model tests, showed that the method accurately predict the bound amplitudes and incident and reflected surface elevations of nonlinear bichromatic waves. The new method is important in order to be able to describe the detailed characteristics of nonlinear bichromatic waves and their reflection.

Keywords: bichromatic waves; reflection separation; bound waves; nonlinear waves

---

1. Introduction

In physical model tests, it is very important to know in detail the wave conditions that are present in the facility. Incident waves hit passive or active absorbers or structures and are partly absorbed and partly reflected. Thereafter, the wave system includes both incident and reflected components. Nonlinear waves might at a given frequency include both bound and free components, making the separation into incident and reflected components more complicated. In regular or bichromatic waves, free incident higher harmonics are usually unwanted, as they mainly exist if the bound components were not correctly generated by the wave maker. Moreover, in regular and bichromatic waves, the bound harmonics at a given frequency usually stem from only one interaction. In irregular waves, this is not the case, and moreover, for wider spectra the primary energy overlaps with the sub- and superharmonics. If the wave height decreases, for example due to wave-breaking or waves propagating into deeper waters, the generation of free harmonics might also occur.

The present paper presents a detailed analysis of two-dimensional model tests with unidirectional bichromatic waves. Bichromatic waves are commonly used for detailed studies on the effect of bound sub- and superharmonics on, for example, the response of beaches and structures. Therefore, being able to accurately separate these waves into incident and reflected as well as bound and free components is of high interest.

Historically, waves have been separated into incident and reflected components based on methods assuming linear waves, cf. Goda and Suzuki [1], Mansard, and Funke [2], and Zelt and Skjelbreia [3]. Today, these methods are standard tools used in most hydraulic laboratories, but these methods do not consider any bound components to be present and also disregard amplitude dispersion. The few existing methods for nonlinear waves use questionable assumptions, e.g., the Lin and Huang [4] method, which applies to regular waves but disregards amplitude dispersion and thus is only applicable to mildly nonlinear regular waves. The nonlinear LASA methods (Local Approximation using Simulated Annealing) by Figueres et al. [5] and Figueres and Medina [6] are based on local time domain solutions fitted to theoretical wave profiles, making application of their method to cases where free energy exists questionable. Even when no free energy exists and even for linear waves, Lykke Andersen et al. [7] and Eldrup and Lykke Andersen [8] found that the LASA V method was not accurate, and as the method at the same time is computationally very expensive, it is not practical for real applications.

Recently, a nonlinear frequency domain method to separate regular waves into incident and reflected components was developed, cf. Lykke Andersen et al. [7]. This method is an extension of the Lin and Huang [4] method, but includes amplitude dispersion based on stream function wave theory (Fenton and Rienecker [9]). This nonlinear separation method was further extended by Eldrup and Lykke Andersen [8] to irregular waves by utilizing a narrow-band assumption for the primary spectrum. They showed that even for typical single-peaked primary spectra (JONSWAP with $\gamma = 3.3$), this assumption was leading to accurate estimated incident and reflected waves even for highly nonlinear waves. The reason for this was that the main contribution to the bound superharmonics stems from the interaction of components near the peak frequency of the primary spectrum.

The narrow-band assumption applied by Eldrup and Lykke Andersen [8] for the separation of irregular waves is not necessarily valid for nonlinear bichromatic waves if the frequency interval between the two primary components is not small enough. However, as the bound waves in bichromatic waves normally stem mainly from only one interaction, a more accurate method could be developed. Such a method is presented in the present paper, and the new method is evaluated on new model tests performed in a wave flume at the Atmosphere-Ocean Interaction Flume in Granada, Spain.

2. Third-Order Wave Theory for Bichromatic Waves

Bichromatic waves refer to waves with two primary incident components at cyclic frequencies ω_n and ω_m. The surface elevation of such incident waves can, with first-order theory, be expressed as

$$\begin{aligned}\eta^{(1)}(x,t) &= a_{n,I}^{(1)}cos(k_n x - \omega_n t + \varphi_{n,I}^{(1)}) + a_{m,I}^{(1)}cos(k_m x - \omega_m t + \varphi_{m,I}^{(1)}) \\ &= A_{n,I}^{(1)}cos(k_n x - \omega_n t) + B_{n,I}^{(1)}sin(k_n x - \omega_n t) + \\ &\quad A_{m,I}^{(1)}cos(k_m x - \omega_m t) + B_{m,I}^{(1)}sin(k_m x - \omega_m t).\end{aligned} \tag{1}$$

Second-order theory for a bichromatic group gives additionally three bound superharmonic components (sum-interactions) at the cyclic frequencies $2\omega_n$, $2\omega_m$, and $\omega_n + \omega_m$, and one subharmonic component (difference-interactions) at cyclic frequency $\omega_m - \omega_n$ (assuming $\omega_m > \omega_n$). Complete expressions, including second-order transfer functions (G), are given in, for example, Sharma and Dean [10], Schäffer [11], and Madsen and Fuhrman [12,13]. The second-order contribution due to the interaction of component n and m can be written as:

$$\begin{aligned}\eta_{n\pm m}^{(2)}(x,t) &= G_{nm}^{\pm}(A_{n,I}^{(1)}A_{m,I}^{(1)} \mp B_{n,I}^{(1)}B_{m,I}^{(1)})cos((k_n \pm k_m)x - (\omega_n \pm \omega_m)t) + \\ &\quad G_{nm}^{\pm}(A_{m,I}^{(1)}B_{n,I}^{(1)} \pm A_{n,I}^{(1)}B_{m,I}^{(1)})sin((k_n \pm k_m)x - (\omega_n \pm \omega_m)t) \\ &= a_{n\pm m}cos((k_n \pm k_m)x - (\omega_n \pm \omega_m)t + \varphi_{n\pm m,I}).\end{aligned} \tag{2}$$

This gives one subharmonic contribution (η_{n-m}) and three superharmonic contributions (η_{2n}, η_{2m}, η_{n+m}). The period of the subharmonic component ($T = 2\pi/(\omega_m - \omega_n)$) also corresponds to the duration of the formed wave group. A correction to the linear dispersion equation appears first at third order, and thus in second order theory the linear dispersion equation is valid for the primary components. The wave numbers of the bound superharmonics are calculated as the sum of the wave numbers of the interacting components and the difference for the subharmonics. Thus, the celerity of the bound components due to interaction of component n and m can be calculated as $c_{n-m} = (\omega_m - \omega_n)/(k_m - k_n)$ for the subharmonic and $c_{n+m} = (\omega_m + \omega_n)/(k_m + k_n)$ for the superharmonic component.

Recently, third-order solutions for bichromatic waves as well as for irregular waves were developed, cf. Madsen and Fuhrman [12,13]. Applying that theory to unidirectional bichromatic waves yields additionally four bound superharmonic components at the cyclic frequencies $3\omega_n$, $3\omega_m$, $\omega_m + 2\omega_n$, $\omega_n + 2\omega_m$ and two additional bound subharmonics at the cyclic frequencies $|\omega_n - 2\omega_m|$ and $|\omega_m - 2\omega_n|$. The computational cost of that method for irregular waves is very high, but for bichromatic waves the computational cost is very small. The wave numbers of the bound waves at the above given angular frequencies are found simply by summation and differences of the wave numbers for the primary frequencies (as in the second-order components given in Equation (2)). Thus, for example, the wave number of the bound component at frequency $\omega_m + 2\omega_n$ is given by $k_m + 2k_n$. The celerity of the bound components can afterwards be calculated using the same principle as for the second-order components. However, the celerity of the primary components is different from the first-order values due to amplitude dispersion. The dispersion equation was derived to third order by Madsen and Fuhrman [12,13]. Here, it was shown that the celerity of each of the two primary components depend on the amplitude and frequency of both components.

3. Nonlinear Reflection Separation Method for Bichromatic Waves

Lykke Andersen et al. [7] showed that for regular waves the celerity of the bound superharmonics is higher than a free component at the same frequency, and this makes it possible to separate the bound and free waves. The same applies to bichromatic waves, and thus for bichromatic waves it is also possible to separate the bound and free superharmonic components due to their different celerities. The bound subharmonic component has a lower celerity than a free one at the same frequency. Thus, it is also possible to separate these components based on their differences in celerity. However, for shallow water waves, the celerities of the bound and the free waves are almost identical, and thus separation of the two is not possible.

The mathematical model used for the separation of nonlinear bichromatic waves is chosen as a third-order surface for the incident as well as for the reflected waves. Moreover, free incident and reflected components are included in all of the sub- and superharmonic frequencies. Finally, a noise term (Ω) is included. In the following, it is assumed that at the frequencies of the two primary components, only primary incident and reflected waves exist. In special cases, it might occur for bichromatic waves that the frequency of one or more of the sub- or superharmonic components is coincident with either a primary component or another higher harmonic. This is a special case that for the moment is not covered by the present method. For irregular waves, this occurs to a large extent, and Eldrup and Lykke Andersen [8] used a narrow band assumption for the primary spectrum to overcome this issue.

In order to separate the bichromatic waves, a mathematical model for the surface is needed at every frequency present in the signal. The surface is expressed at each of the wave gauges (x_j) in the frequency domain (Fourier space). The signal include both incident and reflected waves, and due to amplitude dispersion, the wave numbers are different for the incident waves ($k_{n,I}$, $k_{m,I}$) and the reflected ($k_{n,R}$, $k_{m,R}$) primary components. Based on Equation (1) the following mathematical model is obtained at the two primary frequencies (where index i can take the values n and m and thus represent frequencies ω_n and ω_m):

$$\begin{aligned} \hat{\eta}^{(1)}(x_j) &= C_I^{(1)} X_I^{(1)} + C_R^{(1)} X_R^{(1)} + \Omega_{i,j}^{(1)} \\ X_I^{(1)} &= a_{i,I}^{(1)} exp[-i(k_{i,I} x_1 + \varphi_{i,I}^{(1)})] \\ X_R^{(1)} &= a_{i,R}^{(1)} exp[i(k_{i,R} x_1 + \varphi_{i,R}^{(1)})] \\ C_I^{(1)} &= exp(-ik_{i,I}\,\Delta x_j) \\ C_R^{(1)} &= exp(ik_{i,R}\,\Delta x_j), \end{aligned} \tag{3}$$

where $\Delta x_j = x_j - x_1$ is the distance between wave gauge number j and wave gauge number 1, i is the imaginary unit ($i^2 = -1$), and $\Omega_{\mathrm{i,j}}$ is the noise at frequency ω_i at gauge j. Later, an iterative procedure to estimate the wave numbers will be presented. The amplitudes are then calculated by a least-squares approach identical to that used by Lykke Andersen et al. [7] for regular waves.

Seven superharmonic frequencies ($2\omega_n$, $2\omega_m$, $\omega_m + \omega_n$, $3\omega_n$, $3\omega_m$, $\omega_m + 2\omega_n$, $\omega_n + 2\omega_m$) and three subharmonic frequencies ($\omega_m - \omega_n$, $|\omega_n - 2\omega_m|$ and $|\omega_m - 2\omega_n|$) are analyzed. These corresponded to the second- and third-order interaction frequencies given in Section 2. In the following, these frequencies are termed ω_i, where index i can take the values $2n$, $2m$, $m + n$, $3n$, $3m$, $m + 2n$, $m - n$, $n - 2m$, and $m - 2n$. The related wave numbers of the bound wave are termed k_i: For example, $i = m + 2n$ corresponds to the cyclic frequency $\omega_m + 2\omega_n$, and the wave number of the bound wave is $k_i = k_{m+2n} = k_m + 2k_n$. These wave numbers are different for the incident and the reflected waves due to different primary wave numbers due to amplitude dispersion ($k_{n,I}$, $k_{m,I}$ for incident and $k_{n,R}$, $k_{m,R}$ for reflected). For these ten frequencies, the mathematical model consists of both free and bound components in the incident and reflected wave direction. The mathematical model can be written in frequency domain as

$$\begin{aligned} \hat{\eta}_i(x_j) &= C_{i,I,B} X_{i,I,B} + C_{i,R,B} X_{i,R,B} + C_{i,I,F} X_{i,I,F} + C_{R,F} X_{R,F} + \Omega_{i,j} \\ X_{i,I,B} &= a_{i,I,B}\, exp[-i(k_{i,I} x_1 + \varphi_{i,I,B})] \\ X_{i,R,B} &= a_{i,R,B}\, exp[i(k_{i,R} x_1 + \varphi_{i,R,B})] \\ X_{I,F} &= a_{i,I,F}\, exp\left[-i\left(k_f x_1 + \varphi_{i,I,F}\right)\right] \\ X_{R,F} &= a_{i,R,F}\, exp\left[i\left(k_f x_1 + \varphi_{i,R,F}\right)\right] \\ C_{i,I,B}^{(n)} &= exp(-ik_{i,I}\,\Delta x_j) \\ C_{i,R,B}^{(n)} &= exp(ik_{i,R}\,\Delta x_j) \\ C_{I,F} &= exp\left(-ik_f\,\Delta x_j\right) \\ C_{R,F} &= exp\left(ik_f\,\Delta x_j\right), \end{aligned} \tag{4}$$

where k_f is the wave number of the free waves at the cyclic frequency ω_{i} and is calculated by the linear dispersion equation. The amplitudes of the bound harmonics can also be calculated by a least-square method when the wave numbers are known (see solution in Lykke Andersen et al. [7]).

Including terms up to the third order is reasonable for the tests considered in the present paper. Nevertheless, it is straightforward to extend the method to a higher order by applying Equation (4) at these frequencies. However, it is important to stress that it is assumed that only one (significant) bound interaction must occur at each frequency. This becomes more difficult to fulfill the higher order the waves are.

The wave numbers of the two primary components determine the wave numbers of each of the bound sub- and superharmonics analyzed with Equation (4). Thus, a critical part of being able to separate accurately the bound and the free harmonics is to have an accurate estimate of the wave number (or the celerity) of the two incident and reflected primary components that the bichromatic waves consist of. In principle, these four wave numbers could be unknown parameters in the mathematical model to be calculated, which, however, would result in nonlinear equations. Alternatively, they might be estimated by a wave theory, which was the procedure chosen by Lykke Andersen et al. [7] and Eldrup and Lykke Andersen [8] and which was also chosen for the present paper. In the present case, the wave numbers of the free sub- and superharmonic components (k_f) were calculated with the linear dispersion equation. The wave numbers for the primary may be calculated with the third-order theory of Madsen and Fuhrman [12,13], which is valid for all of the present cases except one test (see Chapter 5). Initial tests showed that the presence of a second primary component did not influence significantly the celerity of the other component. Therefore, another possibility is to calculate the celerity of each primary component based on the stream function theory for regular waves by Fenton and Rienecker [9], with the parameters of each of the primary waves. This might increase the range of applicability to more shallow water waves. For the present test cases, the results were almost identical for the two methods, and thus more nonlinear waves in shallow water are needed to clarify this. The iterative procedure used is as follows:

1. Calculate the wave number for each primary frequency (n,m) based on linear dispersion, i.e., for infinitesimally small amplitudes of the primary components. Use this estimate as a starting guess for the incident and reflected wave numbers ($k_{n,I}$, $k_{m,I}$, $k_{n,R}$, and $k_{m,R}$);
2. Calculate first- to third-order incident and reflected components by applying the Lykke Andersen et al. [7] method on Equation (3) for the two primary frequencies and Equation (4) for the frequencies of the ten bound second and third harmonics. For bound components, use the latest estimated incident and reflected wave numbers ($k_{n,I}$, $k_{m,I}$, $k_{n,R}$, and $k_{m,R}$) calculated in either step 1 for the first iteration or step 3 for the following iterations. Free components are normally of much smaller height, and linear assumption is assumed valid for those;
3. Calculate updated values of $k_{n,I}$, $k_{m,I}$, $k_{n,R}$, and $k_{m,R}$ either based on either a) third-order wave theory using the found amplitudes of the primary components, or b) ignoring the interaction of the two primary components (n and m), and thus a method for highly nonlinear regular waves can be applied for each component (e.g., stream function theory from Fenton and Rienecker (1980));
4. Repeat steps 2 and 3 until convergence is obtained for the incident and reflected wave numbers.

In principle, the calculations of the wave numbers should take into account the generated return currents from the incident and reflected waves. The third-order dispersion equation given by Madsen and Fuhrman [12,13] makes it possible to calculate the influence of the return currents on the celerities. This was done by calculating the celerities when no return currents were considered ($U = 0$) and when a closed basin was considered and thus a return current from the incident waves existed ($q = 0$). As the reflected waves generate a return current in the opposite direction, these calculations show the maximum influence of the return currents. For the test cases considered in the present paper, the difference on the celerities between using $U = 0$ and $q = 0$ was found to be from 0.2% to 0.9%, and thus the return current was not very important to include for the present tests. Therefore, it was decided to use $U = 0$ in the calculations, which was in line with what was used by Lykke Andersen et al. [7] for regular waves.

Moreover, the interaction of the incident and reflected waves was ignored, which from second-order wave theory is known to be a reasonable approximation (Eldrup and Lykke Andersen [8]). When both free and bound waves are present at a given frequency, this causes a modulation of the amplitude of the other components. This effect is mainly relevant when the bound and the free waves are high and have amplitudes of the same order of magnitude. This effect was not included in the mathematical model by Lykke Andersen et al. [7] and Eldrup and Lykke Andersen [8], and it was also ignored in the present paper.

Equations (3) and (4) are of similar form to the equations provided by Lykke Andersen et al. [7] for regular waves. Their solution can thus be applied also to bichromatic waves. Cases where the celerity of the free and the bound waves are almost identical can also be handled in a similar way to that presented in that paper. In bichromatic waves, the two primary frequencies in the signal would normally be known from the input to the wave maker. If this is not the case, a frequency domain analysis can be carried out to identify the frequency of the two highest peaks.

4. Validation on Synthetic Data

In order to validate the method and its implementation, bichromatic synthetic data were generated using the third-order theory of Madsen and Fuhrman [12,13]. The test case corresponds to "Test No. 2" in the later presented physical model tests (see Section 5). The data does not include any reflection or free harmonics. The wave gauge array used is also identical to that used in the physical model tests. The chosen phases of the two primary components are zero at $x = 0$.

Table 1 presents the actual amplitudes and the estimated ones. Note that there is almost full agreement, and errors on the amplitudes are typically less than 1%. Moreover, the estimated reflected free and bound components as well as the incident free components are close to zero, as used in the synthetic data. Minor errors are present mainly in the second-order components and are expected to be caused by small errors on the estimated celerities of the bound waves due to using third-order theory with $q = 0$ in the synthetic data and using stream function theory with $U = 0$ for each primary wave in the analysis method.

Table 1. Actual and calculated amplitudes in millimeters for the synthetic test case.

Component	Actual Incident Bound	Estimated Incident Bound	Estimated Reflected Bound	Estimated Incident Free	Estimated Reflected Free
a_n	50.0	50.1	0.2	-	-
a_m	50.0	49.9	0.2	-	-
a_{2n}	5.8	5.8	0.0	0.0	0.0
a_{2m}	4.7	4.7	0.0	0.1	0.1
a_{n+m}	10.5	10.5	0.0	0.2	0.1
a_{m-n}	7.4	7.6	0.1	0.3	0.1
a_{3n}	0.7	0.7	0.0	0.0	0.0
a_{3m}	0.5	0.5	0.0	0.0	0.0
a_{n+2m}	1.6	1.6	0.0	0.0	0.0
a_{2n+m}	1.8	1.8	0.0	0.0	0.0
a_{n-2m}	7.7	7.7	0.0	0.0	0.0
a_{m-2n}	3.4	3.2	0.0	0.2	0.0

Additionally, the time series are plotted in Figure 1, which demonstrates that phases are also correctly calculated. Moreover, the figure gives an overall impression of the accuracy, as the estimated incident time series including only bound higher harmonics is almost identical to the target. Moreover, the estimated incident free higher harmonics, as well as the reflected components, are insignificant. Therefore, the implemented method was validated for data, fulfilling the assumptions of the method.

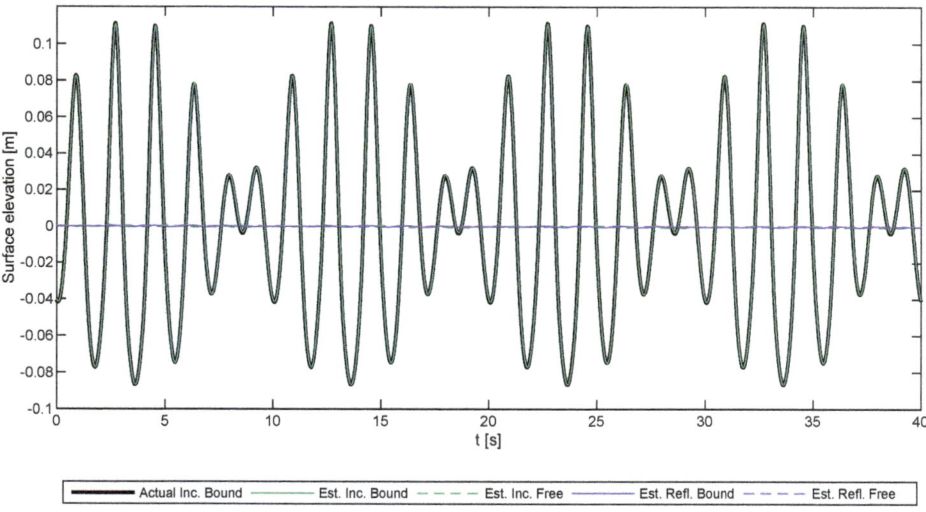

Figure 1. Comparison of measured total and predicted incident and reflected time series at WG1 for the synthetic test case.

5. Physical Model Test Setup and Methodology

In order to evaluate the method, 2-D model tests were performed in a wave flume at the Atmosphere-Ocean Interaction Flume in Granada, Spain. The flume has a piston wave maker in each end, where one wave maker (WM1) is generating and absorbing waves, and the other wave maker (WM2) is only absorbing. The active absorption method used was the one presented by Lykke Andersen et al. [14], and the model test setup was also identical. An array of eight wave gauges (WG1–WG8) were used to test the above given reflection separation method. The method uses data from all gauges to calculate incident and reflected wave train components (amplitudes and phases). Based on the calculated components, the incident and reflected wave trains can be calculated at any point, but any errors in the mathematical model (for example, celerities) will amplify with distance from the central part of the wave gauge array. In the present paper, the time series will only be presented for $x = 5.95$ m (at the location of WG1). The water depth was constant throughout the length of the flume and equal to 0.65 m in all tests. The model test setup is shown in Figure 2.

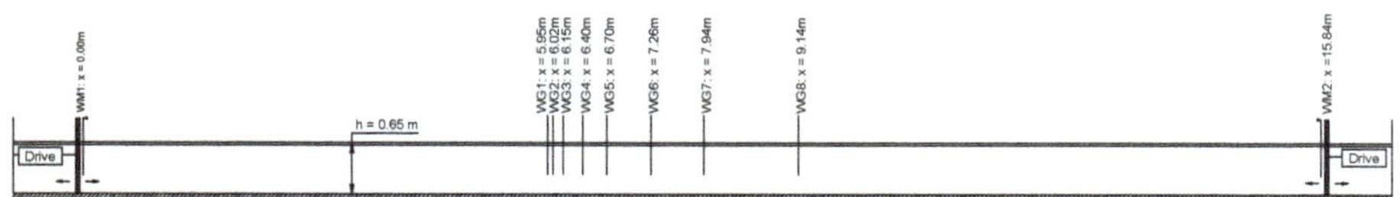

Figure 2. Model test setup.

Bichromatic waves were generated using the second-order wave maker theory of Schäffer [11], which correctly generates the bound sub- and superharmonics to second-order and thus minimizes free second-order incident energy. The test was performed for 240 s with a 10-s ramp up and down in each end. The analyzed part of the tests was from 40 s to 230 s. Table 2 presents the test conditions. With the chosen test conditions, the assumption in the mathematical model of all the primary and the interaction terms occurring at independent frequencies up to the third order was fulfilled. After Test Nos. 3, 8, and 9, some minor cross-modes were observed.

Table 2. Target test conditions.

Test No.	f_n [Hz]	f_m [Hz]	a_n [mm]	a_m [mm]
1	0.30	0.40	50	50
2	0.50	0.60	50	50
3	0.60	0.70	50	50
4	0.60	0.80	50	50
5	0.50	0.55	50	50
6	0.50	0.52	50	50
7	0.30	0.40	25	25
8	0.50	0.60	25	25
9	0.60	0.70	25	25
10	0.60	0.80	25	25
11	0.50	0.55	25	25
12	0.50	0.52	25	25

Figure 3 presents the test conditions in a Le Méhauté [15] diagram, where the data are plotted with the highest period of the two primary components and with the sum of their wave heights. In that way, the diagram should be slightly on the safe side with respect to the valid theory. It appears that third-order theory should be valid for all cases except for Test No. 1.

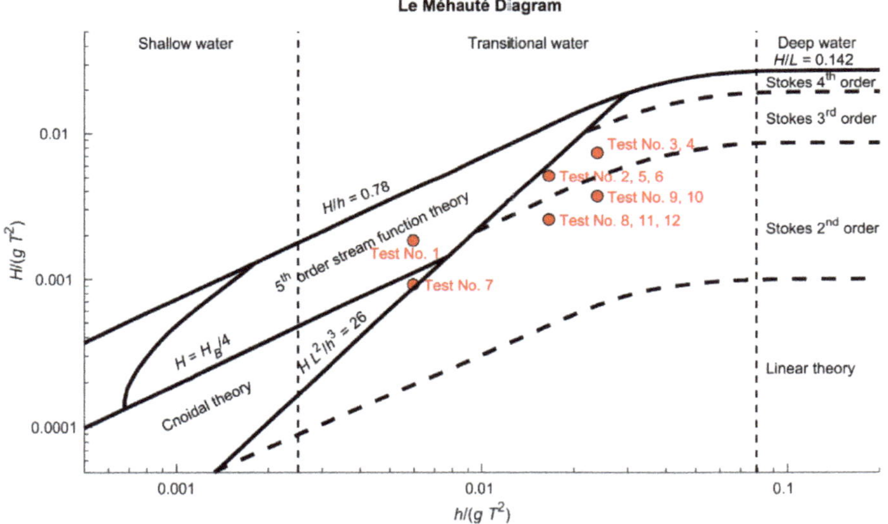

Figure 3. Applicability of different wave theories for the test conditions.

6. Physical Model Test Results

Figures 4 and 5 show the incident and reflected time series for Test Nos. 2 and 5, respectively. The time series for the bound waves also include the two primary components. It appears that reflection is limited and in both cases at the order of 5%–10% of the incident wave. In the following, the focus is on analyzing the incident waves. An in-depth analysis of the reflection of the bound and free waves and a comparison to the linear absorption curve given by the Lykke Andersen et al. [14] method, is covered in detail by Lykke Andersen et al. [16].

Incident free waves of small amplitude are present, indicating that the bound harmonics were not generated accurately. This was maybe expected, as second-order generation was used while third-order theory was needed, according to Figure 3.

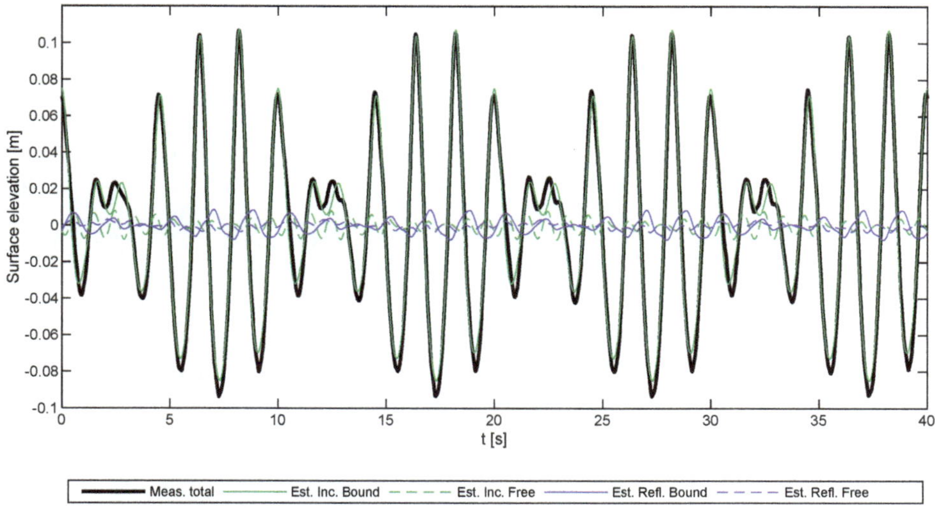

Figure 4. Comparison of measured total and predicted incident and reflected time series at WG1 (x = 5.95 m) for Test No. 2.

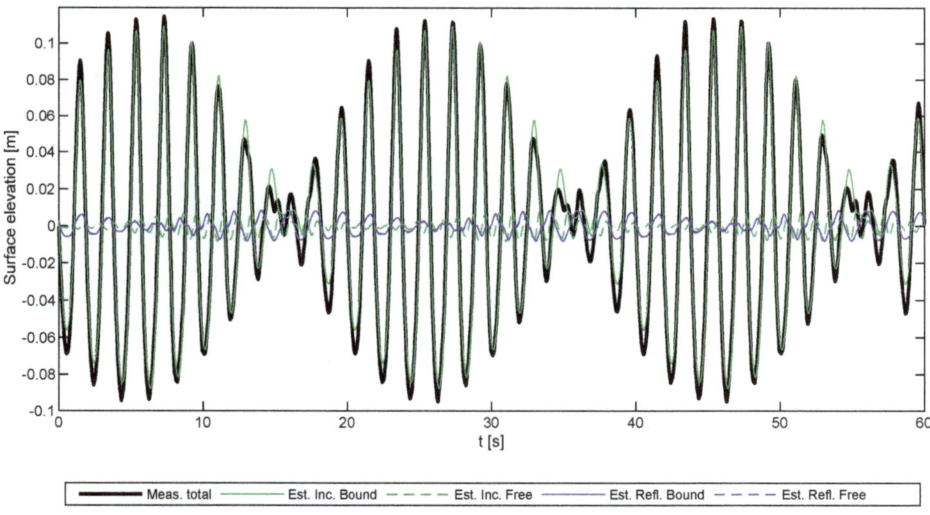

Figure 5. Comparison of measured total and predicted incident and reflected time series at WG1 (x = 5.95 m) for Test No. 5.

Table 3 presents the measured components for all the test cases using the present method. It appears that the two primary amplitudes are usually slightly lower than the target. This might be partly caused by energy needed to be stolen for the third-order components that were not generated and partly due to other issues such as leakage around the paddle.

Table 3. Estimated amplitudes in millimeters of all the incident bound components.

Test	a_n	a_m	a_{2n}	a_{2m}	a_{n+m}	a_{m-n}	a_{3n}	a_{3m}	a_{n+2m}	a_{2n+m}	a_{n-2m}	a_{m-2n}
1	47.0	44.3	12.2	5.7	16.2	13.7	1.7	0.8	3.8	5.5	3.9	1.3
2	47.5	46.8	5.4	4.2	9.8	7.2	0.7	0.6	1.5	1.4	2.9	1.6
3	46.4	47.6	4.6	4.0	8.2	4.9	0.8	0.5	0.9	1.5	0.6	1.4
4	45.8	46.5	4.7	5.2	7.9	2.1	1.2	0.9	2.3	2.5	2.7	1.1
5	50.5	44.5	6.3	4.2	9.6	9.3	0.9	0.5	1.1	1.2	5.5	4.4
6	52.7	40.7	7.1	2.4	9.2	5.4	1.3	0.1	0.8	1.6	6.5	6.0
7	24.2	23.7	3.2	1.7	4.4	4.6	0.2	0.1	0.4	0.5	0.8	0.2
8	23.6	23.9	1.2	1.0	2.3	1.8	0.1	0.0	0.1	0.2	0.4	0.3
9	23.6	24.3	1.0	1.0	1.9	1.3	0.1	0.1	0.2	0.3	0.0	0.4
10	23.6	24.2	1.0	1.1	2.1	0.7	0.1	0.1	0.2	0.2	0.3	0.1
11	24.2	23.4	1.3	1.1	2.3	2.1	0.1	0.0	0.1	0.1	0.8	0.7
12	24.5	23.1	1.4	1.0	2.4	1.3	0.0	0.0	0.1	0.1	0.7	0.9

In Figure 6, the measured amplitudes from Table 3 are compared to theoretical predictions based on third-order wave theory (Madsen and Fuhrman [12,13]) using the two estimated primary amplitudes. The estimated amplitudes of the bound second-order superharmonics by the present method match accurately the ones predicted by the third-order theory. The same is the case for the subharmonic component, but more scatter is observed.

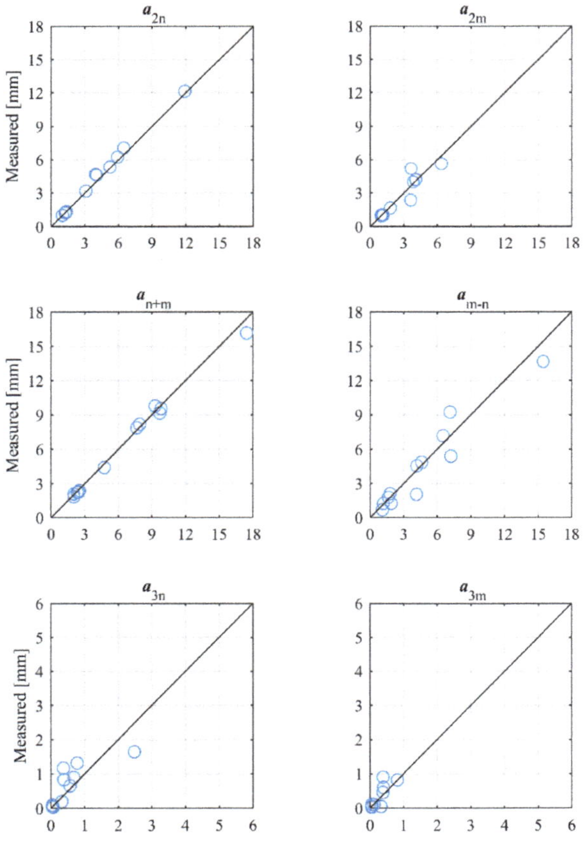

Figure 6. *Cont.*

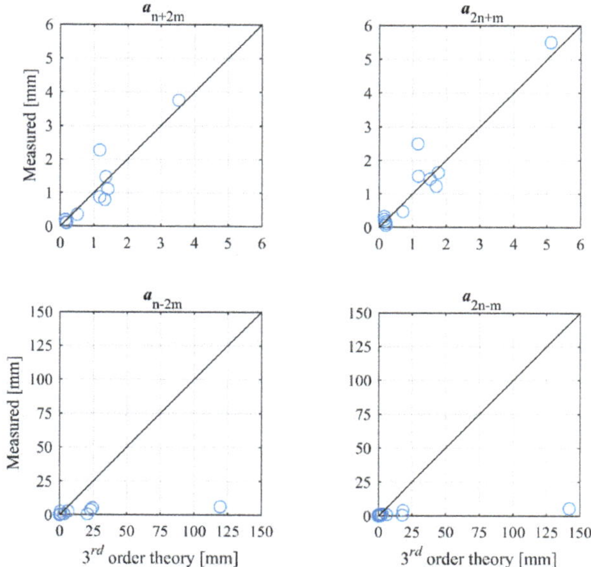

Figure 6. Measured amplitudes of bound second- and third-order components compared to Madsen and Fuhrman [12,13].

With respect to the third-order components, the agreement with third-order wave theory is fair except for the subharmonics, which presented strong deviations. The predicted amplitudes are significantly larger than the second-order components, which is not physically correct and is due to singularities in the transfer functions, as also found by Madsen and Fuhrman [13].

7. Discussion

In the present paper, a new method to separate nonlinear bichromatic waves into incident and reflected surface elevation time series was presented. Most existing methods only consider free energy to be present, and their range of applicability is thus limited to linear waves. The new method includes both free and bound higher harmonics, which is necessary in order to separate nonlinear waves accurately. The method is an extension of the method for nonlinear regular waves presented recently by Lykke Andersen et al. [7]. That method was extended by Eldrup and Lykke Andersen [8] to fully irregular waves by utilizing a narrow band approximation for the primary spectrum. The new method for bichromatic waves includes all interactions correctly to the third order. As expected, the results showed that the new method accurately predicted the incident and reflected waves in synthetically generated third-order bichromatic waves.

Moreover, the method was tested on physical model tests performed in a wave flume with wave makers in both ends. The new method seems to predict accurately the incident and reflected wave trains, and the estimated amplitudes of the second- and third-order bound components matched reasonably with third-order wave theory for irregular waves. For third-order subharmonics, a significant deviation from third-order wave theory was, however, found in some tests, but this could be attributed to singularities in the transfer functions provided by Madsen and Fuhrman [12,13], and thus their method was not valid for these cases.

Author Contributions: T.L.A. and M.R.E. defined the new analysis method, and M.R.E. prepared the synthetic and numerical model data. M.C. provided access to the wave flume and prepared the test setup for the physical model tests. T.L.A. prepared the test methodology and performed the model tests and the analysis. T.L.A. wrote the paper. All authors have read and approved the final manuscript.

Funding: This research received no external funding.

Conflicts of Interest: The authors declare no conflicts of interest.

References

1. Goda, Y.; Suzuki, Y. Estimation of incident and reflected waves in random wave experiments. In Proceedings of the 15th International Conference on Coastal Engineering, Honolulu, HI, USA, 11–17 July 1976; pp. 828–845. [CrossRef]
2. Mansard, E.P.D.; Funke, E.R. The measurement of incident and reflected spectra using a least squares method. In Proceedings of the 17th International Conference on Coastal Engineering, Sydney, Australia, 23–29 March 1980; pp. 154–172. [CrossRef]
3. Zelt, J.A.; Skjelbreia, J.E. Estimating incident and reflected wave fields using an arbitrary number of wave gauges. In Proceedings of the 23rd International Conference on Coastal Engineering, Venice, Italy, 4–9 October 1992; pp. 777–788. [CrossRef]
4. Lin, C.-Y.; Huang, C.-J. Decomposition of incident and reflected higher harmonic waves using four wave gauges. *Coast. Eng.* **2004**, *51*, 395–406. [CrossRef]
5. Figueres, M.; Garrido, J.M.; Medina, J.R. *Cristalización Simulada para el Análisis de Oleaje Incidente y Reflejado con un Modelo de Onda Stokes-V*; VII Jornadas de Puertos y Costas: Almeria, Spain, 2003.
6. Figueres, M.; Medina, J.R. Estimating incident and reflected waves using a fully nonlinear wave model. In Proceedings of the 29th International Conference on Coastal Engineering, Lisbon, Portugal, 19–24 September 2004; pp. 594–603. [CrossRef]
7. Lykke Andersen, T.; Eldrup, M.R.; Frigaard, P. Estimation of Incident and Reflected Components in Highly Nonlinear Regular Waves. *Coast. Eng.* **2017**, *119*, 51–64. [CrossRef]
8. Eldrup, M.R.; Lykke Andersen, T. Estimation of Incident and Reflected Wave Trains in Highly Nonlinear Two-Dimensional Irregular Waves. *J. Waterw. Port Coast. Ocean Eng.* **2019**, 145. [CrossRef]
9. Fenton, J.D.; Rienecker, M.M. Accurate numerical solution for nonlinear waves. In Proceedings of the 17th Conference of Coastal Engineering, Sydney, Australia, 23–29 March 1980; Volume 1, pp. 50–69. [CrossRef]
10. Sharma, J.; Dean, R. Second-order directional seas and associated wave forces. *Soc. Pet. Eng. J.* **1981**, *21*, 129–140. [CrossRef]
11. Schäffer, H.A. Laboratory Wave Generation Correct to Second Order. In Proceedings of the Conference Wave Kinematics and Environmental Forces, London, UK, 24–25 March 1993; Soc Underwater Tech; Volume 29, pp. 115–139.
12. Madsen, P.A.; Fuhrman, D.R. Third-order theory for bicromatic bi-directional water waves. *J. Fluid Mech.* **2006**, *557*, 369–397. [CrossRef]
13. Madsen, P.A.; Fuhrman, D.R. Third-order theory for multi-directional irregular waves. *J. Fluid Mech.* **2012**, *698*, 304–334. [CrossRef]
14. Lykke Andersen, T.; Clavero, M.; Frigaard, P.; Losada, M.; Puyol, J.I. A new active absorption system and its performance to linear and non-linear waves. *Coast. Eng.* **2016**, *114*, 47–60. [CrossRef]
15. Le Mehauté, B. *An Introduction to Hydrodynamics and Water Waves*; Springer: Berlin/Heidelberg, Germany, 1976. [CrossRef]
16. Lykke Andersen, T.; Clavero, M.; Eldrup, M.R.; Frigaard, P.; Losada, M. Active Absorption of Nonlinear Irregular Wave. In Proceedings of the 36th International Conference on Coastal Engineering, Baltimore, MD, USA, 30 July–3 August 2018. submitted.

Journal of
Marine Science
and Engineering

Article

Hydraulic Stability of the Armor Layer of Overtopped Breakwaters

Gloria Argente *, M. Esther Gómez-Martín and Josep R. Medina

Laboratory of Ports and Coasts, Institute of Transport and Territory, Universitat Politècnica de València, 46022 Valencia, Spain; mgomar00@upv.es (M.E.G.-M.); jrmedina@upv.es (J.R.M.)
* Correspondence: gloargar@upv.es; Tel.: +34-963-877-375

Received: 18 October 2018; Accepted: 23 November 2018; Published: 27 November 2018

Abstract: Mound breakwaters with significant overtopping rates in depth-limited conditions are common in practice due to social concern about the visual impact of coastal structures and sea level rise due to climatic change. For overtopped mound breakwaters, the highest waves pass over the crest producing armor damage, not only to the front slope, but also to the crest and the rear slope. To guarantee the breakwater stability, it is necessary to limit the armor damage in the three parts of the structure: Front slope, crest, and rear slope. This paper describes the hydraulic stability of the armor layer of medium and low-crested structures in wave breaking conditions. Small-scale physical model tests were carried out with different relative crest freeboards and three armor units: Rocks, cubes, and Cubipods. The armor damage progression in the front slope, crest, and rear slope was analyzed using the Virtual Net method to consider the heterogeneous packing and porosity evolution along the armor slope. A comparison is provided between the hydraulic stability of the different armors and their relationship with the measured overtopping volumes.

Keywords: hydraulic stability; breaking wave conditions; low-crested structures; mound breakwaters; armor layer

---

1. Introduction

The crest height of a mound breakwater relative to the water level is defined as the crest freeboard, R_c (see EurOtop [1]). This parameter is one of the keys in the structural design as it affects the economic cost, the energy footprint, the overtopping hazards, and the visual impact. Growing social concern regarding the environmental and visual impacts associated with coastal structures and sea level rise due to climatic change is leading to a reduction in crest freeboards and an increase in overtopping hazards. Moreover, structures with a reduced freeboard are usually built in shallow water where the highest waves are depth limited (see Kramer and Burcharth [2]).

Medium and low-crested breakwaters are defined in this paper as emergent structures within the range $0.5 < R_c/H_s < 1$ where H_s is the significant wave height. These types of structures are frequently overtopped by waves (see Burcharth et al. [3]), so wave energy is allowed to pass through or over the structure; consequently, the design of these structures must be different than a conventional type [2]. These overtopping events have a direct impact on the hydraulic stability of the crest and rear slope armors producing several armor damage in these parts of the structure because of the wave energy dissipation (see CIRIA [4]). Therefore, the hydraulic stability of the armor layer of the medium and low-crested structures may be higher on the front slope than for non-overtopped structures because some wave energy passes over the breakwater crest.

The armor layer of an overtopped structure can be divided in three parts (see Figure 1): Front slope (I), crest (II), and rear slope (III). This study focused on the analysis of the hydraulic stability of the armor layer in the three parts of a conventional medium and low-crested mound breakwater protected with three different armors: (1) Double-layer rocks, (2) double-layer randomly-placed cubes, and (3) single-layer Cubipods. A wide range of dimensionless freeboards, with and without overtopping events, were tested ($0.3 < R_c/H_s < 2.6$) in wave breaking conditions ($H_s > 0.4\ h_s$), where h_s is the water depth at the toe of the structure. A total of 144 small-scale tests were carried out in the wave flume at the Laboratory of Ports and Coasts at the Universitat Politècnica de València (LPC-UPV). A detailed quantitative analysis was conducted using photographs to determine the armor damage progression in each part of the structure (front slope, crest, and rear slope).

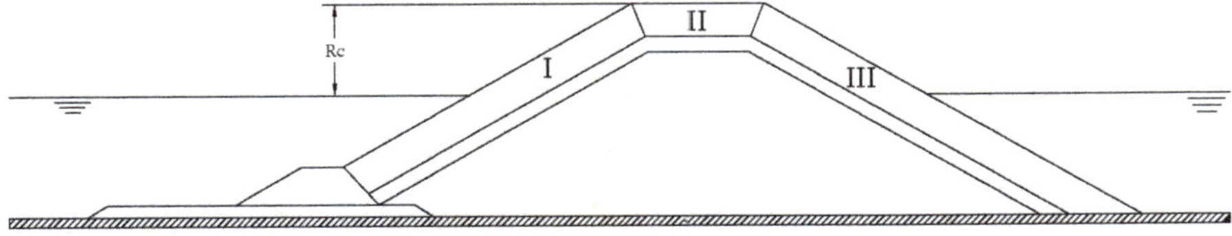

Figure 1. Division of armor layer in three parts.

2. Literature Review

2.1. Hydraulic Stability of the Armor Layer

While most mound breakwaters are constructed in the depth-limited zone, they are usually designed with empirical formulas based on small-scale tests of non-overtopped models in non-breaking conditions such as Hudson's formula [5] or Van der Meer's formula [6]. Armor design in wave breaking conditions involves estimating the incident characteristic wave height at the breakwater toe, but the standard stability formulas found in the literature have rarely taken into account the wave height distribution changes due to wave breaking. Some empirical modifications have been proposed by Herrera et al. [7] to estimate rock armor damage in breaking wave conditions. However, the hydraulic stability formula developed by Herrera et al. [7] is only valid for frontal slope armor with zero or low overtopping rates.

In the case of non-overtopped structures, waves mainly affect hydraulic stability on the front slope, while in the case of overtopped structures, waves do not only affect the hydraulic stability on the front slope, but also the stability of the crest and rear slope. The aim of this paper was to characterize the hydraulic stability of medium and low-crested structures subjected to frequent overtopping events in breaking wave conditions.

The hydraulic stability and performance of low-crested structures ($R_c/H_s < 1$) have been studied in European project DELOS (Environmental Design of Low Crested Coastal Defence Structures) and other research projects reported in the literature. Van der Meer and Daemen [8] compared the hydraulic stability of the armor layer of overtopped and non-overtopped rubble mound breakwaters, concluding that the required stone size for an overtopped rubble mound breakwater can be estimated by applying a reduction factor to the size calculated for a non-overtopped structure using the hydraulic stability formulas given in the literature. Vidal et al. [9] performed model tests with low-crested structures to analyze the different sections of the trunk in order to determine the distribution of damage. For the front armor slope, the results showed a linear relationship between the crest freeboard and the stability number, with lower hydraulic stability corresponding to the case of a non-overtopped rubble mound breakwater. For the armor on the crest, the hydraulic stability increases with the crest freeboard, and the opposite is true for the armor on the rear slope. Burger [10] re-analyzed existing tests and described the hydraulic stability to the initiation of damage of the front, crest, and rear slope, concluding that the damage to the front slope almost always determined the stability of the structure.

Vidal et al. [11] proposed the following hydraulic stability formula for rock-armored low-crested structures:

$$N_s = A + BF_d + CF_d^2 \quad (1)$$

where $F_d = R_c/D_{n50}$ is the non-dimensional crest freeboard; $N_s = H_s/(\Delta D_{n50})$ is the stability number; $D_{n50} = (M_{50}/\rho_r)^{1/3}$ is the nominal diameter of the armor unit; M_{50} is the armor unit mass corresponding to the D_{n50}; $\Delta = (\rho_r - \rho_w)/\rho_w$ is the relative submerged mass density; ρ_r is the mass density of the armor unit; ρ_w is the mass density of the sea water; and H_s is the significant wave height. Coefficients A, B, and C depend on the section of the breakwater and the damage level as specified in Vidal et al. [11]; these coefficients are valid for the experimental range $2.01 < F_d < 2.41$. Kramer and Burcharth [2] calibrated coefficients A, B, and C from Equation (1), based on the least stable section of the structure. Vidal et al. [12] added additional data corresponding to low-crested structures by re-calibrating the coefficients A, B, and C from Equation (1) and formulating a stability formula to design rubble-mound breakwaters in the range $-4 < F_d < 4$.

There have been numerous hydraulic stability studies for the frontal armor slope of non-overtopped structures. Low-crested rubble mound breakwaters with the crest near the still water level have also been well studied. However, the transitional zone between a non-overtopped structure and a low-crested mound breakwater has not been as well studied. In this research, experiments of overtopped structures with medium and low-crest freeboards were carried out to analyze the hydraulic stability in the three categories of the armor layer (frontal, crest, and rear slope) with three different armors (double-layer rock, cube, and single-layer Cubipod®) under wave breaking conditions.

2.2. Armor Damage Measurement

Different methods to characterize armor damage have been described in the literature [13]. The traditional visual counting method [12] assumes a constant porosity along the armor layer, so the heterogeneous packing (HeP) failure mode is not considered. The HeP, defined by Gómez-Martín and Medina [13], is an armor-damaging process without armor unit extractions, but with a reduction of the porosity in the lower area of the armor and a higher porosity in the upper area. The Virtual Net method developed by Gómez-Martín and Medina [13] considered armor unit extractions and changes in the porosity due to HeP. In this paper, the Virtual Net method was used to calculate the damage to single-layer Cubipod® armor, the double-layer rock, and to cube armors on the front slope, the crest, and the rear slope. The Virtual Net method divides the armor into individual strips with a constant width (a) and length (b), allowing for the measurements of the dimensionless damage in each strip, S_i. Integrating S_i over the slope, the equivalent dimensionless armor damage parameter, S_e, is obtained. This method allows for measuring the armor layer porosity in time and space.

$$S_i = k\left(1 - \frac{1 - n_{vi}}{1 - n_{v0i}}\right) \quad (2)$$

$$S_e = \sum_{i=1}^{l} S_i \forall S_i \geq 0 \quad (3)$$

where k is the number of rows in each strip; $n_{vi} = 1 - (N_i D_{n50}^2 / a \cdot b)$ is the porosity of the strip; N_i is the number of armor units whose center of gravity is within the strip; D_{n50} is the nominal diameter of the armor unit; n_{v0i} is the initial porosity of each strip; and l is the number of strips.

The criteria given by Losada et al. [14] and Vidal et al. [15] was followed for double-layer armors: Initiation of Damage (IDa) occurs when the upper armor layer loses some units and gaps in the size of an armor unit are visible. In the case of single-layer armors, the criterion defined by Gómez-Martín [16] was followed: IDa occurs when the upper armor layer has lost one or more units and gaps the size of an armor unit are visible in the armor. For qualitative analysis, only the first level of armor-damage (IDa) was considered in the study.

3. Experimental Methodology

Two-dimensional physical model tests were conducted in the wind and wave test facility (30 m × 1.2 m × 1.2 m) of the Laboratory of Ports and Coasts at the Universitat Politècnica de València (LPC-UPV), with a bottom slope m = 1/50. Figure 2 shows the longitudinal cross-section of the LPC-UPV wave flume.

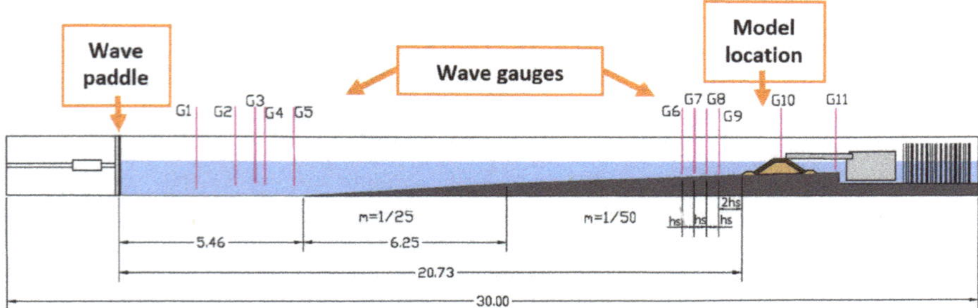

Figure 2. Longitudinal cross-section of the LPC-UPV wave flume (dimensions in meters).

Figure 3 shows the cross-section of the physical model, a low-crested conventional mound breakwater with armor slope cotα = 1.5 in the front slope and in the rear slope, and a toe berm to support them. Using the same core and filter layer, three armors were tested: Rocks (2-L), cubes (2-L), and Cubipods (1-L). A single-layer Cubipod® armor was tested with an initial packing density of $\Phi = 1 - p = 60\%$, and double-layer randomly placed rock and cube armors were tested with the initial packing density coefficients of $\Phi = 1 - p = 63\%$ and 59%, respectively, where p is the armor porosity. A summary of the characteristics of the materials used in the physical models of this paper is presented in Table 1.

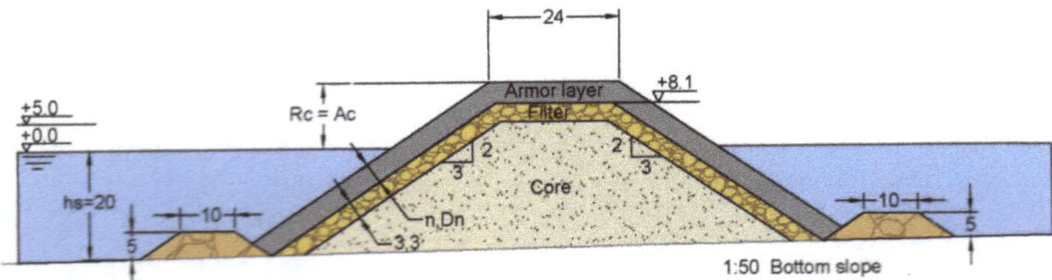

Figure 3. Cross-section of the of the breakwater model (dimensions in centimeters).

Table 1. Characteristics of the materials used in the experiments.

Layer	M_{50} (g)	ρ_r (g/cm^3)	D_{n50} (cm)
Core	0.86	2.72	0.68
Filter	15.40	2.73	1.78
Rocks	86.77	2.68	3.18
Cubes	141.51	2.27	3.97
Cubipods	121.25	2.22	3.79

Tests with runs of 1000 irregular waves were generated following the JONSWAP (γ = 3.3) ·trum in deep water. The AWACS Active Wave Absorption System of the wavemaker was ·ated to avoid multi-reflections in the wave flume. Tests were grouped in a series of constant · of water depth at the toe berm, h_s[cm] = 20, 25, and 30, and constant Iribarren's number, ·an $\alpha/(2\pi H_{m0}/(gT_p^2))^{0.5}$ = three and five, where tan α = 2/3, T_p is the peak period and H_{m0}

is the spectral significant wave height, $H_{m0} = 4(m_0)^{1/2}$. For each series, tests were run by increasing the H_{m0} progressively in steps of 1 cm in the range $8 \leq H_{m0}[cm] \leq 24$ from zero damage until severe damage occurred or the limit of use of the wavemaker was reached. Table 2 summarizes the test characteristics considering waves in the generating zone, where $s_{0p} = 2\pi H_{m0}/(gT_p^2)$ is the wave stepness, N_t is the number of tests, and N_{tw} is the total number of waves in the series.

Table 2. Test matrix.

Series	Armor Layer	h_s (cm)	Ir_p	S_{0p}	H_{m0} (cm)	T_p (s)	R_c (cm)	N_t	N_{tw}
1	Cubipods 1L	20	3	0.049	8–24	1.02–1.76	12	17	17,000
2	Cubipods 1L	20	5	0.018	8–20	1.70–2.68	12	13	13,000
3	Cubipods 1L	25	3	0.049	8–24	1.02–1.76	7	17	17,000
4	Cubipods 1L	25	5	0.018	8–20	1.70–2.68	7	13	13,000
5	Rocks 2L	20	3	0.049	8–16	1.02–1.44	15	9	9000
6	Rocks 2L	20	5	0.018	8–13	1.70–2.08	15	6	6000
7	Rocks 2L	25	3	0.049	8–16	1.02–1.44	10	9	9000
8	Rocks 2L	25	5	0.018	8–13	1.70–2.08	10	6	6000
9	Cubes 2L	25	3	0.049	8–24	1.02–1.76	11	17	17,000
10	Cubes 2L	25	5	0.018	8–20	1.70–2.68	11	13	13,000
11	Cubes 2L	30	3	0.049	8–24	1.02–1.76	6	17	17,000
12	Cubes 2L	30	5	0.018	8–14	1.70–2.25	6	7	7000

Two groups of capacitive wave gauges were placed along the flume to measure the water elevation at different points. One group of five gauges (G1 to G5) was placed near the wavemaker and the other four gauges (G6 to G9) were installed at a distance of 2 h_s, 3 h_s, 4 h_s, and 5 h_s seaward from the structure toe. One wave gauge (G10) was placed on the crest of the structure, and the last one (G13) was placed behind the model to control the water level behind the structure (see Figure 2).

Wave gauges near the wavemaker were distanced to select the combination of gauges needed following the criterion given by Mansard and Funke [17], depending on the wave-length of the test. Using these selected gauges, incident and reflected waves were separated using the LASA-V method proposed by Figueres and Medina [18] allows the separation of non-linear and non-stationary waves. However, neither the LASA-V method, nor other existing methods, are reliable to separate incident and reflected waves in the breaking zone. For this reason, the methodology, validated by Herrera et al. [7], to estimate incident waves in breaking conditions was used. Considering the same bottom profile of the flume and wave characteristics in deep waters as in the physical experiments, numerical simulations using SwanOne software were carried out to estimate the incident wave parameters near the breakwater model. SwanOne is a 1D numerical model that is appropriate to estimate wave propagation and simulate the depth-induced breaking phenomena (see Verhagen et al. [19]). Virtual wave measurements were obtained in the same location as the gauges used in the physical test. To validate the methodology, the numerical SwanOne estimations were compared with measurements in the wave flume without any structure, assuming no reflections. Figure 4 shows the comparison between the incident spectral significant wave height, $H_{m0,i}$, measured without structure in the model zone and estimations given by SwanOne at the same point.

In order to measure the goodness of fit, the relative mean squared error, rMSE, was calculated.

$$rMSE = \frac{MSE}{Var} = \frac{1}{N}\sum_{i=1}^{N}\frac{(o_i - e_i)^2}{Var(o_i)} \tag{4}$$

where MSE is the mean squared error; Var is the variance of the observed values; N is the number of observations; o_i is the observed value; e_i is the estimated value; $\overline{o}$ is the average of the observed values; and $\overline{e}$ is the average of the estimated values. $0 \leq rMSE \leq 1$ estimates the proportion of variance not explained by the model; the lower the rMSE, the better the estimations.

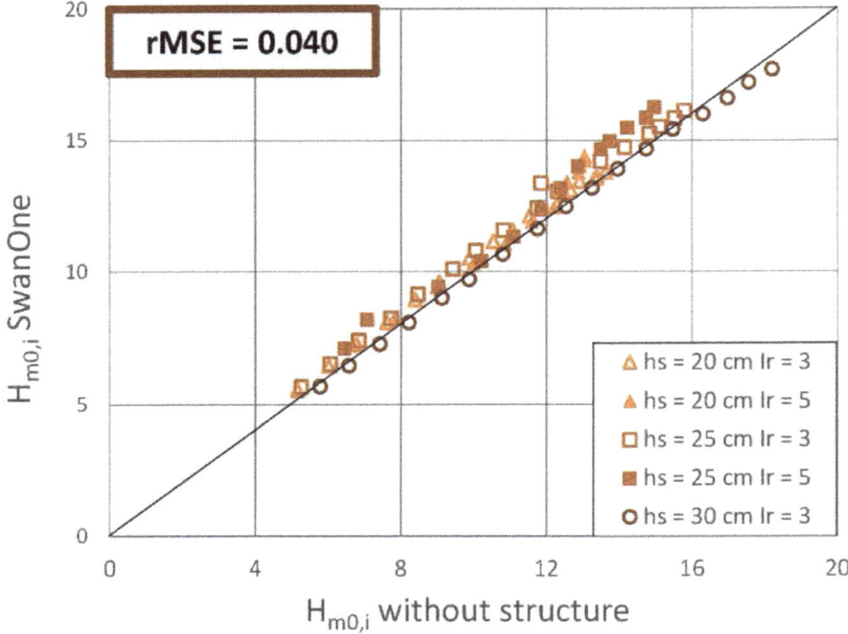

Figure 4. Comparison of $H_{m0,i}$ measured without structure and estimations given by SwanOne in the model zone.

Due to the good results obtained, H_{m0} estimated with SwanOne at a distance of three times the water depth seaward from the structure toe was used in this research to characterize incident waves and analyze armor damage in wave breaking conditions [7].

To measure armor damage, the Virtual Net method was used, taking photographs perpendicular to the front slope, crest, and rear slope armors before and after each test run. A virtual net was projected over each photograph, dividing the armor into individual strips. Three strips of $3D_n$ (strip A, B, and C) and one strip of $4D_n$ (strip D) were used on the front slope and four strips of $3D_n$ (strips A, B, C, and D) on the rear slope. For the crest armor, only one strip of $6D_n$ was considered. Dimensionless armor damage was calculated for each strip (S_i); after integrating this dimensionless armor damage over the slope, the equivalent dimensionless armor damage (S_e) was obtained. Figure 5 shows three photographs with the virtual net used for the single-layer Cubipod® armor during the experiments. Overtopping measurements were taken with a weighing instrument, which registered the overtopped volume in time. Mean overtopping rate, $q[m^3/s/m]$, was routinely calculated for each test.

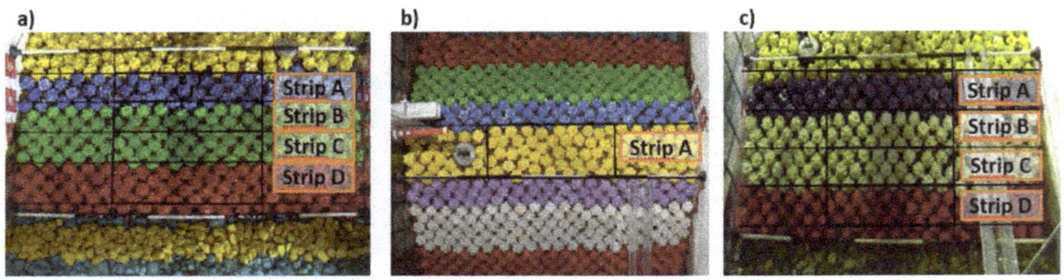

Figure 5. Application of the Virtual Net method to measure armor damage to the front slope (**a**), crest (**b**), and rear slope (**c**).

4. Analysis of Hydraulic Stability Test Results

The stability number $N_s = H_{m0}/(\Delta D_{n50})$ was used to characterize the hydraulic stability performance of the armor layers. For the wave height, H_{m0}, estimated by SwanOne at a distance of three times the water depth seaward from the structure toe, was used in this study to estimate the wave characteristics at the toe of the structure [7]. Three different armor layers were tested in this study: Double-layer rock and cube armors, and a single-layer Cubipod® armor.

4.1. Damage to Double-Layer Armors

Following Medina et al. [20], the failure function of double-layer rock armors follows a 5-power relationship with the stability number, so in this paper the failure function was represented with the linearized dimensionless armor damage ($S_e^{1/5}$).

Figures 6–8 show the linearized equivalent dimensionless armor damage observed during the experiments as a function of the stability number (N_s) for the front slope, crest, and rear slope of the double-layer rock and cube armors. Tests where the Initiation of Damage (IDa) was qualitatively observed are represented in black. Horizontal blue and red broken lines represent the average quantitative damage for IDa corresponding to rocks and cubes, respectively.

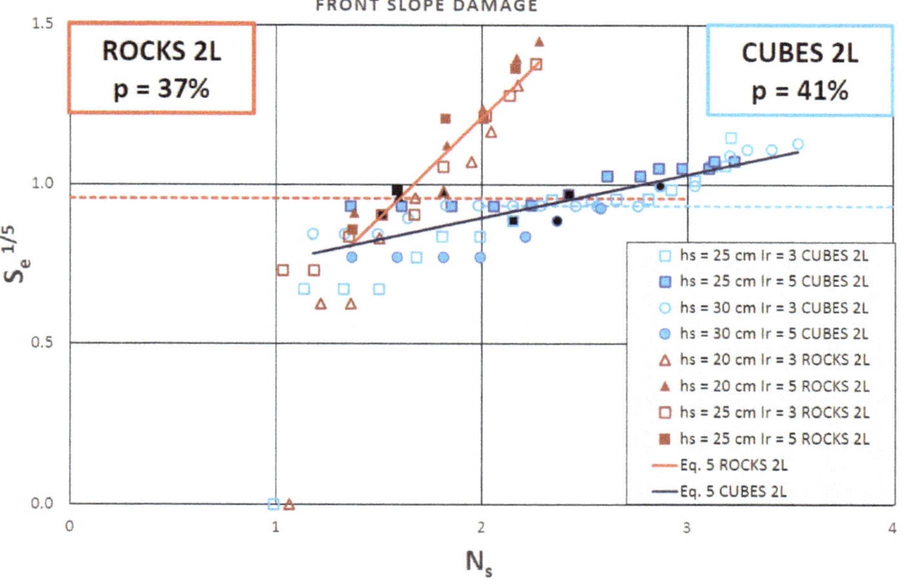

Figure 6. Measured armor damage, S_e, to the front slope as a function of the stability number (rock and cube armors).

Figure 6 shows a higher hydraulic stability of cube armors compared to rock armors on the frontal slope. Using only tests with a minimum of equivalent dimensionless damage ($S_e > 0.25$), a lineal model was developed to estimate armor damage following the expression:

$$S_e^{1/5} = k_1 N_s + k_2 \quad (5)$$

where k_1 and k_2 are fitting parameters. Besides the rMSE, the correlation coefficient, r, was calculated to measure the goodness of fit.

$$r = \sum_{i=1}^{N} \frac{(o_i - \bar{o})(e_i - \bar{e})}{\sqrt{\sum_{i=1}^{N} (o_i - \bar{o})^2 \sum_{i=1}^{N} (e_i - \bar{e})^2}} \quad (6)$$

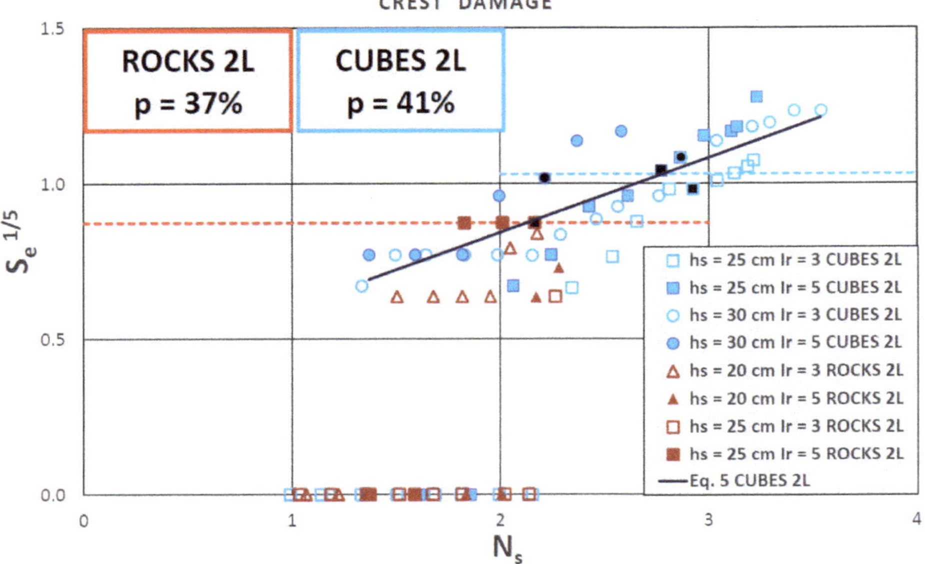

Figure 7. Measured armor damage, S_e, to the crest as a function of the stability number (rock and cube armors).

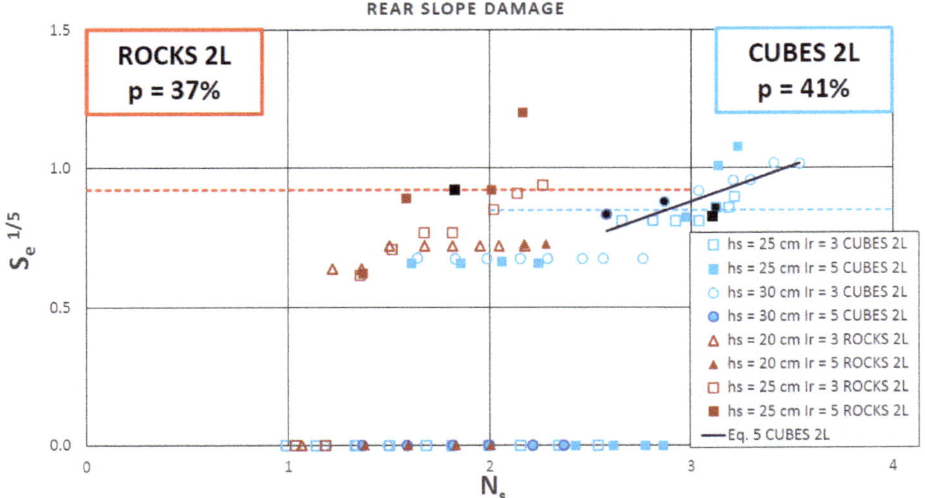

Figure 8. Measured armor damage, S_e, in the rear slope as a function of the stability number (rock and cube armors).

Table 3 summarizes the results and shows the calibrated values k_1 and k_2 for Equation (5) and the rMSE and r values between the measured and estimated armor damage.

Table 3. Calibrated values of k_1 and k_2 for Equation (5) and rMSE and r values.

Armor Layer	Sector	k_1 (Equation (5))	k_2 (Equation (5))	rMSE	r
Rocks 2L	Front slope	0.633	−0.056	0.095	0.949
Cubes 2L	Front slope	0.137	0.621	0.253	0.861
Cubes 2L	Crest	0.240	0.362	0.290	0.838
Cubes 2L	Rear slope	0.255	0.113	0.461	0.716

Figures 7 and 8 also show a higher hydraulic stability of cubes than rocks for the crest and rear slope armors. Nevertheless, in these cases, the values of the rock armor damage obtained were lower due to the high damage that occurred in the front slope armor. For this reason, Equation (5) was only obtained for cube armors in the crest and rear slope.

Equation (5) estimates the equivalent dimensionless armor damage, S_e, for the rocks and cubes armors in the corresponding breakwater sector, within the ranges summarized in Table 4.

Table 4. Ranges of validation for Equation (5).

Armor Layer	Sector	s_{op}	N_s	$h_s/\Delta D_{n50}$
Rocks 2L	Front slope	0.018–0.049	1.36–2.28	3.73–4.66
Cubes 2L	Front slope	0.018–0.049	1.18–3.54	4.96–5.95
Cubes 2L	Crest	0.018–0.049	1.37–3.54	4.96–5.95
Cubes 2L	Rear slope	0.018–0.049	2.58–3.54	4.96–5.95

4.2. Damage in Single-Layer Armors

The same methodology was used to represent the results of damage to single-layer Cubipod® armors. Figures 9–11 show the linearized equivalent dimensionless armor damage measured on the front slope, crest, and rear slope of the breakwater. As observed, the hydraulic stability of single-layer Cubipod® armors was higher than for the double-layer cube or rock armors on the front slope, crest, and rear slope; IDa was not observed in any of the tests.

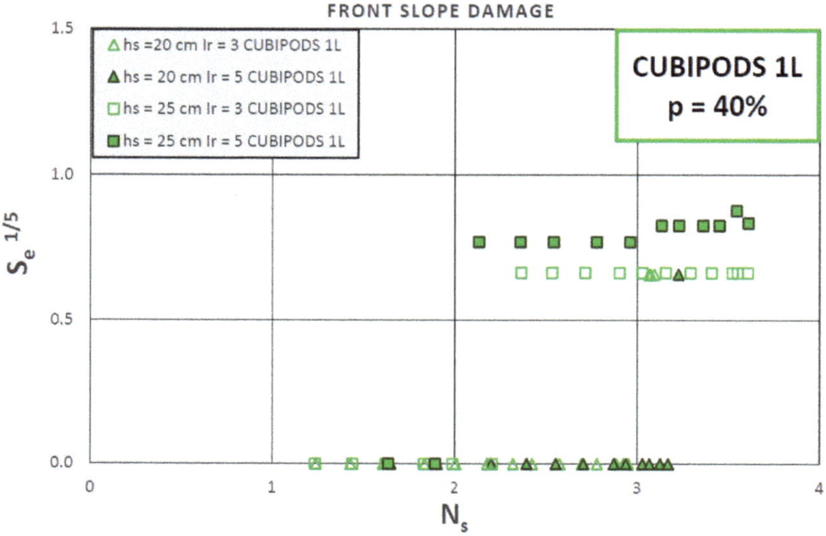

Figure 9. Measured armor damage, S_e, to the front slope as a function of the stability number (Cubipod® armor).

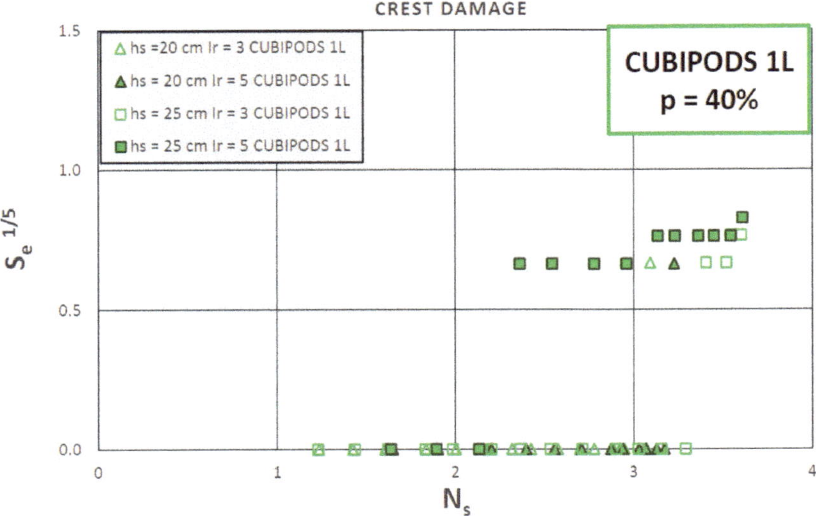

Figure 10. Measured armor damage, S_e, to the crest as a function of the stability number (Cubipod® armor).

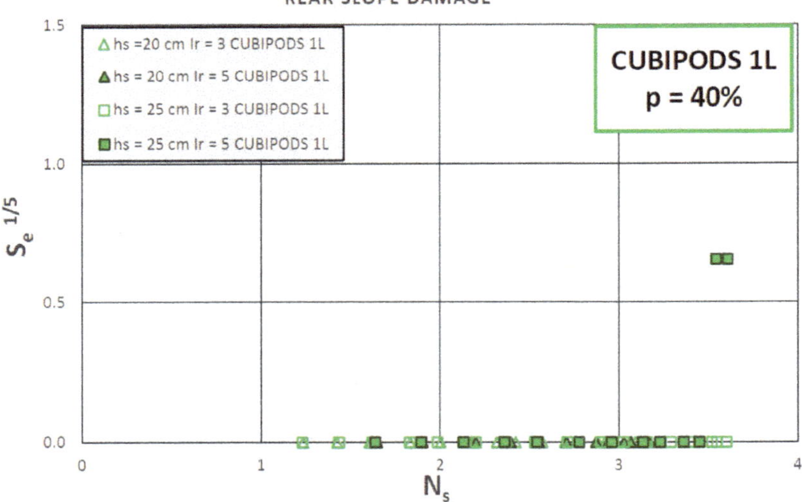

Figure 11. Measured armor damage, S_e, to the rear slope as a function of the stability number (Cubipod® armor).

4.3. Armor Damage and Overtopping Events

In this paper, a wide range of dimensionless crest freeboards was tested ($0.3 < R_c/H_s < 2.6$), and the experiments indicated that the armor damage measured in each section of the structure was clearly related to the mean overtopping rate. Figure 12 shows the dimensionless overtopping rate, $Q = \frac{q}{\sqrt{g \cdot H_{m0}^3}}$, as a function of the dimensionless crest freeboard. The test, where the IDa in the crest was qualitatively observed, is represented in black, while the orange test represents the IDa in the rear slope.

It was observed that the minimum value of Q from which the IDa was detected in the crest or rear slope for the double-layer armors was $Q = 10^{-3}$. This means that tests with $Q > 10^{-3}$ caused significant damage to the crest and rear slope, as was the case of the cube and rock armors with

h_s[cm] = 25 and Ir = 5. However, for Cubipod® armored breakwaters, although the threshold $Q > 10^{-3}$ was exceeded, no significant damage (IDa) was observed in any part of the armor layer. Figure 13 shows the relation between the dimensionless armor damage and the dimensionless overtopping discharged. Test series with a low overtopping rate led to higher armor damage to the front slope, as almost all the energy must be dissipated by this part of the structure. In contrast, when the overtopping rate was high, the crest and the rear slope armors have to dissipate a significant part of the wave energy, so that the armor damage was higher in the crest and rear slope when $Q > 10^{-3}$.

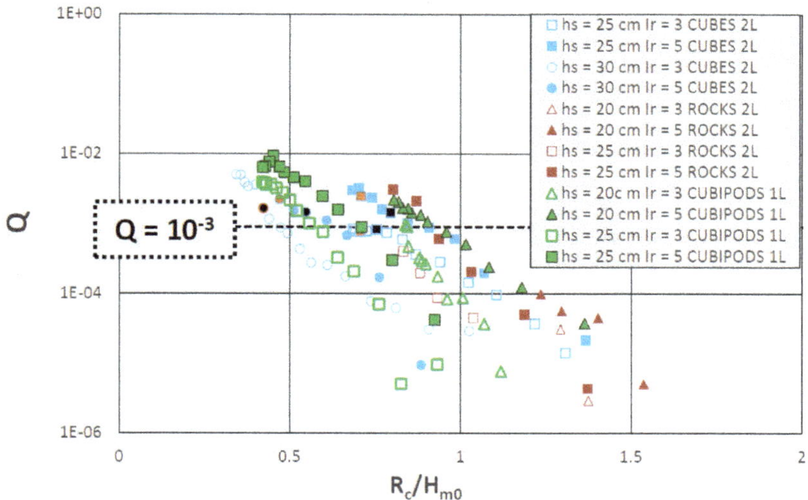

Figure 12. Measured overtopping discharge in terms of normalized freeboard (cube, rock, and Cubipod® armors).

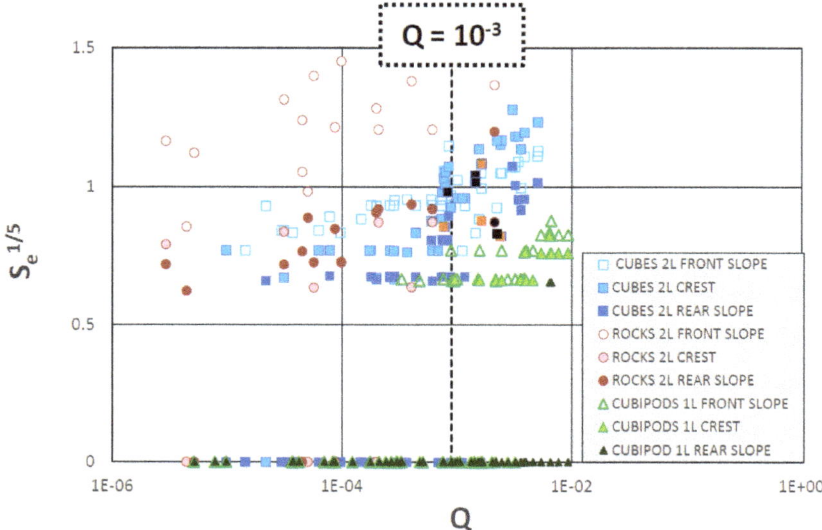

Figure 13. Measured armor damage, S_e, in the front slope, crest and rear slope as function of the Q (cube, rock, and Cubipod® armors).

5. Conclusions

Medium and low-crested mound breakwaters are frequently overtopped by waves, which may cause damage not only to the front slope, but also to the crest and rear slope. To design mound breakwaters subjected to intense overtopping conditions, it is necessary to design front, crest, and rear armors that consider the overtopping rates to withstand throughout the structure's lifetime. Medium and low-crested mound breakwaters with occasional large overtopping events under breaking wave conditions have not been well studied in the literature.

A wide range of dimensionless crest freeboards ($0.3 < R_c/H_s < 2.6$) were tested at the LPC-UPV with single-layer Cubipod® and double-layer rock and cube armors. The Virtual Net method, which takes into account the HeP failure mode, was used to measure the armor damage to the front slope, crest, and rear slope. Results showed a higher hydraulic stability for the double-layer cube armors when compared with the double-layer rock armors. The Equation (5) parameters were calibrated to estimate the equivalent dimensionless damage, S_e, for double-layer rock and cube frontal slope armors and double-layer cube crest and rear slope armors within the ranges of the study. When the overtopping rate exceeded a threshold value ($Q > 10^{-3}$) on the rock and cube armored breakwaters, the damage to the crest and rear slope was higher than that to the front slope, and Initiation of Damage (IDa) was observed in the crest and rear slope.

Single-layer Cubipod® armors showed a higher hydraulic stability for the front slope, crest, and rear slope; the damage was below the Initiation of Damage level (IDa) in all tests. Overtopping rates exceeded the threshold limit for the rock and cube armors ($Q > 10^{-3}$), but did not cause significant damage to the Cubipod® armor. The hydraulic stability of the single-layer Cubipod® armor was higher than that of the double-layer randomly-placed cube armor. These conclusions are valid for overtopped ($m = 1/50$ and $\cot\alpha = 1.5$) structures in the front slope, crest, and rear slope.

Author Contributions: G.A. wrote the original draft and was responsible for the conceptualization, experimental methodology, and analysis. M.E.G.-M. and J.R.M. supervise the investigation, review and approved the manuscript.

Funding: This research was funded by *Ministerio de Economía y Competitividad* and the *FondoEuropeo de Desarrollo Regional* (*FEDER*) under grant BIA2015-70436-R.

Acknowledgments: The authors acknowledge the financial support from the *Ministerio de Economía y Competitividad* and the *FondoEuropeo de Desarrollo Regional* (*FEDER*) under grant BIA2015-70436-R. The authors thank Debra Westall for revising the manuscript. Moreover, the authors acknowledge financial support from the Conselleria d'Educació, Investigació, Cultura i Esport (Generalitat Valenciana) under grant GV/2017/031.

Conflicts of Interest: The authors declare no conflict of interest.

References

1. Pullen, T.; Allsop, N.W.H.; Bruce, T.; Kortenhaus, A.; Schüttrumpf, H.; van der Meer, J.W. EurOtop. Wave Overtopping of Sea Defences and Related Structures: Assessment Manual. Available online: www.overtopping-manual.com (accessed on 20 October 2018).
2. Kramer, M.; Burcharth, H.F. Stability of low-crested breakwaters in shallow water short crested waves. *Proc. Coast. Struct.* **2003**, 137–149. [CrossRef]
3. Burcharth, H.F.; Kramer, M.; Lamberti, A.; Zanuttigh, B. Structural stability of detached low crested breakwaters. *Coast. Eng.* **2006**, *53*, 381–394. [CrossRef]
4. CIRIA/CUR/CETMEF. *The Rock Manual. The Use of Rock in Hydraulic Engineering*, 2nd ed.; CIRIA: London, UK, 2007; 1267p.
5. Hudson, R.Y. Laboratory investigations on rubble mound breakwaters. *J. Waterw. Harb. Div. ASCE* **1959**, *85*, 93–121.
6. Van der Meer, J.W. Rock Slopes and Gravel Beaches under Wave Attack. Ph.D Thesis, Delft Technical University, Delft, The Netherlands, 1988.
7. Herrera, M.P.; Gómez-Martín, M.E.; Medina, J.R. Hydraulic stability of rock armors in breaking conditions. *Coast. Eng.* **2017**, *127*, 55–67. [CrossRef]

8. Van der Meer, J.W.; Daemen, I.F.R. Stability and wave transmission at low-crested rubble mound structures. *J. Waterw. Port Coast. Ocean Eng.* **1994**, *120*, 1–19. [CrossRef]
9. Vidal, C.; Losada, M.A.; Medina, R.; Mansard, E.P.D.; Gómez-Pina, G. A universal analysis for the stability of both low-crested and submerged breakwaters. In Proceedings of the 23rd Conference on Coastal Engineering, ASCE, Venice, Italy, 4–9 October 1992; pp. 1679–1692.
10. Burger, G. Stability of low-crested breakwaters. Final Proceedings 1995. EU research project Rubble mound breakwater failure modes, MAST 2 contract MAS2-CT92-0042. Delft Hydraulic Report H1878/H2415. *Coast. Eng.* **2006**, *53*, 381–394.
11. Vidal, C.; Medina, R.; Martín, F.L. A methodology to assess the armor stability of low-crested and submerged breakwaters. *Coast. Struct.* **1999**, *2*, 721–725.
12. Vidal, C.; López, F.; Losada, I. Stability of low crested and submerged rubble mound breakwaters. *Proc. Coast. Struct.* **2007**, *2*, 939–950.
13. Gómez-Martín, M.E.; Medina, J.R. Heterogeneous packing and hydraulic stability of cube and Cubipod armor units. *J. Waterw. Port Coast. Ocean Eng.* **2014**, *140*, 100–108. [CrossRef]
14. Losada, M.A.; Desiré, J.M.; Alejo, L.M. Stability of blocks as breakwater armor units. *J. Struct. Eng.* **1986**, *112*, 2392–2401. [CrossRef]
15. Vidal, C.; Losada, M.A.; Medina, R. Stability of mound breakwaters' head and trunk. *J. Waterw. Port Coast. Ocean Eng.* **1991**, *117*, 570–587. [CrossRef]
16. Gómez-Martín, M.E. Análisis de la Evolución de Averías en el Manto Principal de Diques en Talud Formado por Escolleras Cubos y Cubípodos. Ph.D. Thesis, Universitat Politècnica de València, Valencia, Spain, 2015.
17. Mansard, E.P.D.; Funke, E.R. The measurement of incident and reflected spectra using a least squares method. In Proceedings of the 17th International Conference on Coastal Engineering, ASCE, Sydney, Australia, 23–28 March 1980; pp. 154–172.
18. Figueres, M.; Medina, J.R. Estimation of incident and reflected waves using a fully non-linear wave model. In Proceedings of the 29th International Conference on Coastal Engineering, Lisbon, Portugal, 19–24 September 2004; pp. 594–603.
19. Verhagen, H.J.; Van Vledder, G.; Eslami Arab, S. A practical method for design of coastal structures in shallow water. In Proceedings of the 31st International Conference on Coastal Engineering, Hamburg, Germany, 31 August–5 September 2008; Volume 4, pp. 2912–2922.
20. Medina, J.R.; Hudspeth, R.T.; Fassardi, C. Breakwater Armor Damage due to wave groups. *J. Waterw. Port Coast. Ocean Eng.* **1994**, *120*, 179–198. [CrossRef]

Journal of
Marine Science and Engineering

Article

Experimental and Numerical Analysis of the Hydrodynamics around a Vertical Cylinder in Waves

Sara Corvaro [1,*], Andrea Crivellini [2], Francesco Marini [1], Andrea Cimarelli [3], Loris Capitanelli [2] and Alessandro Mancinelli [1]

1 Dipartimento di Ingegneria Civile, Edile e dell'Architettura, Università Politecnica delle Marche, Via Brecce Bianche 12, 60131 Ancona, Italy; f.marini@pm.univpm.it (F.M.); a.mancinelli@univpm.it (A.M.)
2 Dipartimento di Ingegneria Industriale e Scienze Matematiche, Università Politecnica delle Marche, Via Brecce Bianche 12, 60131 Ancona, Italy ; a.crivellini@univpm.it (A.C.); loris.capitanelli@gmail.com (L.C.)
3 Dipartimento di Ingegneria "Enzo Ferrari", Università di Modena e Reggio Emilia, Via Pietro Vivarelli 10, 41125 Modena, Italy; andrea.cimarelli@unimore.it
* Correspondence: s.corvaro@univpm.it; Tel.: +39-071-220-4521

Received: 18 November 2019; Accepted: 6 December 2019; Published: 10 December 2019

Abstract: The present study provides an extensive analysis on the hydrodynamics induced by a vertical slender pile under wave action. The authors carried out the study both experimentally and numerically, thus enabling a deep understanding of the flow physics. The experiments took place at a wave flume of the Università Politecnica delle Marche. Two different experimental campaigns were performed: In the former one, a mobile bed model was realized with the aims to study both the scour process and the hydrodynamics around the cylinder; in the latter one, the seabed was rigid in order to make undisturbed optical measurements, providing a deeper analysis of the hydrodynamics. The numerical investigation was made by performing a direct numerical simulation. A second order numerical discretization, both in time and in space, was used to solve the Navier–Stokes equations while a volume of fluid (VOF) approach was adopted for tracking the free surface. The comparison between experimental and numerical results is provided in terms of velocity, pressure distributions around the cylinder, and total force on it. The analysis of the pressure gradient was used to evaluate the generation and evolution of vortices around the cylinder. Finally, the relation between scour and bed shear stresses due to the structure of the vortex pattern around the pile was assessed. It is worth noting that the physical understanding of this last analysis was enabled by the combined use of experimental data on scour and numerical data on the flow pattern.

Keywords: vertical cylinder; DNS model; pressure gradient; wave force; scour and shear stress

1. Introduction

The knowledge of the hydro- and morphodynamics induced by a slender vertical pile is very important for the design of marine structures, such as offshore platforms or wind farms. In particular, the evaluation of the scour at the base of the cylinder and the force acting on it is fundamental for a correct design of a marine structure. The present topics have been experimentally and numerically studied in the literature. The pioneering work of Morison [1] started an extensive series of studies on the forces over vertical piles under waves. Suddenly, Keulegan and Carpenter [2] found that those processes are highly influenced by a nondimensional parameter (named the Keulegan-Carpenter number *KC*) that depends on, among the others, the particle velocity and on the pile diameter. Scour process in river hydraulics has been extensively studied because of its importance on bridge failure; however, this process in coastal and offshore engineering has not received the same attention. The work of Herbich [3] was the first important contribution to the knowledge of the hydrodynamic processes in marine environment. From the 1990s, several studies on the hydrodynamics have been realized [4–7], mostly focusing on the analysis of the scour process and the formation of vortices. Sumer et al. [4,5] observed that the vortex flow regimes for a free cylinder subjected to an oscillatory flow and the lee-wake vortex flow are governed primarily by *KC*. The bed shear stress under the horseshoe vortex and the bed shear stress in the lee wake are also influenced by *KC*. Olsen and Kjellesvig [6] simulated the scour process in clear water regime by employing the k-ε turbulence model. Roulund et al. [7] studied numerically the live-bed scour for cohesionless sediment by making use of Reynolds-averaged Navier–Stokes (RANS) equations with a k-ω closure model. Subsequently, Baykal et al. [8] incorporated the effects of lee-wake vortices, as well as suspended sediment transport. In the majority of the works the forcing was either current or combined wave and current; however, in a number of instances, the tidal range was significantly smaller than the wave height at breaking [9]. So waves became significant with respect to current and, hence, the flow forcing was only due to waves. In the present paper, the focus was on the analysis of the hydrodynamics around a vertical slender cylinder subjected to wave forcing only, by making use of both the experimental and numerical models. In particular, the direct numerical simulation (DNS) approach was used. The numerical results were compared with the experimental data. The combined analysis focused on the pressure distributions around the pile and on the evaluation of the total force on it at different wave phases. The total force acting on the pile was also compared with the widely used Morison formula [1] and a discussion on the applicability of such formula for nonlinear waves has been provided. Moreover, the wall shear stress induced by the waves around the cylinder and the analysis of the streamlines connected with the vortex pattern evolution around the cylinder allowed us to explain the movement of sand particles and the seabed morphology. In conclusion, in the present paper the authors exploited the different information provided by the numerical and experimental approaches in order to shed light on relevant physical processes which would be otherwise difficult to understand by using solely experimental or numerical data.

2. Experimental Setup

The experiments were carried out at the wave flume of the Hydraulics and Maritime Construction Laboratory of the Università Politecnica delle Marche (Ancona, Italy). The channel is 50-m long, 1-m wide, and 1.3-m high. The flow motion was forced by regular waves (Stokes I order theory) generated by a piston-type wavemaker. An adsorbing mildly sloping beach (slope 1:20) made of coarse gravel was used to reduce the wave reflection and to guarantee an undisturbed flow in the measuring area. The reflection was studied and considered negligible (<5%). The side walls of the flume were glassed for the central 36 m and enabled one to video record and carry out optical measurements.

Two different experimental campaigns were performed. The former experimental campaign aimed to investigate the bottom morphology induced by the presence of a vertical slender pile under regular and random wave action in live-bed regime. Several wave tests were performed as a combination of different wave heights H, wave periods T, and water depths h. The detailed description of the mobile-bed experiments and the analysis of the results were reported in Corvaro et al. [10]. Recently, a second experimental campaign was realized for a deeper analysis of the hydrodynamics around the pile induced by regular waves. Even for this experimental campaign, several wave tests, characterized by different combinations of the wave heights, periods, and water depths, were studied.

Since the purpose of the present analysis was that of characterizing the hydrodynamics around a vertical slender cylinder, we here analyzed one specific regular wave that evolved over a rigid bed with a water depth h = 0.35 m, a wave height H of 0.10 m, a wave period T of 1.77 s, and a wave length L of 3.0 m. The Keulegan–Carpenter parameter, defined as $KC = U_m T/D$ (U_m is the maximum velocity at the bottom just outside of the boundary layer and D is the diameter of the pile), is $KC \approx 6$ and the Reynolds number, defined as $Re = U_m A/\nu$ (A is the semi-amplitude of the horizontal motion of the water particles just outside of the bottom boundary layer, ν is the kinematic viscosity), is $Re \approx 3 \times 10^4$. Due to the significant computational cost, we focused on one specific forcing condition characterized by vorticity generation and scour formation, as found in the first experimental campaign [10] for waves with similar values of a Keulegan–Carpenter parameter. Being the vortex flow regimes governed primarily by KC for a cylinder subjected to an oscillatory flow [5], the wave here analyzed was representative of waves within the vortex shedding flow regime. The scour pattern map obtained in the previous experiments [10] was here used in order to relate it with the vortical structures.

The water level evolutions, and thus wave heights, at different locations of the flume were measured and monitored by five electro-sensitive elevation gauges (see Figure 1). They were located as follows: S1 just downstream of the wave paddle at a distance of 8.43 m, S2 at 12.43 m, S3 at 16.43 m, S4 at 16.83 m, and S5 at 17.15 m, in correspondence to the cylinder.

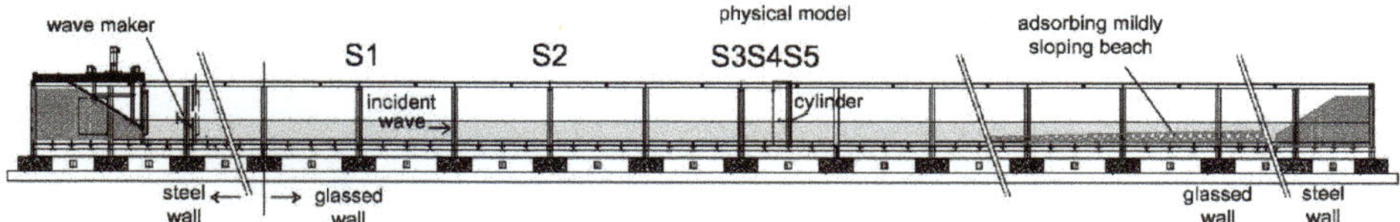

Figure 1. Wave flume and location of the elevation gauges and of the cylinder (rigid-bed model).

The sketch of the physical model (rigid bed) is shown in Figure 2a. A Cartesian coordinate system x, y, z was used. The origin was placed in the centre of the pile at the bottom of the flume, the x-axis was in the direction of the flow with the wave propagating from negative to positive values, the y-axis was horizontal and transversal to the incoming wave direction (see Figure 2b for the plan view of the coordinate system) and the vertical z-axis, coinciding with the pile's axis, pointed upward.

In the mobile-bed experiments a steel pile with a dimeter $D = 0.10$ m was fitted in the sandy seabed. In order to measure the pressure distribution around the cylinder at different water depths, in the rigid-bed experiments a PVC pile (PolyVinyl Chloride) was used because the housing of the pressure sensors along the vertical was easier by using a plastic material with respect to the steel. The pile was connected at the bottom by means of a hinge that allowed the complete rotation of the pile around its axis. The cylinder, being moment free around the hinge, was ensured in the equilibrium condition at the top of the pile by means of a biaxial load cell (P500.BIAX-S/400N, Deltatech, Sogliano al Rubicone, FC, Italy) for the total force measurement. Four pressure sensors Keller 23Y were embedded in the pile along the same vertical ($z = 0.01$ m, $z = 0.08$ m, $z = 0.16$ m, and $z = 0.24$ m). In order to obtain a good estimate of the pressure distribution around the cylinder, the pile was rotated by incremental angles $\Delta\varphi$ around its vertical axis (see Figure 2b) allowing it to collect the data with a good discretization of the pile surface. Hence, each wave test was repeated more times from $\varphi = 0°$ to $\varphi = 180°$ with an angle $\Delta\varphi = 45°$. The locations where the pressure was monitored in the numerical model are at $\varphi = 0°$, $\varphi = 90°$, $\varphi = 180°$, and $\varphi = 270°$ and vertical depths $z = 0.00$–0.32 m (every 0.08 m). Furthermore, the local water particle velocity was measured by means of an acoustic Doppler velocimetry (ADV) placed in front of the pile ($\varphi = 0°$), in the lateral side ($\varphi = 90°$) and behind it ($\varphi = 180°$) at different depths ($z = 0.01$ m, $z = 0.08$ m, $z = 0.16$ m, $z = 0.24$ m). In the numerical model, the velocity was monitored at $\varphi = 0°$, $\varphi = 90°$, $\varphi = 180°$, and $\varphi = 270°$ and at vertical depths $z = 0.00$–0.32 m (every 0.08 m). The elevation gauge S5 was placed in correspondence with the pile in order to measure the water level in its section and to synchronize the pressure recordings over different tests (Figure 2).

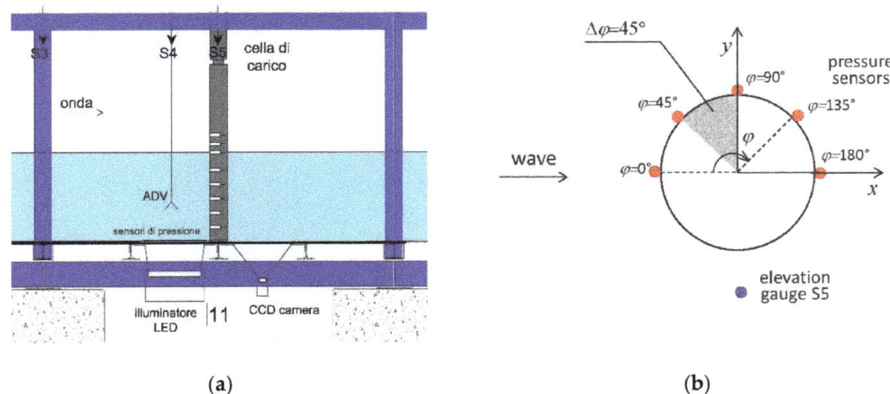

Figure 2. Sketch of the rigid-bed model setup with the position of the elevation gauges S3, S4, S5, and of the acoustic Doppler velocimetry (ADV) at $\varphi = 0°$ (**a**) and of the planar coordinate system, with the position of pressure sensors (**b**).

3. Numerical Setup

The approach used in this study to solve the Navier–Stokes equations consisted of in a direct numerical simulation (DNS) modelling conducted by using the commercial software ANSYS Fluent. Space and time were discretized and the resulting equations were numerically solved without any turbulence model. Increasing the Reynolds number Re, the turbulent scales became smaller and smaller both in space and time; therefore, fine grids and small time steps were needed. The DNS solved all the scales of the motion, from the integral scale to the smallest dissipative one (Kolmogorov's scale). Therefore, the accuracy obtained with this model was the highest possible (compared to large eddy simulations or to RANS equations closed by a turbulence model) with the disadvantage of a huge computational cost, even at low Reynolds numbers like those considered in this work (as above reported, the wave was characterized by a Reynold number Re $\approx 3 \times 10^4$). Note that here the Re number was defined according to the maximum velocity U_m (as usually done for flow induced by waves) and not according to the mean velocity value, which was almost null in an oscillatory flow. From the practical point of view, we found that the computational resources required by DNS are large but significantly lower than those usually adopted for solving a different flow problem (e.g., current) characterized by a similar Re value which was obtained by using the mean velocity. The employed discretization scheme adopted a finite volume method in space and an implicit algorithm in time. More in detail, it used a second order backward time integration scheme, while a second order reconstruction of the numerical fluxes at faces yields a second order accuracy also in space. The volume of fluid (VOF) method was used for locating and tracking the free surface. It consisted of the assumption that the two-phase flow was a mixed fluid whose density and viscosity were weighted functions of the volume fraction which evaluated the fraction of the volume occupied by the two phases and was computed by an advection equation.

The solved wave flume was 1-m wide, 6–7.5 m long, and 0.8-m height. The cylinder had a diameter of 0.1 m and it was placed 3 m downstream the inlet boundary condition. A no-slip boundary condition was applied in the bottom and lateral boundaries of the domain, while at the top and outer boundaries the hydrostatic pressure was imposed. Close to the outlet of the bigger, 7.5-m-long domain, a 1.5-m-long absorption zone, called "numerical beach", was employed. The approach used the last portion of the numerical domain to damp the downstream propagation of the waves by adding a sink term in the vertical component of the momentum equations. This term depended on the vertical velocity and its strength rose with the value of the distance from both the free surface and the interface between the true domain of interest and the absorption zone [11]. Finally, at the inlet, a time-dependent Dirichlet condition was imposed which enforced the creation of waves of different types, according to the Stokes theory, and intensity. A regular wave was forced with a Stokes I order theory with the same characteristics of the one reproduced in the experimental test (H = 0.10 m, T = 1.77 s, h = 0.35 m, and L = 3.0 m). The numerical simulation of the regular wave was performed with a fixed dt = 0.01 s, a coupled solver for the discretized equations and with different grid densities. In the zone surrounding the pile (located at 3 m from the inlet), at the bottom and close to the free surface, the grid was clustered to adequately resolve the boundary layer and the position of the interface between the two fluids, see Figure 3.

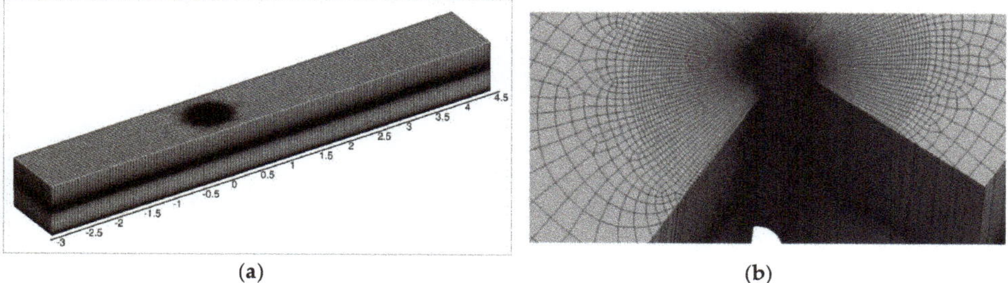

(a) (b)

Figure 3. Numerical model domain (**a**) and detail of the mesh in the surroundings of the pile (**b**).

3.1. Sensitivity Analysis

The sensitivity of the numerical solution with respect to the mesh resolution was analyzed by means of 6 different grids, two of them 6-m-long and the remaining 7.5-m-long. The characteristics of the grids are summarized in Table 1. To give an idea of the computational costs, the simulation with the largest number of volumes required more than a month by using a small Linux cluster consisting of 32 Intel Xeon E5 2620 v4 cores running at 2.1 GHz.

Table 1. Characteristics of the different meshes.

Name	Num. of Volumes	Length (m)	Numerical Beach
0	432'800	6.0	No
0a	492'300	7.5	Yes
1	750'660	6.0	No
1a	784'500	7.5	Yes
2a	880'900	7.5	Yes
3a	1'965'600	7.5	Yes

In order to analyze the effectiveness of the adoption of the numerical beach, in Figure 4 we report the behavior of the force over the pile and of the wet surface around it for simulations 0 and 0a (continuous lines) and 1 and 1a (dashed lines) where the suffix "*a*" represents the use of the numerical beach. The use of the absorption zone drastically changed the results, thus highlighting the effectiveness of the approach. From one side, a significant reduction and regularization of the force acting on the pile was observed. From the other, the wet surface did not show a superposition of different types of waves by using the absorption zone. Indeed, these smaller and irregular waves we a result of wave reflection at the end of the finite numerical domain and an increase in terms of resolution did not lead to a more stable result when no absorption zone was used.

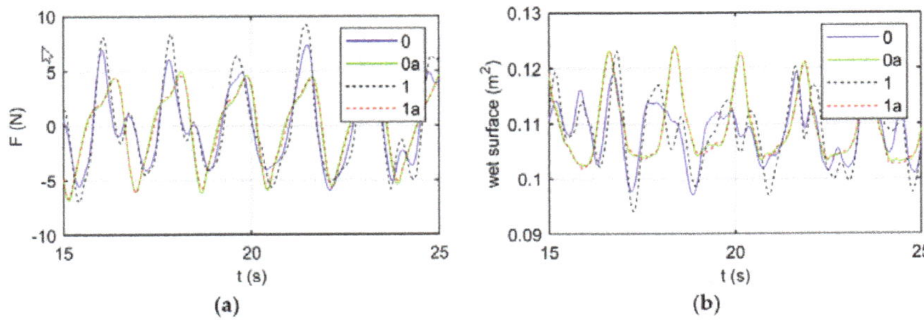

Figure 4. Force (**a**) and water level (**b**) comparisons in correspondence to the pile to evaluate the effectiveness of the numerical beach. Simulations 0 and 0a (continuous lines), 1 and 1a (dashed lines).

To perform an analysis of sensibility of the grid resolution, the force over the cylinder in the direction of wave propagation was analyzed. As shown in Figure 5a, the solved oscillation of the streamwise force over the pile changed by increasing the mesh resolution from the case 1a to 3a. Note that even if the number of elements of the grid 2a was only slightly larger than those of the grid 1a, the first was significantly refined in the boundary layer around the pile and at the still water level. On the other hand, no relevant differences were observed by comparing cases 2a (green continuous lines) with 3a (black dashed lines), thus highlighting that a numerical convergence of the results as a function of the grid resolution was finally achieved. The same behavior was observed also for other observables such as the wet surface of the pile reported in Figure 5b.

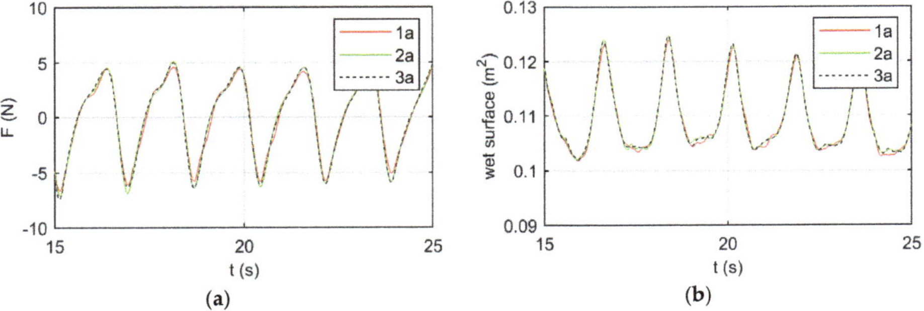

Figure 5. Comparison of the force over the pile (**a**) and the wet surface around it (**b**) among different mesh (1a–3a) with numerical beach.

The mesh resolutions used for the following analyses was the finest one (3a), which consisted of about 2 M elements corresponding to a near-wall resolution $\Delta h/D = 5.0 \cdot \times 10^{-3}$ around the pile and $\Delta h/D = 1.2\times 10^{-2}$ above the bottom wall, where Δh was the normal to the wall grid size.

3.2. Spectral Analysis

To highlight multiscale features of the flow, we considered the frequency spectra of kinetic energy defined as:

$$k = \frac{1}{2}(\hat{u}\hat{u}^* + \hat{v}\hat{v}^* + \hat{w}\hat{w}^*) \tag{1}$$

where $\hat{u}, \hat{v}, \hat{w}$ are the Fourier transform of the velocity components along x, y, and z axes, respectively, and the superscript * denotes their conjugate.

Hence, the kinetic energy K can be defined as $K = \int k df$, where f is the frequency. Figure 6 shows the spectra of the kinetic energy K evaluated in the near-bed region (z = 0.003 m) and around the cylinder at $\varphi = 0°$ (panel a) and $\varphi = 90°$ (panel b). A clear peak was observed corresponding to the frequency of the oscillating flow: $f_p = 1/T = 0.56$ Hz. The frequency spectrum showed a decrease of the spectral energy content typical of turbulent fluctuations up to a frequency (f_u), where viscosity canceled out all the fluctuating motions. The developed range of scales was very tiny, being characterized by a few decades (the ratio f_u/f_p was about equal to 30), suggesting that the turbulent flow emerging from the interaction of the oscillating flow with the vertical cylinder was very weak. Indeed, no evident inertial range, typical of fully developed turbulence where the spectrum was supposed to follow the $f^{-5/3}$ power law, was observed.

The above analysis of the fluctuating field had a strong consequence on the modelling approach. The observed weak behavior of turbulence challenged turbulence models, which are essentially conceived for fully developed turbulence. Furthermore, the simultaneous presence of laminar flow regions away from the pile and turbulence close to it, was again a challenge for turbulence closure, which should be able to recognize laminar oscillations induced by the wave motion where no model was needed from turbulence fluctuations where the model should be active. In conclusion, taking into account these turbulence modelling issues and by considering that the computational costs for a direct numerical simulation were relatively small due to the small separation of scales f_u/f_p, we argue that the DNS approach is strongly recommended for the solution of the wave action on a pile with no current.

In closing this section, let us further support the sensitivity analysis of the numerical approach reported in the previous section. Indeed, in Figure 6 spectra are shown for both simulations 2a and 3a. The quality of the spectral convergence of results at different spatial resolutions is a clear indication of the accuracy of the numerical solution as already assessed in the previous section.

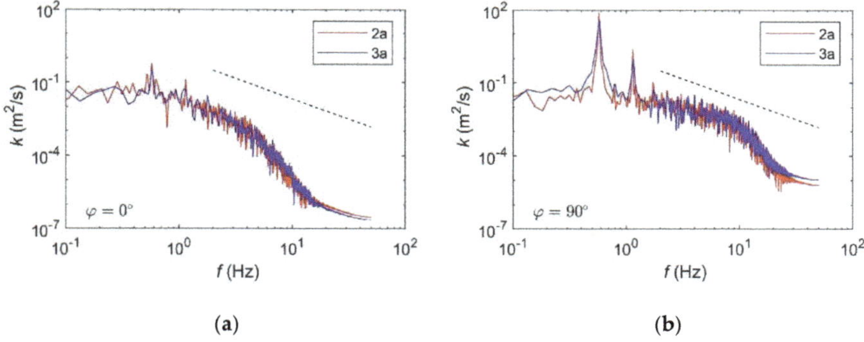

Figure 6. Frequency spectra evaluated in the near-bed (z = 0.003m) at $\varphi = 0°$ (**a**) and $\varphi = 90°$ (**b**). The inertial subrange $f^{-5/3}$ (black dotted line). The red and blue lines report the behavior for mesh 2a and 3a, respectively.

4. Results

In this section the results obtained in the two experimental campaigns and those of the numerical simulation are presented and compared. The validation was performed by the analysis of the main hydrodynamic characteristics, such as the water surface elevation, water particle velocity, pressure distributions and gradients, and total force over the pile. This section is finally closed by the analysis of the vortex pattern and of the related bed-shear stresses which are known to be the main cause of the scour formation. Performing direct experimental measurements of the shear stress was very complicated, hence, the use of a numerical simulation (fully validated) proved to be useful to get more information that can be very difficult to obtain in an experimental way.

The following subsections show the detail of the comparisons. In Sections 4.1 and 4.2 are reported the comparison of the temporal evolution of the water level η and the horizontal velocity u. In Sections 4.3 and 4.4, we make use of the so-called phase average of the analyzed data (pressure and force) in order to obtain values that express correctly the statistics at the wave phases of interest (ωt). The phase average consisted of fixing the phase (ωt) and averaging arithmetically all the "waves" of each realization. The reference wave phases ωt were defined according to the water level evolution η measured at elevation gauge S5, by assuming that $\omega t = 0°$ corresponded to the up-crossing of the water level signal. Finally, in Section 4.5 the results on the vortex pattern, the time-averaged streamlines, and shear stress linked with the equilibrium seabed morphology are reported.

4.1. Water Surface Elevation

The comparison in terms of water surface elevation was realized in correspondence to the three water level gauges placed inside the numerical domain at the same relative distances from the pile of the water level gauges (S3, S4, and S5) of the experimental model. The results are shown in Figure 7. Although the wave was run with the same characteristics at generation both in the experimental and in the numerical models, the wave generation and propagation processes in intermediate water depths led to nonlinear effects such as changes in the wave shape.

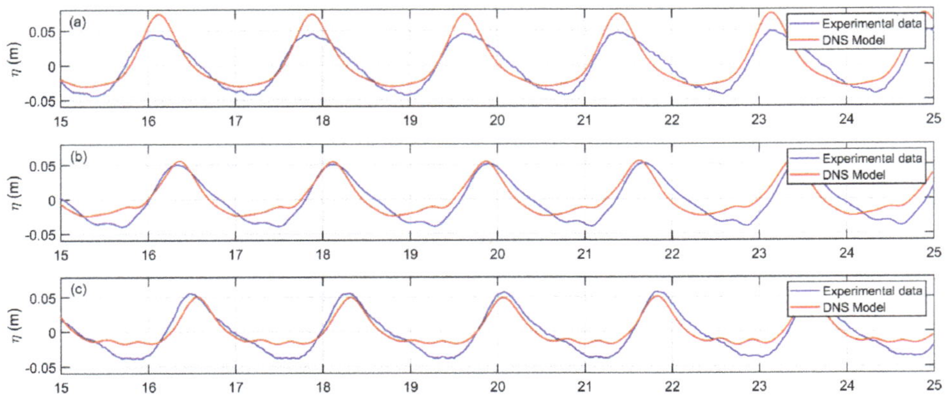

Figure 7. Comparison between numerical and experimental data of water surface elevation in correspondence of water gauges S3 (**a**), S4 (**b**), and S5 (**c**).

In the experimental model, the wave was generated at about 17 m from the pile, while in the numerical model at about 3 m. The nonlinear effects acted differently during the wave propagation over the flume, being more significant at the beginning of the wave propagation, closer to the wave generation. Therefore, such nonlinear effects were more evident in the numerical data (see Figure 7) because of the relative smaller distances between the water level gauges (S3, S4, and S5) and the wave generation boundary. Such behavior explained the small discrepancy between the water surface elevations of the numerical and experimental models observed in Figure 7. Even if a scatter was observed, the water elevation time evolution in correspondence to the pile (elevation gauge S5) showed a good agreement, especially for the evaluation of the maximum values of the wave crest. Even if the wave was rather nonlinear, due to the processes of generation and propagation along the flume in intermediate water depth condition, the nonlinearity of the water level elevation did not reflect to a high nonlinearity of the associated velocities [12,13], which can be well represented by the linear wave theory, as stated also by Le Méhauté [14].

4.2. Particle Velocity

The reliability of the mesh quality of the model has been fully explained in Section 3. The validation of the model was performed by, among the others, the analysis of the horizontal velocity of the water particles in specific points close to the cylinder (at $\varphi = 0°, 90°$, and $180°$, as shown in Figure 2b).

As an example, in Figure 8 the comparison between the experimental (red line) and numerical data (blue line) yielded in front of the pile at different heights ($\varphi = 0°$, z = 0.01–0.24 m) is reported. Even if the nonlinearity of the water surface was not reproduced in the same way (as showed in Figure 7), the DNS model well represented the horizontal velocity magnitude and the shape of the velocity under the crest and trough phases, although the maximum values were slightly over estimated.

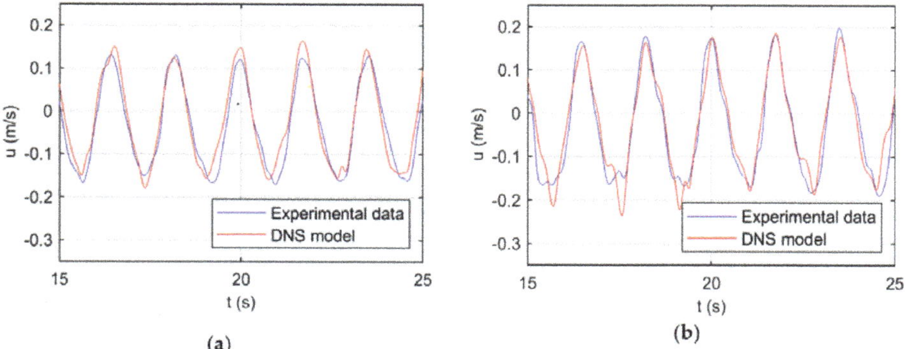

Figure 8. Comparison between numerical and experimental data of water particle velocity in position $\varphi = 0°$ at z = 0.01 m (**a**) and $\varphi = 0°$ at z = 0.24 m (**b**).

4.3. Pressure Distribution

In the present section the comparison of the vertical pressure distributions and the pressure gradients are analyzed. All the pressure P refers to the atmospheric pressure, hence it is null at the free water surface.

The experimental pressure distributions were compared with the numerical data. In Figure 9, the comparison of the vertical distribution of the pressure P evaluated in front of the pile ($\varphi = 0°$) and behind it ($\varphi = 180°$) is reported. By analyzing the pressure distributions for each wave phase, it was observed that the discrepancy was mainly due to the small difference in terms of water level elevation between the numerical and experimental data. The phase-averaged analysis of the water elevation η revealed that the wave crest occurred at wave phases $\omega t = 65°$ and $\omega t = 63°$, respectively, for the experimental and numerical models, while the down-crossing occurred at a larger wave phase for the experimental model with respect to the numerical one. Moreover, as argued in Section 4.1, a lower wave trough and larger nonlinearities were observed in the numerical model. Despite the small difference in the water level elevation, the pressure results showed a quite good agreement. The lower panels of Figure 9 show the vertical distributions of the dynamic pressure P_d evaluated in front of the pile ($\varphi = 0°$) and behind it ($\varphi = 180°$) for the wave phases $\omega t = 45°$ and $\omega t = 270°$. The dynamic pressure P_d is given by the difference between the pressure P and the hydrostatic pressure $P_h = \rho g h$ (ρ is the water density and $\underline{g}$ the gravity acceleration). The analysis of the dynamic pressure distribution around the pile was very important because its asymmetry with respect to the vertical plane (yz-plane) produced a horizontal force on the cylinder acting along the x-axis. During the growth of the positive water surface elevation (e.g., $\omega t = 45°$), a larger dynamic pressure P_d was observed in front of the pile with respect to the area behind it; hence, a positive horizontal force was expected. During the negative water surface elevation (e.g., $\omega t = 270°$) the pressure distributions in front of and behind the pile were quite similar; hence, the horizontal force along the x-axis was expected to be very small (in absolute value). Moreover, the dynamic pressure data at $\varphi = 90°$ and $\varphi = 270°$, here not shown, were similar, confirming that the horizontal force along the transversal direction y was negligible, as found in Section 4.4.

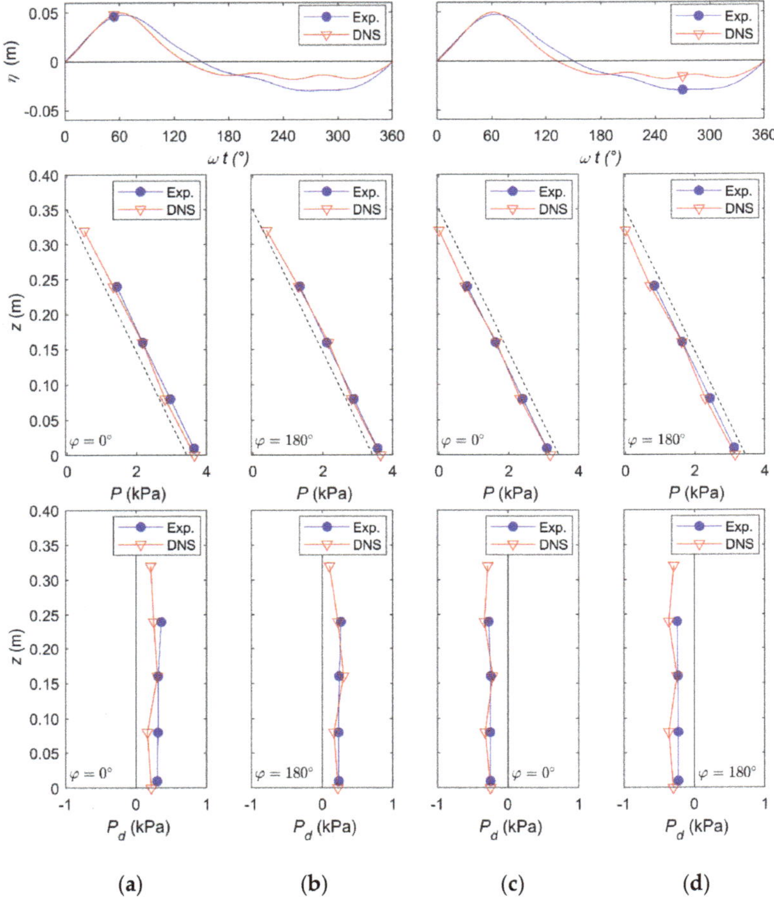

Figure 9. Vertical distributions of the pressure P and of the dynamic pressure P_d for the wave phase $\omega t = 45°$ (**a**,**b**) and $\omega t = 270°$ (**c**,**d**) in front of the pile at $\varphi = 0°$ (**a**,**c**) and behind it at $\varphi = 180°$ (**b**,**d**). Experimental data: Filled blue circle. Numerical data: Red empty triangle. The dashed line is the hydrostatic pressure.

The measurements of the dynamic pressure around the pile also enabled us to confirm the presence, for specific wave phases, of positive pressure variations which induced the separation of the flow within the boundary layer and the subsequent formation of vortices. In Figure 10, the plan view (xy-plane) of the dynamic pressure distribution around the pile is shown at $z = 0.01$ m. The pressure gradients around the pile dP/dx' (dx' is the length of the arc of the cylinder circumference formed by the angle $\Delta\varphi = 45°$) were studied only experimentally, because the pressure sensors were located at $\Delta\varphi = 45°$, while in the numerical model $\Delta\varphi$ was too large ($\Delta\varphi = 90°$).

The dynamic pressure distribution was reported for specific wave phases: Two wave phases ($\omega t = 36°$ and $\omega t = 54°$) representing the rising (Figure 10a,b) and other two wave phases ($\omega t = 72°$ and $\omega t = 90°$), representing the decaying of the positive water surface elevation (Figure 10c,d). Only wave phases corresponding to the positive water surface elevation are shown, in order to emphasize the pressure variations and, hence, support the subsequent analysis of the physical process. The pressure distribution showed an asymmetric shape with respect to the y-axis, which changed with the wave phases. During the rising of the positive water surface elevation, a favorable pressure gradient ($dP/dx' <$

0) was observed around the pile. Such condition was needed for the transport of the horseshoe vortices generated in front of the pile, as found during the mobile-bed experiments [10,15]. From the wave phase $\omega t = 36°$, the pressure gradient dP/dx' switched back to positive values (adverse pressure gradient) within a small area ($\varphi = 135°–180°$). Such area became increasingly larger. After the occurrence of the wave crest, the adverse pressure gradient affected the area $\varphi = 90°–180°$. Such conditions led to the separation of the flow within the boundary layer and, hence, to the formation of lee-wake vortices, which were the main properties responsible for the scour under the wave action. Such findings were in agreement with the results of the vortex pattern reported in Section 4.5, confirming that, for specific wave phases, the coherent structures were formed, developed, and transported during the interaction of the flow with the cylinder.

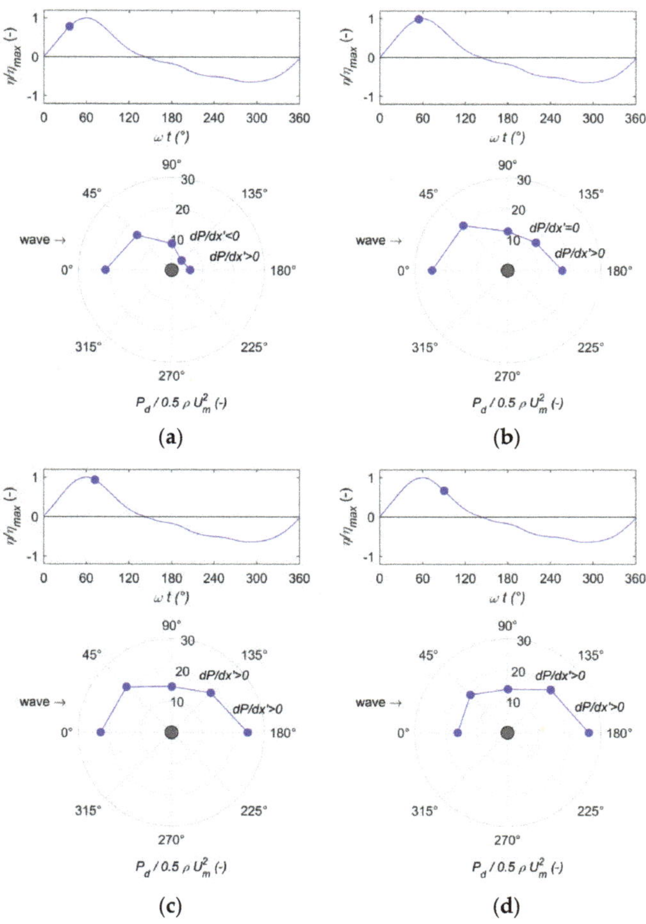

Figure 10. Plan view (*xy-plane*) of the nondimensional dynamic pressure *Pd*, for the wave phase $\omega t = 36°$ (**a**), $\omega t = 54°$ (**b**), $\omega t = 72°$ (**c**), and $\omega t = 90°$ (**d**). Pressure gradients dP/dx' values (positive, null, or negative) in the area $\varphi = 90°–180°$.

4.4. Total Force

In this section the total horizontal force acting over the surface of the pile is analyzed. The forces acting on slender cylinders were often evaluated by the Morison equation [1] for linear small amplitude waves. Here, Equation (2), the total wave force is computed as the sum of two different components, a drag force (F_D), and an inertial component (F_I).

$$F = F_M + F_D = \int_0^h C_M \rho \frac{\pi D^2}{4} \frac{\partial u}{\partial t} dz + \int_0^h C_D \frac{1}{2} \rho D |u| u dz, \tag{2}$$

where C_D and C_M are, respectively, the drag and inertia coefficients.

The drag force, which depends on the square of the velocity, is given by the asymmetric pressure distribution around the pile. Indeed, in the wake zone, the pressure tended to a constant lower value. The cylinder geometry dove the fluid to go around it, modifying all the local particle velocities and accelerations, thus originating the inertia force. This equation proved to be a good approximation for the estimation of the force, although it has some weak points. Indeed, the integration was done considering the linear superposition of the two components which reached the maximum values at different wave phases. The wave was considered linear, and the run-up and the impact force were not considered. Furthermore, the estimation depended on two experimental coefficients, which vary with the geometry, the roughness, and the characteristics of the flow and, for simplicity, were considered constant along the pile and for all the wave phases even if they varied in an oscillatory flow because of their dependence on the flow. In the DNS model, the total force acting over the cylinder was computed at every time step as the integration of the pressure of the wet points over the whole surface.

The pile of the experimental model was equipped with a hinge at the bottom and it was fixed with the load cell on the other end at the upper side. According to this setup, the measured horizontal force at the load cell was elaborated to obtain the wave force over the pile that depended on the lever arm between the line of application of the force and the hinge (b_W). Knowing the distance between the load cell and the hinge (b_C = 1.04 m) and estimating for each time the distance b_W, the force on the pile was obtained. These assumptions were verified by means of a calibration process of the load cell, applying different known horizontal forces at different locations.

The force obtained by means of the load cell measurements was compared with that obtained by means of the integration of the pressure distribution recorded around the pile (Figure 11). The results show a very good agreement between the two sets of data, especially during the crest phase. The underestimation of the minimum value of the force can be due to the discretization of the pressure distribution around the pile. Therefore, although the pressure sensor discretization was not tiny ($\Delta\varphi$ = 45°, $dz \approx$ 8 cm), the comparison of the force obtained by two different types of measurements confirmed the validity of the data.

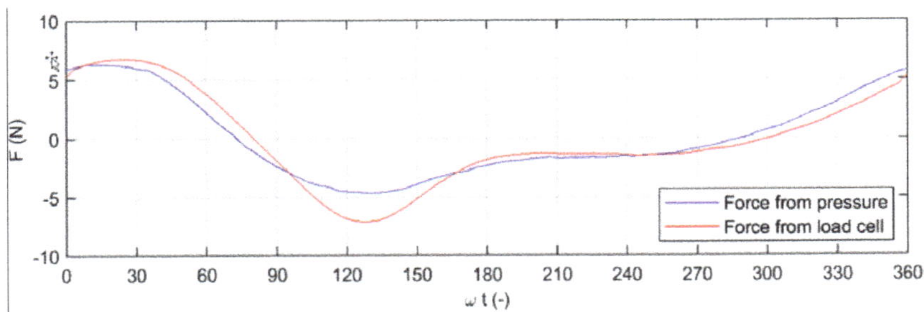

Figure 11. Comparison between the force obtained from the load cell measurements, and as integral of the pressure measurements.

Figures 12 and 13 show the phase-averaged force, along the x direction, evaluated both numerically and experimentally. The wave forces along the transversal direction (y-axis) were found to be negligible.

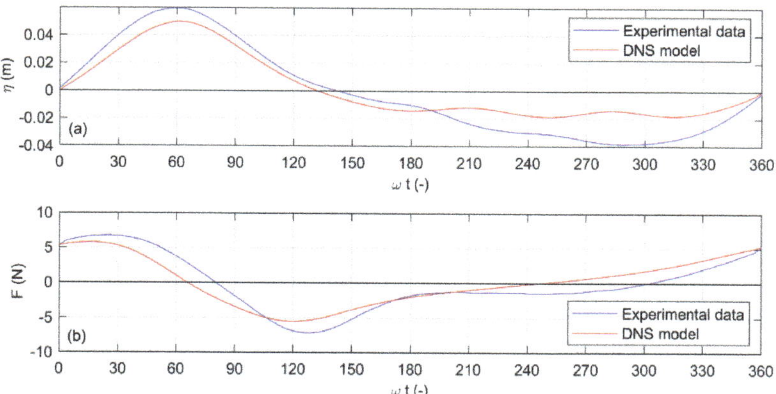

Figure 12. Comparison between experimental and numerical data of water level (**a**) and force over the pile (**b**).

Although the shape of the wave of the numerical model is steeper and reveals a larger nonlinearity than that of the experimental test, a very good agreement over the evaluation of the force was achieved. The maximum numerical value was slightly lower, according to the lower wave crest elevation. The shape of the force evolution along the wave phases was very similar; the slight shift of the force signals was only due to the difference in the occurrence of the wave crest between the numerical and experimental data, as already pointed out in Section 4.3. Therefore, the dynamic response of the wave over the cylinder was well reproduced and the model can be considered fully validated.

The theoretical evaluation of the wave force over a vertical cylinder can be made by using Equation (2). The problems of its application for design purposes are the correct evaluation of the water particle velocities and accelerations and the choice of the right drag and inertia coefficients (usually obtained for smaller Reynolds numbers with respect to prototype conditions [16,17]). The direct integration can be performed with the hypothesis that the coefficients are constant, the wave is linear (Airy's theory), and its amplitude can be considered small in comparison with the water depth (Morison formula [1]). These assumptions cannot be considered reliable for real water waves, which show nonlinearities, with a more peaked wave crest and a more flattened trough.

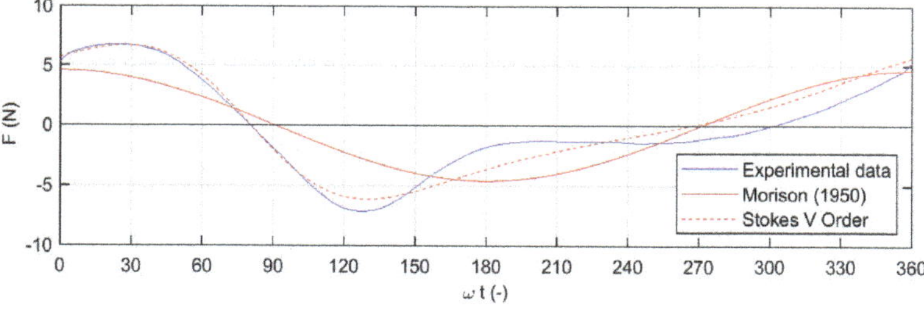

Figure 13. Comparison between the experimental force and the force evaluated by applying Morison's formula [1] with the linear wave theory (original formulation) and with Stokes V order wave theory (modified formulation).

Furthermore, the right choice of the coefficients of Equation (2) plays a significant role on the evaluation of the force acting on the pile. For the present laboratory conditions, according to the work of Sarpkaya [16], a pair of drag and inertia coefficients are chosen (C_D = 0.9 and C_M = 2.0). The application of the linear theory as for the evaluation of particle velocities and accelerations could lead to a significant underestimation of the maximum value of the force (here about –30%). The use of higher order theories in Equation (2) (e.g., Stokes V order theory) improved significantly the estimation of the maximum value of the wave force, as shown in Figure 12, and allowed us to better reproduce its trend over the whole wave period (the velocities and the accelerations of water particles were computed according to the methodology proposed by Fenton [18]). Therefore, the use of the Morison equation with higher order wave theories is recommended for the prediction of the wave force on the pile.

4.5. Flow Pattern, Shear Stress, and Scour

In this section we make use of the different information provided by the numerical and experimental approach in order to highlight the relevant physical processes responsible for the scour at the base of the pile, which would be otherwise difficult to understand by solely using experimental or numerical data.

It is well known that the process of scour in the wave regime significantly depends on the wave characteristics and, hence, by the Keulegan–Carpenter number. As shown by Corvaro et al. [10], the formation of the scour is mainly due to the presence of phenomenon of shedding of vortices at the base of the pile. Accordingly, we made use of the direct numerical simulation data to analyze the flow pattern around the pile. More in detail, we made use of the second invariant Q of the velocity gradient tensor $\overline{D} = D_{ij} = \frac{\partial u_i}{\partial x_j}$ (the velocity gradient D_{ij} can be decomposed into a symmetric and a skew-symmetric part $D_{ij} = S_{ij} + \Omega_{ij}$, where $S_{ij} = \frac{1}{2}\left(\frac{\partial u_i}{\partial x_j} + \frac{\partial u_j}{\partial x_i}\right)$ is known as the rate-of-strain tensor and $\Omega_{ij} = \frac{1}{2}\left(\frac{\partial u_i}{\partial x_j} - \frac{\partial u_j}{\partial x_i}\right)$ as the vorticity tensor), defined as:

$$Q = \frac{1}{2}\left[tr\left(\overline{D}\right)^2 - tr\left(\overline{D}^2\right)\right] = \frac{1}{2}\left[\,\|\overline{\Omega}\|^2 - \|\overline{S}\|^2\right]. \tag{3}$$

Indeed, it is well known that connected flow regions characterized by a positive value of the second invariant, $Q > 0$, allow the identification of vortical structures [19]. Q represents the local balance between shear-strain rate and vorticity magnitude, defining vortices as areas where the vorticity magnitude is greater than the magnitude of rate-of-strain [19,20]. In Figure 14, the isosurface $Q = 10$ colored by enstrophy (the square of the vorticity vector) is shown.

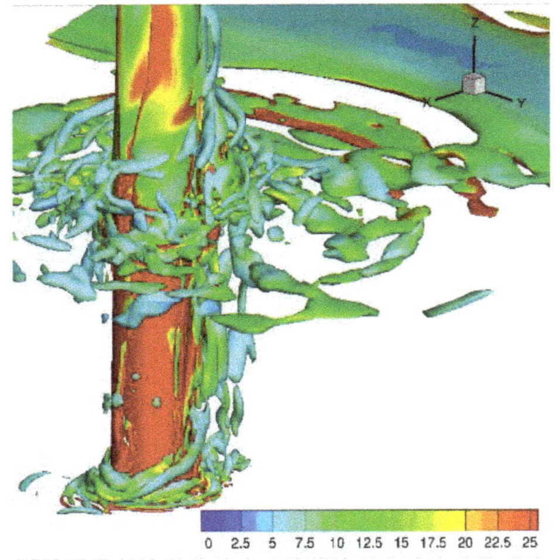

Figure 14. Isosurface of $Q = 10$ colored with enstrophy.

As can be seen, the wave motion and, consequently, the alternating flows gave rise to a complex flow pattern consisting of shedding of structures from the pile. In particular, vortices were found to be essentially aligned in the vertical direction along the pile, while, at its base, flow motion organized in structures that lie in the horizontal direction forming ring shapes around the vertical cylinder. The vertical-aligned and ring-like vortex structures of the flow are known to play an important role on the fluid dynamic loads on the vertical cylinder and on the erosion phenomena at its base. These two flow structures are commonly referred as lee-wake and horseshoe vortices, respectively, see Sumer et al. [5].

As extensively argued by Sumer et al. [5], one of the process related to the vortex formation is an increase of the bed-shear stress which, in turn, is related to the scour process. Again, here we made use of the combined information from the numerical and experimental models to shed light on this phenomenon. The vortex-flow regimes, the bed-shear stress, and the scour are all related together and governed primarily by KC [4,5]. Figure 15 shows the measured scour pattern obtained in a previous experimental campaign with a mobile bed (Corvaro et al. [10]) for a wave characterized by a value of KC similar to that of the wave here analyzed. In particular, Figure 15 plots the nondimensional scour S/D, where S is the scour depth and D is the pile diameter. The analysis of the seabed morphology revealed that the area subjected to the maximum scour was located behind the pile ($\varphi = 180°$) at a distance of about 10 cm (1.0 D) and the maximum value of S was equal to 1.4 cm ($S/D = 0.14$). The most widely used formula for the scour prediction under regular waves is given by Sumer et al. [4]. Other zones where scour occurred were located at $\varphi = \pm 45°$, while deposit zones were observed at $\varphi = \pm 135°$ and in front of the pile ($\varphi = 0°$).

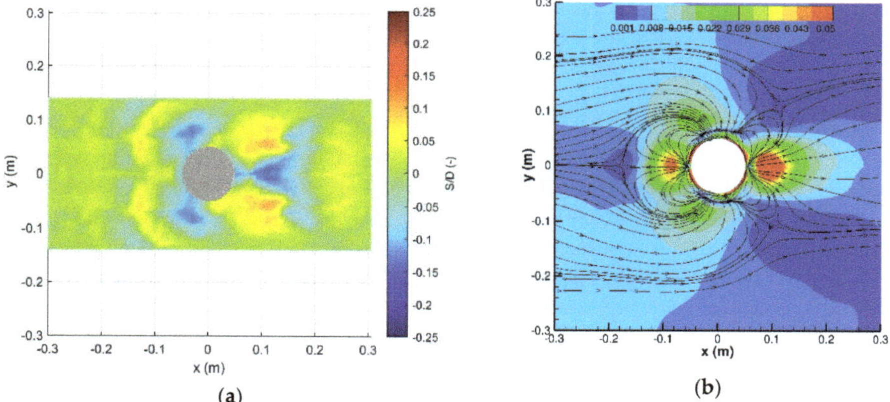

Figure 15. Contour map of the measured scour pattern evaluated in the mobile-bed experiments (**a**) and of the bed-shear stress magnitude computed from simulation (**b**). The streamlines in (**b**) report the behavior of the bed-shear stress vector field ($\mu\partial u/\partial z$, $\mu\partial v/\partial z$). Wave direction from left to right.

In Figure 15b, the computed bed-shear stress from the numerical simulation is reported. Note that an increase in terms of bed-shear stress and, thus, of turbulence levels, resulted in a tremendous increase in the sediment transport leading to the formation of a scour hole around the pile in a fairly short period of time [21].

The comparison of the two plots proves that zones of high scour are associated with those of intense shear stress. However, the intensity of the shear stress alone does not allow us to distinguish zones of sediment deposition from those where erosion occurs. To overcome this limitation, in Figure 15b the streamlines of the bed-shear stress vector field ($\mu\partial u/\partial z$, $\mu\partial v/\partial z$) are also shown. The presence of the pile was found to strongly modify the shear-stress pattern that in free conditions would lead to a transport of sediment in the direction of propagation of the waves. Particularly, we observed four singularity points for the field lines. The first one is right in front of the pile and represents the origin of shear stresses transporting sediment in the upstream direction before symmetrically bending in the lateral directions. This lateral bending was induced by the interaction with the shear-stress field lines emerging further upstream and transporting sediment in the wave direction. This interaction gave rise to the second singularity point at $(x,y) = (-0.17\text{ m}, 0.00\text{ m})$ and to two lateral arcs of convergence of the field lines. Hence, sediment was transported from the front of the pile towards the second singularity point and towards the two lateral arcs. In accordance with this picture, Figure 15a shows a deposition of sediments conforming with these two paths and an erosion in the embedded region.

The third singularity point for the shear-stress field lines was located right in the back of the pile. From this point, two branches of field lines emerged. A central portion of the field lines transporting sediment was directed downstream. Hence, the rear central region of the pile was characterized by shear stresses whose action was solely of extracting sediment and transporting it downstream. Accordingly, the measured scour in Figure 15a highlights the presence of erosion. On the other hand, the other portion of field lines was found to symmetrically bend in the lateral directions first and then to rise in the upstream direction. By moving upstream, the field lines interacted with those emerging from the upstream regions. The region of convergence of these two branches of field lines became thinner and thinner by moving further upstream and formed an arch shape which was finally closed by the last (fourth) singularity point where all these field lines converged. Overall, this pattern conformed with two lateral symmetrical regions of convergence of field lines transporting sediment from the rear of the pile and from the upstream regions whose size reduced by moving upstream, thus forming an arc shape. In accordance with this picture, the measured scour in Figure 15a shows two symmetrical regions of deposition which became thinner by moving in the upstream direction. Finally, very strong shear stresses were found around the pile in proximity of it, confirming that in such area the bed forms typical of a mobile bed were destroyed.

5. Discussion and Conclusions

The combined use of direct numerical simulation and experimental data proved to be fundamental for a deep understanding of the hydrodynamics around a vertical cylinder under wave action. The convergence of the numerical simulation with experiments was achieved and its validation was performed by means of all the major hydrodynamic quantities. A very good agreement was found by the comparison of the water surface elevation, the water particle horizontal velocity, the pressure distribution around the pile, and the total wave force. The combination between experimental and numerical data is always the best procedure to guarantee the reliability of the results. Furthermore, once the numerical simulation was fully validated, it was possible to extract a huge amount of information that it is very difficult to obtain experimentally.

The analysis of the pressure distribution at different wave phases confirmed the presence of pressure gradients around the pile responsible for the formation, development, or transport of vortices. The asymmetric pressure distribution between the front and the wake region of the pile also produced a total force on it. The maximum force occurred during the rising of the positive water elevation ($\omega t > 0°$) and before the wave crest phase ($\omega t \approx 65°$). Such result suggests that both the drag and inertial components of the total force are important. It is well known that the total force is maximum at $\omega t = 0°$ when the inertial force is dominant, while it is maximum at the crest wave phase when the drag force is dominant. The application of the Morison's equation [1] to estimate the maximum force can lead to a significant difference with the experimental results for nonlinear waves. The problem of using this equation is on the assumptions made to simplify the calculations (linear theory, constant coefficients, small amplitude wave). Furthermore, the use of empirical coefficients is needed for its application and their determination from literature (e.g., [16]) as a function of KC and Re is not easy. Several authors ([22,23]) addressed how the determination of the drag and inertia coefficients is not unique and depends on, among the others, the technique used for their estimation. As shown in Figure 13, the simple application of the formula can lead to an underestimation of the maximum values of the force for the analyzed values of Re ($Re \approx 10^4$). The use of the Morison's equation with higher order wave theories improved the prediction of the maximum force and, hence, is recommended.

Field conditions were associated with higher values of Re ($Re \approx 10^6 \div 10^7$) and, thus, to correctly design a structure, a safety factor for the total force of 1.5 ÷ 2.0 was needed, due to the uncertainties on the determination of the coefficients in such conditions. Full scale experiments at higher Reynolds numbers were useful to optimize the design process.

The DNS model was able to reproduce correctly the main hydrodynamic characteristics. The analysis of the bottom shear stress gave useful information to understand the scour process. The wake region where the maximum shear stress was obtained corresponded with that of maximum scour for a wave with similar characteristics. The main deposit regions were located at $\varphi \approx 150°$ (according to Figure 2) and about 1.5 D from the axis of the pile. Such results were in agreement with the contour map of Figure 15. Indeed, even if mobile-bed conditions can modify the bottom shear stress, the DNS model results of a rigid bed proved to be consistent with the seabed morphology data, providing qualitative information about the regions in which the bottom shear stresses are more significant. Furthermore, the analysis of the streamlines allowed us to understand the general movement of the sand around the pile and to explain the relation between the erosion and deposition zones. The formation of vortices, in addition to the streamline contraction, removed the sand particles from the surroundings of the pile, especially in the zone where the lee-wake vortices were stronger and, thus, the suspended particles tended to accumulate in the regions where the shear stress decreased.

Finally, by means of a spectral analysis of kinetic energy, it was pointed out that the wave action on a pile was responsible for an oscillating flow whose turbulence fluctuations were weak and confined in the pile region. These aspects challenged for turbulence closures that should be able to distinguish laminar oscillations induced by the wave motion from turbulence fluctuations. Given these modelling issues and by considering the relatively small computational costs of a DNS due to the small separation of scales generated by the wave action on a pile without current, we finally strongly recommend the use of a DNS approach for the solution of these types of problems.

Author Contributions: Conceptualization, S.C., A.C. and A.M.; methodology, S.C.. and A.C.; software, L.C. and A.C.; experimental investigation; F.M., S.C. and A.M.; validation, F.M., A.C. and S.C.; formal analysis, S.C., A.C., F.M. and A.C.; writing—original draft preparation, F.M.., S.C., A.C. and A.C.; writing—review and editing S.C., F.M. and A.C; supervision, S.C., A.C. and A.M..; project administration, S.C

Funding: The present work was partially funded by the Università Politecnica delle Marche (Italy) through the Scientific Research Project "Coastal Areas and Maritime Structures Adaptation to Climate Change (CAMSA).

Conflicts of Interest: The authors declare no conflict of interest.

References

1. Morison, J.R.; O'Brien, M.P.; Johnson, N.P.; Schaaf, S.A. The force exerted by surface waves on piles. *Pet. Trans. AIME* **1950**, *189*, 149–157. [CrossRef]
2. Keulegan, G.H.; Carpenter, L.H. Forces on cylinders and plates in an oscillating fluid. *J. Res. Natl. Bur. Stand.* **1958**, *60*, 423–440. [CrossRef]
3. Herbich, J.B. *Scour around Pipelines and Other Objects, Offshore Pipeline Design Elements*; Marcel Dekker Inc.: New York, NY, USA, 1981.
4. Sumer, B.M.; Fredsøe, J.; Christiansen, N. Scour around vertical pile in waves. *J. Waterw. Port Coast. Ocean Eng.* **1992**, *118*, 15–31. [CrossRef]
5. Sumer, B.M.; Christiansen, N.; Fredsøe, J. The horseshoe vortex and vortex shedding around a vertical wall-mounted cylinder exposed to waves. *J. Fluid Mech.* **1997**, *332*, 41–70. [CrossRef]
6. Olsen, N.R.B.; Kjellesvig, H.M. Three-dimensional numerical flow modeling for estimation of maximum local scour depth. *J. Hydraul. Eng.* **1998**, *36*, 579–590. [CrossRef]
7. Roulund, A.; Sumer, B.M.; Fredsøe, J. Numerical and experimental investigation of flow and scour around a circular pile. *J. Fluid Mech.* **2005**, *534*, 351–401. [CrossRef]
8. Baykal, C.; Sumer, B.M.; Fuhrman, D.R.; Jacobsen, N.G.; Fredsøe, J. Numerical investigation of flow and scour around a vertical circular cylinder. *Philos. Trans. R. Soc. A* **2015**, *373*, 20140104. [CrossRef] [PubMed]
9. Masselink, G.; Short, A.D. The Effect of Tide Range on Beach Morphodynamics and Morphology: A Conceptual Beach Model. *J. Coast. Res.* **1993**, *9*, 785–800.

10. Corvaro, S.; Marini, F.; Mancinelli, A.; Lorenzoni, C.; Brocchini, M. Hydro- and Morpho-dynamics Induced by a Vertical Slender Pile under Regular and Random Waves. *J. Waterw. Port Coast. Ocean Eng.* **2018**, 144. [CrossRef]
11. Park, J.; Kim, M.; Miyata, H. Fully non-linear free-surface simulations by a 3D viscous numerical wave tank. *Int. J. Numer. Methods Fluids* **1999**, *29*, 685–703. [CrossRef]
12. Corvaro, S.; Seta, E.; Mancinelli, A.; Brocchini, M. Flow dynamics on a porous medium. *Coast. Eng.* **2014**, *91*, 280–298. [CrossRef]
13. Miozzi, M.; Postacchini, M.; Corvaro, S.; Brocchini, M. Whole-wavelength description of a wave boundary layer over permeable wall. *Exps. Fluids* **2015**, *56*, 127. [CrossRef]
14. Le Méhauté, B. *An Introduction to Hydrodynamics and Water Waves*; Springer-Verlag: New York, NY, USA, 1976; ISBN 978-3-642-85567-2.
15. Miozzi, M.; Corvaro, S.; Pereira, F.A.; Brocchini, M. Wave-induced morphodynamics and sediment transport around a slender vertical cylinder. *Adv. Water Resour.* **2019**, *129*, 263–280. [CrossRef]
16. Sarpkaya, T. In-line and transverse forces on cylinder in oscillating flow at high Reynolds number. In Proceedings of the Offshore Technology Conference, Houston, TX, USA, 3–6 May 1976; pp. 95–108.
17. Chakrabarti, S.K. Inline forces on fixed vertical cylinder in waves. *J. Waterw. Port Coast. Ocean Div.* **1980**, *106*, 145–155.
18. Fenton, J.D. A Fifth-Order Stokes Theory for Steady Waves. *J. Waterw. Port Coast. Ocean Eng.* **1985**, *111*, 216–234. [CrossRef]
19. Hunt, J.C.R.; Wray, A.; Moin, P. *Eddies, Stream and Convergence Zones in Turbulent Flows*; Center for Turbulence Research Report: CTR-S88; NASA: Washington, DC, USA, 1988.
20. Kolar, V. Vortex identification: New requirements and limitations. *Int. J. Heat Fluid Flow.* **2007**, *28*, 638–652. [CrossRef]
21. Sumer, B.M.; Fredsøe, J. *The Mechanics of Scour in the Marine Environment*; Advanced Series on Ocean Engineering: Hackensack, NJ, USA, 2002; Volume 17, ISBN 10.1142/4942.
22. Isaacson, M.; Baldwin, J.; Niwinski, C. Estimation of Drag and Inertia coefficients from random wave data. *J. Offshore Mech. Arct. Eng.* **1991**, *113*, 128–136. [CrossRef]
23. Dean, R.G. Methodology for evaluating suitability of wave and wave force data for determining drag and inertia coefficient. In *Proceedings of the 1st International Conference on Behaviour of Off-Shore Structures, BOSS'76*; Norwegian Institute of Technology: Trondheim, Norway, 1976; pp. 40–64.

Journal of
Marine Science and Engineering

Article

Damage Characterization of Rock Slopes

Ermano de Almeida [1,2,*], Marcel R. A. van Gent [2] and Bas Hofland [1,2]

1 Department of Hydraulic Engineering, Faculty of Civil Engineering, Delft University of Technology, 2628CN Delft, The Netherlands; B.Hofland@tudelft.nl

2 Department of Coastal Structures and Waves, Deltares, 2629HV Delft, The Netherlands; Marcel.vanGent@deltares.nl

* Correspondence: e.deAlmeida@tudelft.nl; Tel.: +31-015-2789448

Received: 5 December 2018; Accepted: 2 January 2019; Published: 9 January 2019

Abstract: In order to design reliable coastal structures, for present and future scenarios, universal and precise damage assessment methods are required. This study addresses this need, and presents improved damage characterization methods for coastal structures with rock armored slopes. The data used in this study were obtained from a test campaign carried out at Deltares within the European Union (EU) Hydralab+ framework. During these tests, advanced measuring techniques (digital stereo photography) were used, which are able to survey the full extension of the structure and identify local variations of damage. The damage characterization method proposed here is based on three fundamental aspects: clear damage concepts, precise damage parameters, and high resolution measuring techniques. Regarding damage concepts, first, the importance of the characterization width is studied. For damage parameters obtained from the maximum erosion depth observed in a given width ($E_{3D,m}$), the measured damage increases continuously with increased characterization width. However, for damage parameters obtained from width-averaged profiles (S and E_{2D}), the measured damage reduces with increased characterization width. Second, a new definition of damage limits (damage initiation, intermediate damage, and failure) is presented and calibrated. Regarding the damage parameters, the parameter $E_{3D,5}$, which describes the maximum erosion depth within the characterization width, is recommended as a robust damage parameter for conventional and non-conventional configurations based on four main characteristics: its low bias, its low random error, the ability to distinguish damage levels, and its validity and suitability for all types of structures (conventional and non-conventional). In addition, the results from this study show that the damage measured with the damage parameter $E_{3D,5}$ presents an extreme value distribution.

Keywords: rock slopes; damage characterization; damage parameters; physical model tests

---

1. Introduction

This study addresses the required research on the assessment and characterization of damage to coastal structures with rock armored slopes produced by the impact of environmental loads. The development of consistent and accurate damage characterization methods aims to fully describe the response, as well as the remaining strength of the coastal structures after facing a given loading condition. This paper will elaborate on how, with high resolution measurements from digital stereo photography (DSP), more advanced damage parameters can be used. Combined with clear damage definitions (such as characterization width and damage limits), this can offer a precise damage characterization of conventional (e.g., straight slope) and non-conventional (e.g., slopes with a berm and roundheads) structures.

Climate change should also be considered, as this phenomenon increases the environmental loads acting on coastal structures. In such scenarios, the damage characterization methods presented in this paper aim to improve the assessment of conventional and non-conventional structures used as upgrading and adaptation alternatives.

1.1. Background

In the assessment of coastal structures, damage can be defined as "the movement of armor units as consequence of the impact of environmental loads" [1]. Hudson [2] used the percentage of displaced units to characterize damage. Later on, Thompson and Shuttler [3] and Broderick [4] introduced parameters to describe damage as the number of removed units from the slope. The parameter S [4] is widely used and is based on the eroded area (difference between the initial and final width-averaged profiles) divided by the nominal diameter of the armor units.

More recently, Melby and Kobayashi [5] introduced three additional damage parameters including the normalized erosion depth (E), the normalized erosion length (L), and the normalized cover depth (C) obtained for averaged profiles over the width. Hofland et al. [6,7] recently incorporated innovative and more accurate measuring techniques, such as DSP, for the survey of damage in physical modelling tests, and presented a three-dimensional erosion depth parameter ($E_{3D,m}$) to be used in the damage characterization of conventional and non-conventional coastal structures.

1.2. Current Limitations

Firstly, the concepts of "damage initiation", "intermediate damage", and "failure" are not accurately and uniformly described. Different authors established the current limits based on unconsolidated arguments, also influenced by less detailed measurement techniques (such as being obtained from a limited number of profiles). Broderick [4] defined the condition of no-damage as S = 2, "which is the lowest level of damage that can be consistently detected in the survey data", while failure is described as "when enough rip-rap is shifted to expose the filter material", again without a filter exposure extension. Van der Meer [1] defines these limits in a similar manner—start of damage is described as S = 2 for a 1:2 slope and failure is reached when the filter layer becomes visible, without describing an extension threshold.

Secondly, the suitability of innovative damage parameters to accurately characterize the response and remaining strength of rubble mound structures is further investigated. The damage parameter S used by Van der Meer [1] provides limited information about the structure conditions and is only considered for trunk sections. The damage parameters presented by Melby and Kobayashi [5] and by Hofland et al. [6,7] could present a more accurate description of damage for conventional and non-conventional slopes, providing a significant improvement in the characterization of the structure conditions.

Thirdly, damage characterization methods for rubble mound structures can increase their accuracy and reliability as a result of the continuous development of testing and measuring techniques. New measuring techniques such as DSP [6,7] are able to provide model surveys with millimeter resolution and describe the damage in detail.

Following the three previously described limitations of current methods, this study focuses on the following aspects of damage characterization: concepts (demand for unified damage characterization concepts) in Section 3.1, parameters (demand for universal and more accurate damage characterization parameters) in Section 3.2, and measuring techniques (demand for validating the suitability of innovative survey methods) in Section 3.3. The physical model tests are presented in Section 2 and the main conclusions are discussed in Section 4.

2. Physical Model Tests

A physical modelling campaign was carried out at Deltares within the European Hydralab+ framework. These tests were planned with the aim of obtaining better validation data for addressing the research aims. They were carried out in the Scheldt Flume at Deltares, Delft, The Netherlands (see Figure 1). The set-up was based on shallow coastal areas with wave conditions that were depth-limited. Thus, conditions under which the increase in the water level due to climate change would be directly linked to a change in the incident wave height at the toe of the structure were studied.

Figure 1. Scheldt Flume, Deltares.

This set-up consisted of a non-overtopped rubble mound structure with an impermeable core, a filter (D_{n50} = 9.4 mm), a 1:3 two-layer rock armored slope (D_{n50} = 16.3 mm and D_{n85}/D_{n15} = 1.19), and a foreshore (see Figure 2). In addition, the main particularity of these tests for shallow coastal areas is that the generated wave was equal for all test runs, and only the change in the water depth at the toe of the structure defined the incident wave height. This way, the effect of the sea level rise in the damage to the structure was clearly represented. Thus, this set-up provided a large amount of data given the number of tests carried out, which allowed the testing of different slope configurations and a number of repetition tests. The testing conditions included the following:

- Current sea level and sea level rise scenarios (in depth-limited conditions).
- Cumulative and non-cumulative damage.
- Damage variability: repetition tests.

The cross-section of the model set-up used in the Deltares tests is shown in Figure 3. During these tests, only the damage to the front slope was evaluated. In this study, only the straight slope configuration is considered. The test program, including slope configurations with a berm, is described in the works of [8,9]. Incident wave heights were depth-limited given that an 8 m long foreshore was present in this model. The model (considering reference wave climate in the Dutch coast) followed the Froude criterion. Moderate Reynolds ($Re > 1 \times 10^4$) ensured turbulent conditions in the filter and armor (no permeable core) and negligible scale effects. In addition, the wave generation equipment included active compensation for the reflected wave at the wave board.

Figure 2. Deltares shallow water tests set-up.

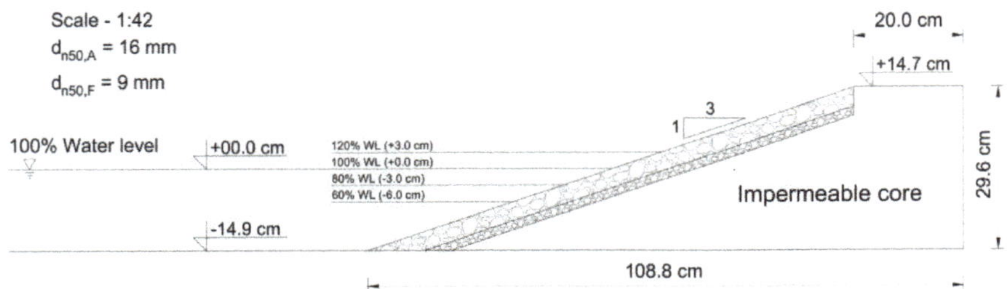

Figure 3. Deltares tests cross-section.

The (straight slope) tests included 25 runs in two series (repeated five times each), described in Table 1, which shows the type of damage (cumulative or non-cumulative), the water depth at the toe of the structure (h_T), the test condition (Cond.), the targeted generated wave height ($H_{s,g}$), the targeted wave height at the toe of the structure ($H_{s,T}$), the targeted peak wave period (T_p), and the number of waves of each test run (N_w).

Table 1. Test conditions.

Series	Run	Damage	h_T (m)	Cond.	$H_{s,g}$ (m)	$H_{s,T}$ (m)	T_p (s)	N_w
Series 1 (5 repetitions)	Run 1	Cum.	0.089	60%	0.087	0.037	1.27	1000
	Run 2	Cum.	0.119	80%	0.087	0.050	1.27	1000
	Run 3	Cum.	0.149	100%	0.087	0.062	1.27	1000
	Run 4	Cum.	0.179	120%	0.087	0.074	1.27	1000
Series 2 (5 repetitions)	Run 1	Non-cum.	0.149	100%	0.087	0.062	1.27	1000

During those tests, the measurement of damage to the armor layer was done using the digital stereo photography (DSP) technique. This method used the following equipment described in Figure 4 and in the list hereafter.

- A—Markers base plate: fixed origin of coordinates for the 3D reconstruction and comparison between test runs, to be kept fixed during the complete testing period.
- B—Additional markers: pairs of reference points used in the 3D reconstruction, to be randomly placed before a measurement and removed after it (if necessary, they can also be fixed).
- C—Cameras: two digital cameras with a fixed distance between each other.
- D—Laptop: used for obtaining the pairs of images and carrying out the post-processing.

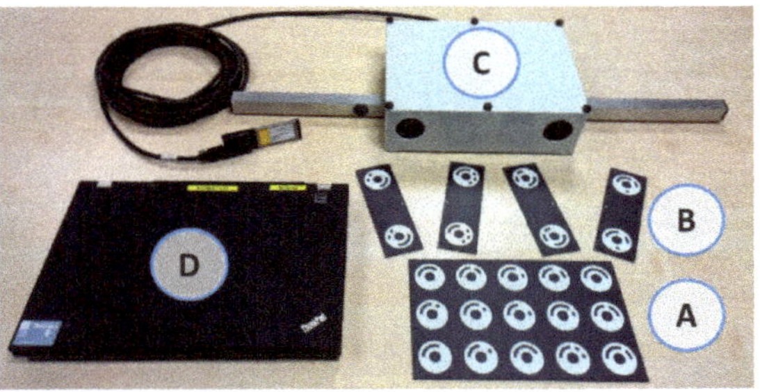

Figure 4. Digital stereo photography (DSP) equipment.

Figure 5 shows the workflow for the characterization of damage to rock slopes using digital stereo photography. The images and measurements shown in Figure 5 are obtained from the measurement campaign carried out during this research (see tested rock slope in Figure 2). From about 10 pairs of images (Figure 5a) before and after a given test run, the 3D reconstructions of the structure are obtained (Figure 5b). Comparing those 3D models from before and after the given tests, the 3D damage reconstruction is obtained (Figure 5c). This method for measuring damage is used during this research. From those measurements, the erosion areas (in yellow-orange-red in Figure 5c) are assessed and characterized (for more details on the DSP technique, see [6,7,10])

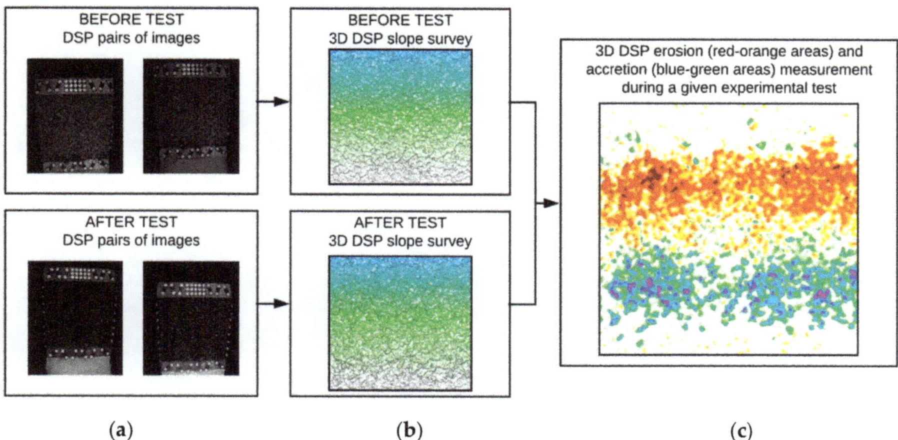

Figure 5. Digital stereo photography (DSP) workflow. (**a**) Pairs or images from the surveyed slope, before and after a test, (**b**) 3D slope survey before and after a test (white color being the lower part of the slope, and blue color being the higher part of the slope), (**c**) erosion and accretion measurement following a given test (red-orange areas being erosion areas, and blue-green areas being accretion areas).

3. Damage Characterization Method

This section has the aim of validating the damage characterization methods, which include the damage concepts (Section 3.1), damage parameters (Section 3.2), and measuring techniques (Section 3.3). For that, the physical modelling tests presented in the previous section are used. As a first step, the damage parameters considered in this study are described in more detail, including the procedure for their calculation in Figure 6, for the S and E_{2D} parameters (6a) and for the $E_{3D,m}$ parameters (6b).

The first parameter considered is S, as defined by Broderick [4] (see Equation (1)). This damage parameter is widely used and describes the damage to the structure as the number of units eroded in the width-averaged profile.

$$S(number\ of\ units) = \frac{\langle Ae \rangle_w}{D_{n50}^2} \tag{1}$$

where $\langle Ae \rangle_w$ (m^2) is the eroded area from the averaged profile obtained over a given characterization width w and D_{n50} (m) is the nominal median diameter.

The parameters E_{2D} [5] and $E_{3D,m}$ [6,7] are shown in Equations (2) and (3), respectively. These two parameters estimate the damage to the structure considering the maximum erosion depth perpendicular to the slope. The distinction between them is that for E_{2D}, this erosion depth is measured in the width-averaged profile, while for $E_{3D,m}$, this erosion depth is obtained as the maximum erosion depth recorded at any point of the structure or defined characterization width. In addition, for $E_{3D,m}$, the initial profile and the profile after the test are averaged with a circular spatial moving average with mD_{n50} diameter.

$$E_{2D}\ (erosion\ depth\ in\ units) = \frac{max(\langle e \rangle_w)}{D_{n50}} \tag{2}$$

$$E_{3D,m}\ (erosion\ depth\ in\ units) = \frac{max(\langle e \rangle_{mDn50})_w}{D_{n50}} \tag{3}$$

where $max(\langle e \rangle_w)$ (m) is the maximum erosion depth from the averaged profile obtained over a given characterization width w, $max(\langle e \rangle_{mDn50})_w$ (m) is the maximum erosion depth averaged over an area of mD_{n50} diameter obtained over a given characterization width w, and D_{n50} (m) is the nominal median diameter.

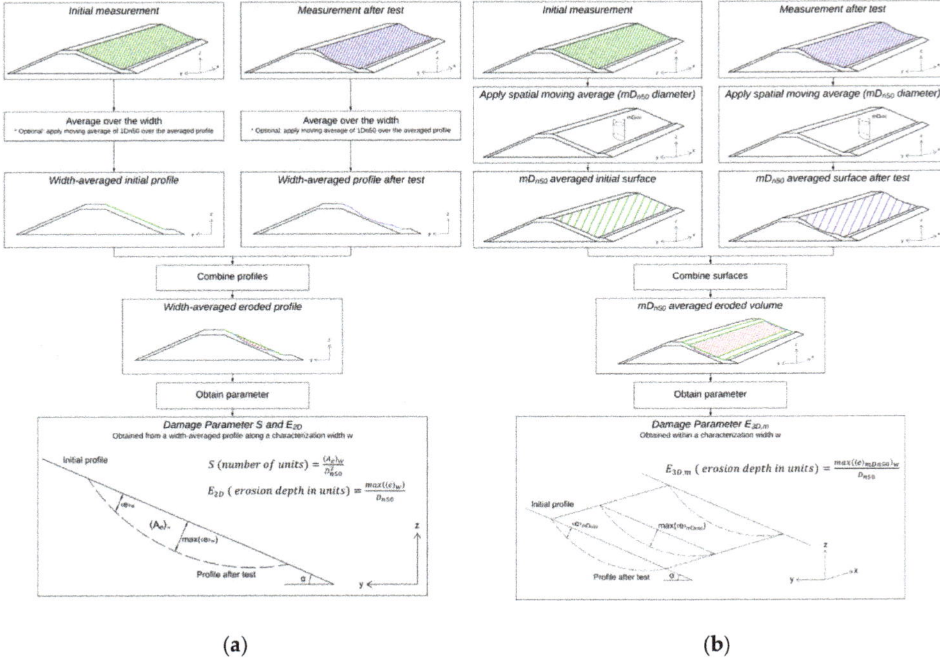

(a) (b)

Figure 6. Damage parameters methodology. (**a**) S and E_{2D} parameters obtained from a width-averaged profile, (**b**) $E_{3D,m}$ parameters obtained from the maximum erosion depth observed within a given characterization width.

3.1. Damage Concepts

Two main damage concepts will be discussed here; namely characterization width (i.e., the width over which the damage is evaluated in a physical model) and the damage limits (i.e., damage initiation, intermediate damage, and failure).

3.1.1. Characterization Width

This is addressed given that using different characterizations widths in the assessment of coastal structures would lead to different damage results for all damage parameters and would influence the definitions of damage limits. Among others, the authors of [11] recommend the use of characterization widths larger than 15–20 rock diameters in order to achieve representative results, but do not discuss the influence it has on the measured damage.

Mean Damage Evolution

The influence of the characterization width on the mean measured damage can be observed in Figure 7, obtained from the survey of a $54D_{n50}$ wide slope in the Deltares tests Series 1 (see Table 1), carried out five times in five identical repetitions. The values for each characterization width are obtained as the average of all the values for that given width, being between 135 values for the 2 D_{n50} width and 5 values for the 54 D_{n50} width. This average damage is then normalized by the damage obtained for the 54 D_{n50} characterization width.

Regarding the parameters obtained from width-averaged profiles (S and E_{2D} in Figure 7a), it can be observed that the measured mean damage reduces with increased characterization width because the damage to certain areas is hidden by the accretion in other areas. Regarding the parameters obtained as the maximum erosion depth observed within a given characterization width ($E_{3D,1}$ and $E_{3D,5}$ in Figure 7b), it can be observed that the measured damage increases with increased characterization width.

The main observation for the damage parameters $E_{3D,1}$ and $E_{3D,5}$ (also referred in general form as $E_{3D,m}$) is that the measured damage continues to increase with increasing characterization widths. This suggests that there is no upper limit for the damage to the structure and that when considering wider structures, the probability of observing a larger extreme damage will continue to increase. Thus, besides considering the characterization width, this length effect for the design and characterization of coastal structures should then be taken into account (see Section 3.2.2 and [9]).

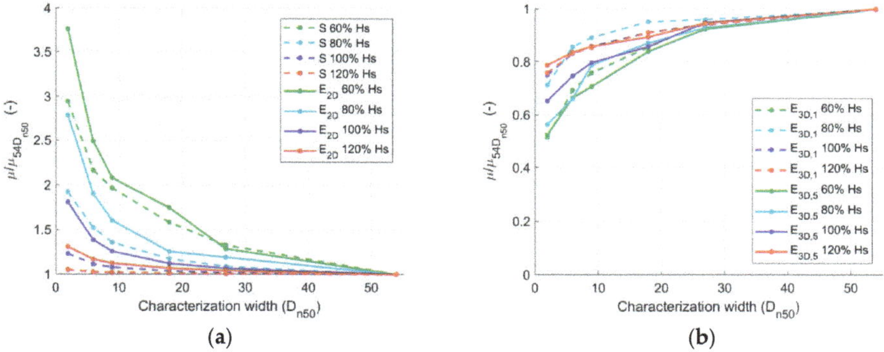

Figure 7. Characterization width. Mean damage variation (bias), calculated as the mean damage from each characterization width (μ) divided by the mean damage from the $54D_{n50}$ characterization width ($\mu_{54D_{n50}}$). **(a)** S and E_{2D} parameters, **(b)** $E_{3D,1}$ and $E_{3D,5}$ parameters.

Damage Variability Evolution

The variability of measured damage is evaluated for different characterization widths. The five identical realizations of Deltares tests Series 1 are considered again, which consists of four test runs each (see Table 1). In Figure 8, the standard deviation normalized by the mean for the four parameters can be observed.

Regarding the parameters obtained from width-averaged profiles (S and E_{2D} in Figure 8a), it can be observed that the increase in the characterization width leads to a general reduction of the variability in the measurements. Regarding the parameters obtained as the maximum erosion depth observed within a given characterization width ($E_{3D,1}$ and $E_{3D,5}$ in Figure 8b), they also show a reduction in the variability (mainly because of an increase in the mean values) with increasing characterization width, with significantly less variability than for parameters S and E_{2D}. For additional details on damage variability, consider the work of [9].

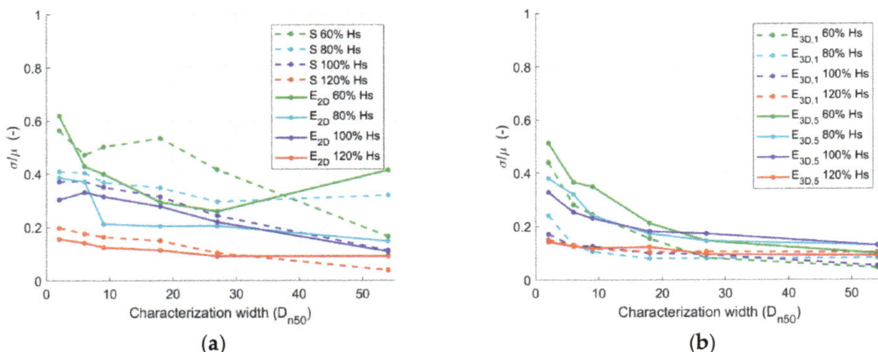

Figure 8. Characterization width. Standard deviation of the damage (σ) divided by the mean of the damage (μ), for each characterization width (random error). (**a**) S and E_{2D} parameters, (**b**) $E_{3D,1}$ and $E_{3D,5}$ parameters.

Conclusions Regarding Characterization Width

Following the previous results, it can be stated that the characterization width is an important factor in the definitions of damage to coastal structures. According to the present results, a characterization width of approximately $25D_{n50}$ is recommended, but the following aspects should be taken into account:

- Mean evolution: as shown in Figure 7, different characterization widths will affect the measured damage (bias). Thus, ideally, a suitable characterization width will be the one that shows a smaller difference in mean measured damage compared with what is obtained when considering different characterization widths.
- Variability evolution: as shown in Figure 8, it should be considered that different characterization widths will present different variability in the results when the same condition is tested several times (random error). Thus, ideally, a suitable characterization width will be the one that shows a smaller variability in the results.
- Laboratory conditions: the definition of the most suitable characterization widths should take into account the limited space available for physical modelling tests and the associated costs that limit the experiment dimensions. In addition, the implication of the scaling constraints that limit the possibility of reducing the units' size in the armor of the structure should be taken into account. Thus, a suitable characterization width will be the one that could fit into general testing facilities without introducing scale effects.

3.1.2. Damage Limits

This study aims to establish unified damage limits for the characterization of coastal structures such as damage initiation, intermediate damage, and failure. On the basis of the contributions and limitations found in literature (see Section 1.2) and the need for establishing unified concepts for damage characterization, the following damage limits are proposed for a nD_{n50} thick rock armored slope:

- Damage initiation: defined as the condition in which a circular hole of $1D_{n50}$ diameter and a depth of $1D_{n50}$ is observed in the armor layer.
- Intermediate damage: defined as the condition in which a circular hole of $1D_{n50}$ diameter and a depth of 1.5 D_{n50} is observed in the armor layer.
- Failure limit: defined as the condition in which a circular hole of $1D_{n50}$ diameter and a depth of nD_{n50} is observed in the armor layer.

In order to characterize a given rock armored slope according to these damage limits, the parameter $E_{3D,1}$ is used as the calibration method. This $E_{3D,1}$ parameter can be described as "the erosion depth measured in D_{n50} perpendicular to the slope averaged over a circular area of $1D_{n50}$", which allows capturing the damage limits described above.

Traditional damage assessment methods were based on slope surveys with profilers with a limited number of profiles and the exposure of the filter layer for the failure limit; $D_{50}/2$ for Thompson and Shuttler (1975), $1D_{n50}$ for Melby and Kobayashi (1998), and to a less clear extent for other authors (see Section 1.2). According to such methods, damage initiation and intermediate damage are based on defined parametric limits, while failure is determined by the visual observation of images or using a cylindrical gauge with a hemispherical foot. In contrast, the use of high resolution surveys such as the ones provided by DSP (see Figure 2) allows a more precise visualization of the state of the structure for all damage levels. In this case, all the damage limits can be quantified according to the depth and extension of the damage area.

Thus, considering all the measurements carried out with a high resolution technique (DSP) and the damage parameter $E_{3D,1}$, the damage limits can be calibrated based on the Deltares tests (characterization width of $27D_{n50}$), see Figure 9 and Table 2.

Table 2. Damage limits (for $27D_{n50}$ characterization width).

Series	$E_{3D,1}$	$E_{3D,5}$	E_{2D}	S
Damage initiation	1.0	0.3	0.2	1
Intermediate damage	1.5	0.7	0.5	4
Failure	2.0	1.1 (90% conf.)	0.9 (90% conf.)	11 (90% conf.)

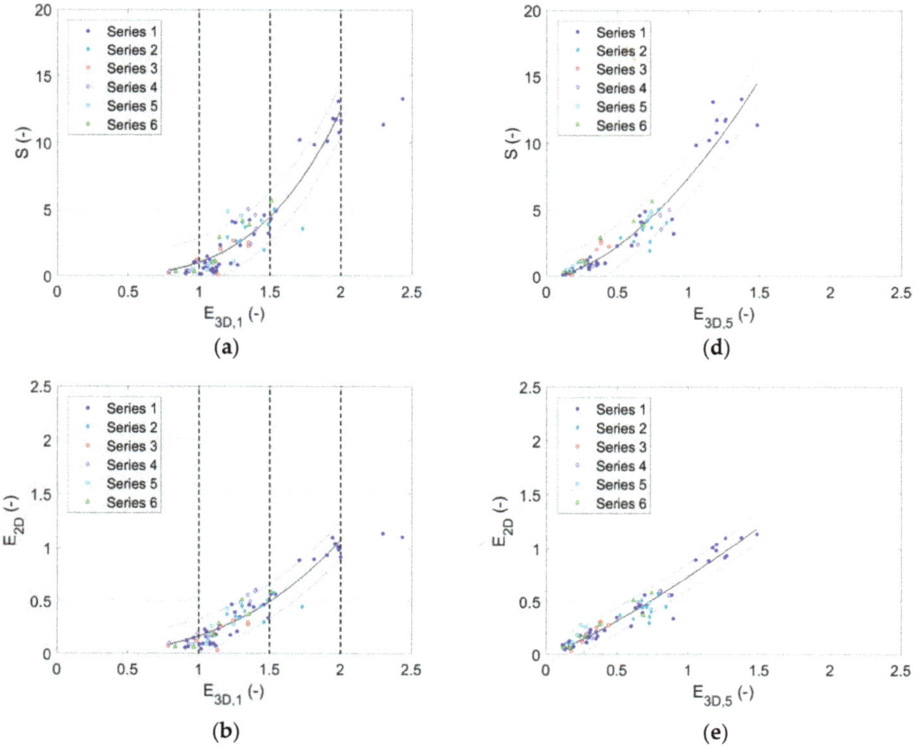

Figure 9. *Cont.*

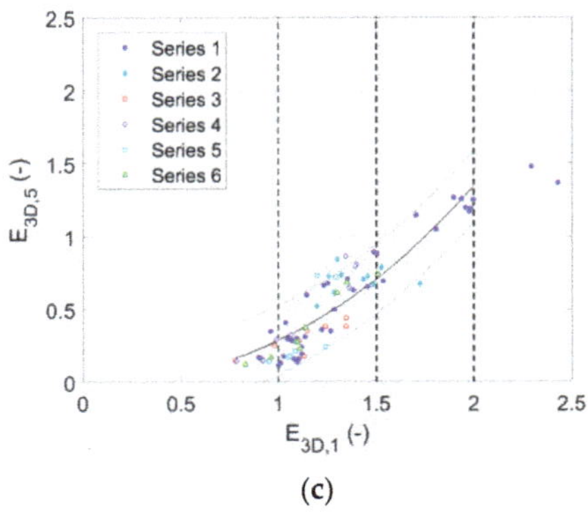

(c)

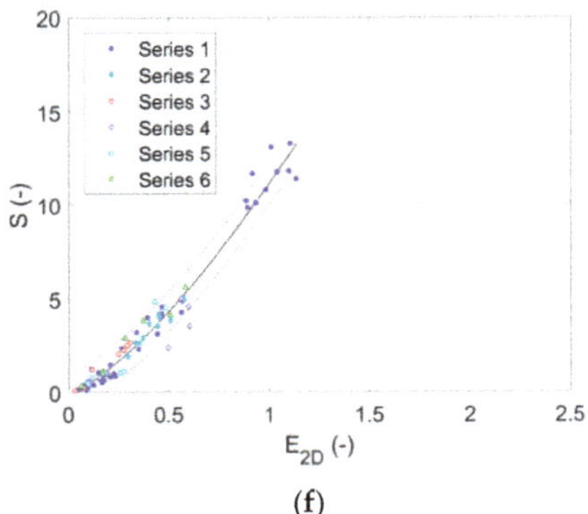

(f)

Figure 9. Comparison of all damage parameters (for $27D_{n50}$ characterization width). (**a**) $E_{3D,1}$ vs. S parameters, (**b**) $E_{3D,1}$ vs. E_{2D} parameters, (**c**) $E_{3D,1}$ vs. $E_{3D,5}$ parameters, (**d**) $E_{3D,5}$ vs. S parameters, (**e**) $E_{3D,5}$ vs. E_{2D} parameters, (**f**) E_{2D} vs. S parameters.

An important remark from the previously defined calibrated damage limits is the fact that, according to the available data, they are constant for the slopes applied (1:3 in Deltares tests). Nevertheless, this should be further investigated considering a larger number of tests and different slope configurations in order to validate these conclusions.

3.2. Damage Parameters

This section addresses the need for establishing and validating a robust damage parameter able to accurately characterize the damage and remaining strength of rubble mound structures. This will be done by analyzing the Deltares shallow water test results in more detail and using a characterization width of 27 D_{n50} and the previously established damage limits. Only parameters S, E_{2D}, and $E_{3D,5}$ will be evaluated ($E_{3D,1}$ calibration parameter will not be considered).

The preference for $E_{3D,5}$ ($E_{3D,m}$ with m = 5, which describes the erosion depth in D_{n50} over an averaged circular area of $5D_{n50}$ diameter) over other values of m is justified given the fact that in the erosion profile from the test results, the damage cavity repeatedly showed a cross-profile extension of approximately $5D_{n50}$.

3.2.1. Damage Parameters Analysis

The analysis of damage parameters is done considering the following four criteria: bias, random error, distinction of damage range, and its values for different structures.

Bias

The first element to be considered, the bias in damage characterization of coastal structures, is related to the ability of the damage parameters to provide the most precise description of the state of the structure. This can be achieved with damage parameters that do not lead to hidden erosion, defined as the conditions in which the damage in one location of the structure is hidden by the accretion in other location when considering width-averaged profiles.

Damage parameters based on width-averaged profiles (S and E_{2D}) are associated with a larger bias because a larger characterization width will lead to an averaged profile, which reduces (averages out) the magnitude of the maximum erosion areas, especially for smaller damage levels (damage initiation). On the contrary, damage parameters that capture the maximum erosion depth observed within a given characterization width ($E_{3D,5}$) have limited bias because they provide an assessment of the complete structure surface and identify all damaged areas.

Random Error

This second aspect, the random error in the measured damage to coastal structures, is related to the ability of damage parameters to describe the state of the structure with the smallest variability and larger confidence in the measured result. This will be analyzed further in order to estimate the random error associated with each of the damage parameters.

The damage data for Series 1 (composed of four test runs of 60%, 80%, 100%, and 120%) and Series 2 (composed of one test run of 100%) is described in the table below. For each test run, 10 observations are considered, obtained from five realizations in a model with a width of $54D_{n50}$ (twice the considered characterization width of 27 D_{n50}). For these 10 observations and the 3 damage parameters, Table 3 and Figure 10 show the 80% confidence prediction interval, applying a Student's *t*-distribution with a critical t-value of 1.383 ($\delta = \pm 1.383 \cdot \sigma \cdot \sqrt{1 + 1/10}$) expressed as a percentage of the mean.

According to these results, it can be observed that damage parameters $E_{3D,5}$ present a lower variability (random error), especially for the lower conditions associated with a lower amount of damage (60% and 80%). Damage parameters S and E_{2D} present low variability in higher damage levels (120% in Series 1 run 4), but much higher variability (random error) in the other damage levels.

Table 3. Eighty percent prediction interval with respect to the mean (for $27D_{n50}$ characterization width).

Parameter	S1 Run 1 60%	S1 Run 2 80%	S1 Run 3 100%	S1 Run 4 120%	S1 Run 2 100%
$E_{3D,5}$	21%	21%	25%	14%	18%
E_{2D}	38%	30%	32%	13%	29%
S	60%	43%	35%	15%	38%

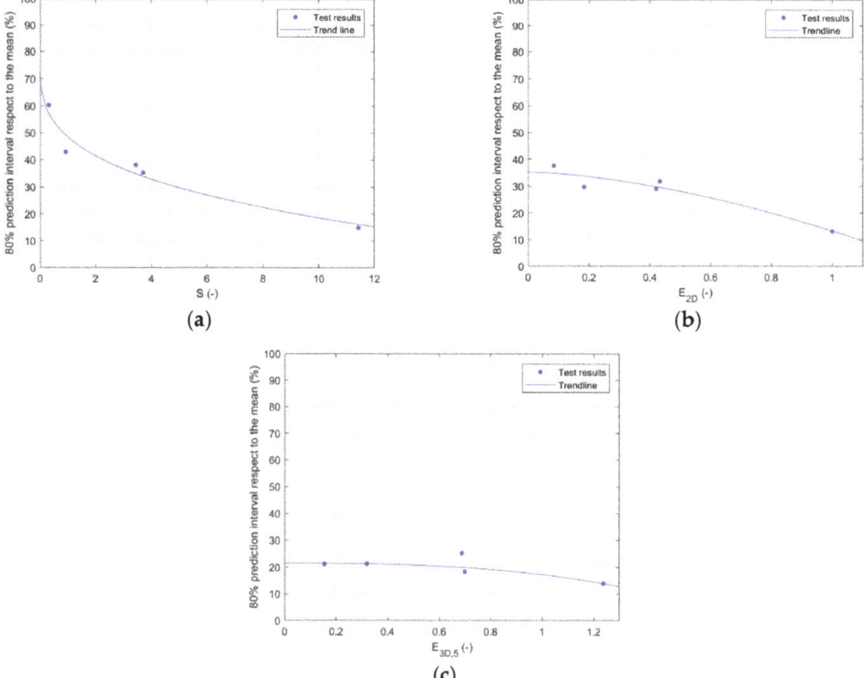

Figure 10. Eighty percent prediction interval with respect to the mean (for $27D_{n50}$ characterization width). **(a)** S parameter, **(b)** E_{2D} parameter, $E_{3D,5}$ parameter.

Distinction of Damage Range

The third element to be considered is the ability of damage parameters to distinguish the range of damage levels between initial damage, intermediate damage, and failure. According to the experimental data from the tests, the S, E_{2D}, and $E_{3D,5}$ parameters were able to distinguish the different damage levels with a similar degree of scatter (for parameter $E_{3D,1}$, a larger scatter was observed). For additional details, consider the work of [12].

Values for Different Structures

The fourth aspect to be considered in this analysis is the suitability of damage parameters to be used in different structures. In this case, different structures do not only include variations in the characteristics of the rock slope, but also 3D features such as roundheads or slopes with a berm. Of the three parameters considered here, only the parameter that captures the maximum erosion depth observed within a given characterization width ($E_{3D,5}$) is suitable for describing damage to coastal structures with any geometry or 3D feature.

Conclusions on Damage Parameters Analysis

According to the four criteria previously discussed, the damage parameter $E_{3D,5}$ is considered as the most suitable damage parameter for the characterization of damage to coastal structures based on the following four reasons:

- Low bias: the damage to the structure is clearly captured, without including hidden erosion present in the parameters that consider width-averaged profiles.
- Low random error: this parameter describes the damage to the structure with low variability also for a relatively low amount of damage, which increases the confidence in the measured and expected damage.
- Distinguish damage range: the different states of damage to the structure can be recognized according to the limits established.
- Constant value for different structures: for all structure configurations, this parameter can be used without the need for any modification.

3.2.2. Extreme Damage Distribution for $E_{3D,5}$

According to the results from this study, the measured damage with parameter $E_{3D,5}$ can be adjusted to an extreme value distribution. This illustrates that for increasingly long structures, a larger damage is expected, in the same way that for an increasingly long time period, a larger wave height is expected, or that for increasingly long dikes, a larger probability exists that one section can fail [13].

In order to define an extreme value distribution for the measured damage with $E_{3D,5}$, the results shown in Table 4 are considered. Each test condition (four for Series 1 and one for Series 2) had five realizations each with a width of $54D_{n50}$, which leads to 10 damage results for a characterization width $27D_{n50}$. For each test condition, the measured damage was normalized by the mean value, so all 50 measurements are combined in one single distribution. These values are adjusted to a Gumbel distribution, with the distribution parameters determined by the least square method.

The exceedance probability for the damage is translated to a return width in a similar form to that used for the study of extreme wave climate (see Equation (4)). The combined extreme distribution for the normalized damage parameter $E^N{}_{3D,5}$ is shown in Figure 11.

$$R_w = \frac{\lambda}{1 - F_X(x)} \tag{4}$$

where R_w (m) is the return width, λ (m) is the characterization width ($27D_{n50}$ in this study), and $1 - F_X(x)$ (-) is the exceedance probability.

Table 4. Parameters for Gumbel distribution for $E^N_{3D,5}$ parameter (for $27D_{n50}$ characterization width).

Parameter	S1 Run 1 60%	S1 Run 2 80%	S1 Run 3 100%	S1 Run 4 120%	S1 Run 2 100%
$E^N_{3D,5} = \frac{E^N_{3D,5}}{E_{3D,5}}$	0.74	0.75	0.72	0.85	0.74
	0.85	0.89	0.87	0.92	0.88
	0.88	0.92	0.91	0.94	0.96
	0.97	0.93	0.95	0.96	0.97
	1.02	0.95	0.96	0.97	1.01
	1.02	0.97	0.98	1.01	1.03
	1.05	1.09	1.00	1.02	1.04
	1.10	1.09	1.03	1.02	1.05
	1.14	1.13	1.28	1.10	1.12
	1.21	1.28	1.29	1.20	1.20
$\alpha E^N_{3D,5}$			0.94		
$\beta E^N_{3D,5}$			0.12		

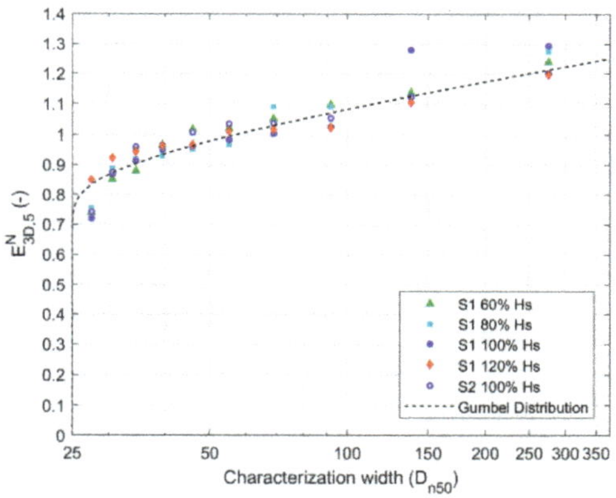

Figure 11. Gumbel extreme distribution for $E^N_{3D,5}$ parameter (for $27D_{n50}$ characterization width).

Figure 11 shows that a single realization of a test run for a characterization width of $27D_{n50}$ is expected to present damage of approximately 0.8 as the mean damage (calculated from 10 realizations), while for a structure 10 times wider (approximately $270D_{n50}$ wide), damage of 1.2 as the mean damage is expected. Thus, according to this study, the damage to a structure $270D_{n50}$ wide is expected to present damage up to 50% (1.2/0.8 = 1.5) larger than that obtained in a single realization in a $27D_{n50}$ characterization width. Detailed guides on how to account for the length effect on rock armored structures are presented in the work of [9].

3.3. Measuring Technique

Significant differences can be observed in the damage characterization outcome when using high resolution and high accuracy measurements techniques compared with low resolution measurements used in traditional surveys. In contrast to low resolution techniques, high resolution techniques (e.g., DSP) are able to measure the state of the whole structure surface and capture all the erosion areas.

These high resolution measuring techniques, such as DSP, present a number of advantages for the damage characterization of coastal structures. First, they allow the introduction of more precise damage parameters, such as $E_{3D,5}$, and allow the improvement in the damage limits and definitions. Thus, these high definition measuring techniques, together with suitable parameters and definitions, can describe the state of the full structure surface in order to identify the presence of weak areas, which threatens the stability of the structure. Second, for parameters obtained from width-averaged profiles such as S and E_{2D}, high resolution surveys that capture the state of the whole structure surface are able to reduce the variability and uncertainty in the damage characterization results, especially for situations with a relatively low amount of damage.

In summary, high resolution measurement techniques, in combination with the use of other damage parameters, are able to reduce the bias related to hidden erosion and random errors related to the position of the measured profiles, which are present in the damage characterization methods carried out with low resolution techniques. Thus, it can be concluded that innovative measuring techniques such as DSP are able to significantly improve the damage characterization of coastal structures.

4. Conclusions

Damage characterization for rock armored slopes was addressed in this paper, with a focus on three key elements: damage concepts, damage parameters, and measuring techniques. It is highlighted that these three elements should be taken into account for a reliable damage assessment. For this research, a physical modelling campaign was carried out within the EU Hydralab+ framework. The Deltares shallow water tests were performed in a wave flume with a foreshore (depth-limited waves), a 1:3 slope, and an impermeable core.

On damage concepts, the characterization width was discussed, because this influences the measured damage and the associated assessment description. The main effect is that for parameters obtained from width-averaged profiles (S and E_{2D}), the measured damage reduces with increased characterization width, because the damage to certain areas will be hidden by the accretion in other areas, while for parameters obtained as the maximum erosion depth observed within a given characterization width ($E_{3D,1}$ and $E_{3D,5}$), the measured damage increases with increased characterization width. Considering this and also the damage variability, it is recommended to carry out laboratory experiments with a characterization width of around $25D_{n50}$.

The calibrated damage limits and damage parameters for rock armored slopes with a characterization width of about $25D_{n50}$ and an armor thickness of $2D_{n50}$ are as follows:

- Damage initiation: defined as the condition in which a circular hole of $1D_{n50}$ diameter and a depth of $1D_{n50}$ is observed in the armor layer: $E_{3D,1} = 1$; $E_{3D,5} = 0.3$; $E_{2D} = 0.2$; $S = 1$.
- Intermediate damage: defined as the condition in which a circular hole of $1D_{n50}$ diameter and a depth of $1.5D_{n50}$ is observed in the armor layer: $E_{3D,1} = 1.5$; $E_{3D,5} = 0.7$; $E_{2D} = 0.5$; $S = 4$.
- Failure limit: defined as the condition in which a circular hole of $1D_{n50}$ diameter and a depth of nD_{n50} is observed in the armor layer: $E_{3D,1} = 2.0$; $E_{3D,5} = 1.1$; $E_{2D} = 0.9$; $S = 11$.

Regarding damage characterization parameters, different parameters were considered. It was found that $E_{3D,5}$ is the parameter that can better describe the damage and remaining strength of conventional and non-conventional structures. The key reasons for defining this parameter as the most suitable damage characterization parameter are the following:

- Low bias: the damage to the structure is clearly captured, without including hidden erosion present in the parameters that consider width-averaged profiles.
- Low random error: this parameter describes the damage to the structure with low variability also for a relatively low amount of damage, which increases the confidence in the measured and expected damage.
- Distinguish damage range: the different states of damage to the structure can be recognized according to the limits established.
- Constant value for different structures: for all structure configurations, this parameter can be used without the need for any modification.

In addition, the results from this study show that the damage measured with the damage parameter $E_{3D,5}$ presents an extreme value distribution. This illustrates that for increasingly long structures, a larger damage is expected.

Regarding damage characterization measuring techniques, the importance of high resolution measuring techniques for delivering accurate damage characterization of coastal structures was demonstrated. These techniques (such as DSP) allow the use of the most precise damage parameter, that is, $E_{3D,5}$.

Author Contributions: Conceptualization, Methodology, Validation and Investigation: E.d.A., M.R.A.v.G. and B.H.; Formal Analysis, Data Curation, Writing-Original Draft Preparation and Visualization: E.d.A.; Resources, Supervision and Funding Acquisition: M.R.A.v.G. and B.H.; Writing-Review & Editing: E.d.A., M.R.A.v.G. and B.H.

Funding: This project received co-funding from the European Union's Horizon 2020 research and innovation programme under grant agreement No 654110, HYDRALAB+.

Acknowledgments: We would like to thank our colleagues from Deltares and Hydralab+ for their contribution to the present work. Most of the research described here has first been presented and published [14] at the Coastlab 2018 conference (22–26 May 2018, Santander, Spain); the conference is acknowledged for providing the platform to present the described research.

Conflicts of Interest: The authors declare no conflict of interest.

References

1. Van der Meer, J.W. Rock Slopes and Gravel Beaches under Wave Attack. Ph.D. Thesis, Delft University of Technology, Delft, The Netherlands, 1988.
2. Hudson, R.Y. Laboratory investigation of rubble-mound breakwaters. *J. Waterw. Harb. Div.* **1959**, *85*, 93–121.
3. Thompson, D.M.; Shuttler, R.M. *Riprap Design for Wind-Wave Attack, a Laboratory Study in Random Waves*; Report EX707; Hydraulic Research Station: Wallingford, Oxford, UK, 1975.
4. Broderick, L.L. Riprap Stability versus Monochromatic and Irregular Waves. Ph.D. Thesis, Oregon State University, Corvallis, OR, USA, 1984.
5. Melby, J.A.; Kobayashi, N. Progression and variability of damage on rubble mound breakwaters. *J. Waterw. Port Coast. Ocean Eng.* **1998**, *124*, 286–294. [CrossRef]
6. Hofland, B.; Van Gent, M.R.A.; Raaijmakers, T.; Liefhebber, T. Damage evolution using the damage depth. In Proceedings of the Coastal Structures 2011, Yokohama, Japan, 5–9 September 2011.
7. Hofland, B.; Disco, M.; Van Gent, M.R.A. Damage characterization of rubble mound roundheads. In Proceedings of the Coastlab 2014, Varna, Bulgaria, 29 September–2 October 2014.
8. Van Gent, M.R.A.; De Almeida, E.; Hofland, B. Statistical analysis of the stability of rock armoured slopes. In Proceedings of the Coastlab 2018, Santander, Spain, 22–26 May 2018.
9. Van Gent, M.R.A.; De Almeida, E.; Hofland, B. Statistical analysis of the stability of rock armoured slopes. *J. Mar. Sci. Eng.* **2019**, submitted.
10. Raaijmakers, T.; Liefhebber, T.; Hofland, B.; Meys, P. Mapping of 3D-bathymetries and structures using stereo photography through air-water interface. In Proceedings of the Coastlab 2012, Ghent, Belgium, 17–20 September 2012.
11. Frostick, L.E.; McLelland, S.J.; Mercer, T.G. *User Guide to Physical Modelling and Experimentation: Experience of the Hydralab Network*; CRC Press/Balkena: Leiden, The Netherlands, 2011.
12. De Almeida, E. Damage Assessment of Coastal Structures in Climate Change Adaptation. Master's Thesis, Delft University of Technology, Delft, The Netherlands, 2017.
13. VNK. *The National Flood Risk Analysis for The Netherlands, Final Report*; Rijkswaterstaat VNK Project Office: Utrecht, The Netherlands, 2014.
14. De Almeida, E.; Van Gent, M.R.A.; Hofland, B. Damage characterisation of rock armoured slopes. In Proceedings of the Coastlab 2018, Santander, Spain, 22–26 May 2018.

Journal of
Marine Science and Engineering

Article

Applicability of Nonlinear Wavemaker Theory

Mads Røge Eldrup * and Thomas Lykke Andersen

Department of Civil Engineering, Aalborg University, 9220 Aalborg, Denmark; tla@civil.aau.dk
* Correspondence: mre@civil.aau.dk

Received: 22 November 2018; Accepted: 8 January 2019; Published: 14 January 2019

Abstract: Generation of high-quality waves is essential when making numerical or physically model tests. When using a wavemaker theory outside the validity area, spurious waves are generated. In order to investigate the validity of different wave generation methods, new model test results are presented where linear and nonlinear wave generation theories are tested on regular and irregular waves. A simple modification to the second-order wavemaker theory is presented, which significantly reduces the generation of spurious waves when used outside its range of applicability. For highly nonlinear regular waves, only the ad-hoc unified wave generation based on stream function wave theory was found acceptable. For irregular waves, similar conclusions are drawn, but the modified second-order wavemaker method is more relevant. This is because the ad-hoc unified generation method for irregular waves requires the wave kinematics to be calculated by a numerical model, which might be quite time-consuming. Finally, a table is presented with the range of applicability for each wavemaker method for regular and irregular waves.

Keywords: linear waves; nonlinear waves; wavemaker theory; wavemaker applicability

1. Introduction

When performing tests in laboratories or numerical models, high-quality waves representing conditions in prototype as close as possible is of highest priority. In the early 20th century, linear wavemaker theory was developed by Havelock [1], which was later extended to a fully second-order irregular wavemaker theory by Schäffer [2,3]. This extension made it possible to generate mildly nonlinear waves without spurious free waves. The spurious free waves contaminate the wave field and can easily be seen for regular waves for cases with low wave reflection, as the wave shape is not constant in space. For example, Orszaghova et al. [4] showed that using first-order wavemaker theory could lead to erroneously wave run-up and wave overtopping results compared to generating the waves with second-order wavemaker theory. Furthermore, Sriram et al. [5] showed that the breaking point of focused waves is different when using first-order and second-order wavemaker theory. This was expected to be caused by the influence from the free spurious long-wave components generated with first-order theory.

Recently, ad-hoc unified wavemaker theories were proposed by Zhang and Schäffer [6] for regular waves and by Zhang et al. [7] for irregular waves. These ad-hoc unified wave generation methods make it possible to generate highly nonlinear waves of high quality in intermediate and shallow water. These methods require a depth-averaged velocity as input to control the motion of the piston wavemaker. For regular waves, a fully nonlinear wave theory is available in the form of the stream function wave theory by Fenton and Rienecker [8], and from this, the depth-averaged velocity can be calculated. At the moment, there exists no analytical model to calculate the kinematics for highly nonlinear irregular waves, but the kinematics can be obtained by numerical models, for example, Boussinesq type wave models. The propagation of waves from deep to shallow water with numerical models can be time-consuming. Therefore, it is more efficient to use the first or second-order wavemaker theory for irregular waves when they are valid.

Unwanted free waves are generated when using a wavemaker theory outside its validity area. Schäffer [3] specified that the second-order wavemaker theory is not valid for regular waves when a secondary crest is produced in the wave trough. This happens when the second-order amplitude is larger than $\frac{1}{4}$ of the first-order amplitude. To describe this, he introduced the nonlinearity parameter, S which must not exceed unity for the second-order wavemaker theory to be valid. For regular waves, Schäffer [3] defined S as four times the ratio between the amplitudes of the second-order and the first-order components in regular waves and is given by:

$$S = 2\left|HG_{\mathrm{nm}}^{\pm}\right| \quad (1)$$

where $G_{\mathrm{nm}}^{\pm}$ is the second-order surface elevation transfer function given for example in Schäffer [3], H is the wave height. For the application of Equation (1) on irregular waves Schäffer [3] proposed to use a characteristic wave height ($H = H_{1/3}$) and $f_{\mathrm{n}} = f_{\mathrm{m}} = f_{\mathrm{p}}$ to calculate $G_{\mathrm{nm}}^{\pm}$.

A more well-known approach to check the applicability of wave maker theories is the diagram by Le Méhauté [9], which described what wave theory was valid depending on the relative water depth and the wave steepness. Figure 1 shows an example of the Le Méhauté diagram and colored areas that illustrates different S ranges calculated with Equation (1). From the figure, it is seen that using the applicability criteria for second-order waves given by Schäffer [3] corresponds to fourth order Stokes waves according to the diagram by Le Méhauté [9]. This clearly shows the need of testing the applicability range of the existing methods and to provide some recommendations that can easily be followed.

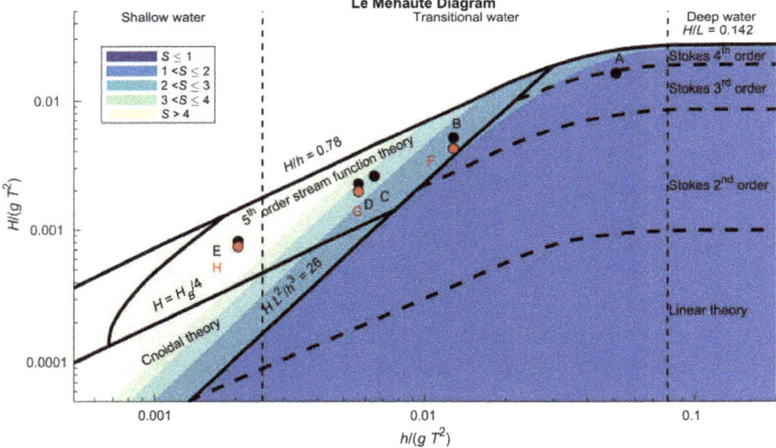

Figure 1. Tested sea states, where black dots are for the regular waves and red dots are for irregular waves. The colored areas show different ranges for the nonlinearity parameter S calculated with Equation (1). (Adapted from Le Méhauté [9]).

The present paper presents new physical model test results to study the applicability of the different wavemaker methods for regular and irregular waves. A simple modification to the second-order wavemaker theory by Schäffer [3] is proposed to extend its range of applicability and to significantly reduce errors for highly nonlinear cases. Furthermore, recommendations are given for the validity of each of the tested wavemaker theories.

2. Present Study

New model tests covering linear to highly nonlinear waves have been performed in the new wave flume at Aalborg University for regular (5 tests) and irregular (3 tests) waves, see Figure 1.

The purpose of the tests is to study the applicability of each wavemaker theory and to provide a clear definition for the applicability range of each wavemaker theory. Furthermore, is it investigated if the applicability range of the second-order wavemaker theory can be extended with a simple modification. The second-order wavemaker theory is modified in the present paper by introducing a maximum allowed relation between the second-order amplitude and the first-order amplitude. This limitation is controlled by a S_{max} value, so for example, $S_{max} = 1$ means that the second-order amplitude will be reduced to $\frac{1}{4}$ of the first-order amplitude if it was predicted higher. Without this upper limit, the second-order amplitude might be even larger than the first-order amplitude which is physically incorrect. Instead the second-order component should saturate at a level smaller than the primary component and third and higher order components should increase instead.

The second-order components generated by two interacting components were scaled by introducing a more generalized form of S. This is done by rewriting Equation (1) to consider the two interacting components:

$$S(f_n, f_m) = HG^{\pm}_{nm}/\delta_{nm} \tag{2}$$

where f_n and f_m are the frequencies of the two interacting components. δ_{nm} is 0.5 for $f_n = f_m$ and 1 for $f_n \neq f_m$. In the present paper the characteristic wave height is on the safe side taken as the maximum wave height ($H = 2H_{m0} \approx H_{max}$).

The modification to the second-order wavemaker theory is performed by the scaling factor, λ, given by Equation (3).

$$\lambda_{nm} = \min\left(1, \frac{S_{max}}{S(f_n, f_m)}\right) \tag{3}$$

The scaling factor should be calculated for all interacting frequencies in the first-order wave spectra and be multiplied to both second-order transfer functions $G^{\pm}_{nm}$ and $F^{\pm}_{nm}$. $G^{\pm}_{nm}$ is the transfer function for the second-order surface elevation and $F^{\pm}_{nm}$ is the transfer function of the second-order paddle movement. This reduces the amount of second-order energy so no secondary crest is calculated for the interaction of the individual frequencies. The optimal value of S_{max} is expected in this interval ($1 \leq S_{max} \leq 4$) which corresponds to a second-order amplitude of 25–100% of the first-order amplitude. Although the criteria of $S \leq 1$ given by Schäffer [3] is for superharmonics in regular waves the scaling is applied also to irregular waves. The present paper focus only on superharmonics (G^{+}_{nm} and F^{+}_{nm}), but the scaling is expected necessary also to subharmonics (G^{-}_{nm} and F^{-}_{nm}). Correct generation of the subharmonics is very important for response of many structures, but the present paper focus only on the superharmonics for which free energy is much easier to observe in the measured time series.

3. Theoretical Optimal S_{max}

The nonlinear wave theory by Fenton and Rienecker [8] can estimate the correct amount of second-order energy that exists for a given regular wave over a horizontal sea bed. With the use of nonlinear wave theory, it is thus possible to estimate what artificial limit (S_{max} in Equation (3)) should be used in the second-order wave theory to obtain the best results. In Figure 2, the calculated amplitude of the second-order component by second-order wavemaker theory is compared with the calculated by stream function wave theory by Fenton and Rienecker [8]. This is done for different values of S_{max}, where $S_{max} = \infty$ corresponds to no limit on the second-order energy (unmodified Schäffer [3] method).

The given applicability range by Schäffer [3] ($S \leq 1$) shows errors smaller than 10% when using the second-order wavemaker theory by Schäffer [3], see Figure 2. The figure shows that the recommendations given by Le Méhauté [9] based on this analysis might be on the safe side. However, this comparison though only considers the amplitude of the second-order component, but higher order components might be relevant.

By limiting the second-order energy to $S_{max} = 1$ the modified second-order wave generation gives errors smaller than 10% for wave conditions up to $S \approx 1.5$ as illustrated by the dotted line in the figure. Using values of S_{max} = 1.5–1.75 gives a larger area where the error is below 10%. Even though $S_{\max}$ = 1.5–1.75 gives a larger area where the second-order amplitude is calculated with a small error, the second-order wavemaker theory is not necessarily valid as the third, and higher order energy might have a significant contribution. Therefore, the relation between the second-order and third order components must additionally be small. In Figure 3, the third order amplitude is compared to the first-order amplitude calculated by stream function wave theory. The figure shows that for sea states with $S \leq 1.5$ the amplitude of the third order component is below 10% of the first-order component. From these results, an optimum of $S_{\max} = 1$ is found, and the expected range of applicability is extended from $S = 1$ to $S = 1.5$ with the modified second-order wavemaker theory.

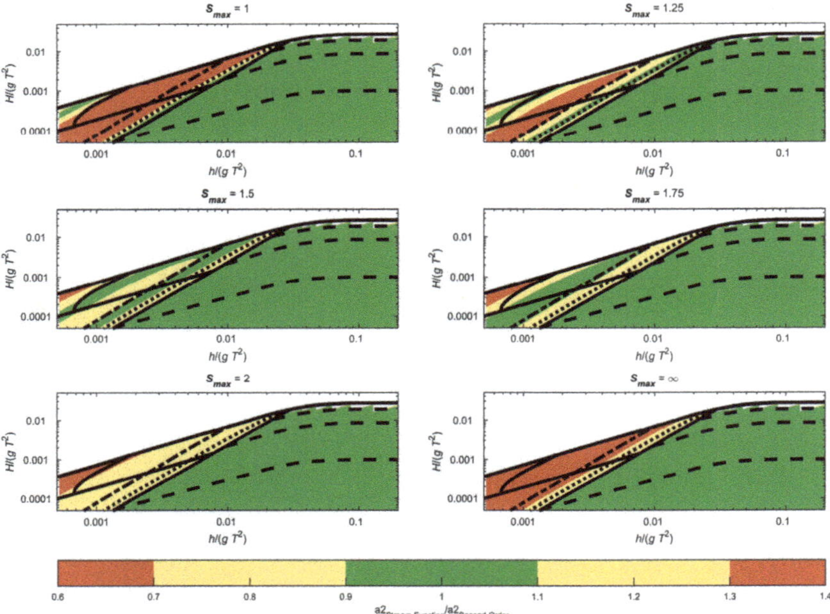

Figure 2. Comparison between the second-order amplitude calculated with second-order and stream function wave theory. $S = 1.5$ for the dotted line and $S = 3$ for the dashed-dotted line, which can also be seen in Figure 1 by the colored areas.

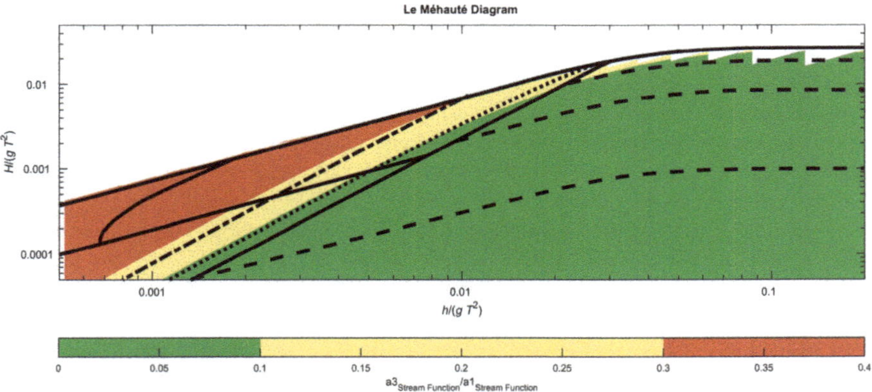

Figure 3. Comparison between the third order and first-order amplitude calculated with stream function wave theory. $S = 1.5$ for the dotted line and $S = 3$ for the dashed-dotted line.

4. Model Test Setup and Methodology

To verify the results from Section 3, physically experiments have been performed where the surface elevation was measured in different locations in the wave flume, see Figure 4.

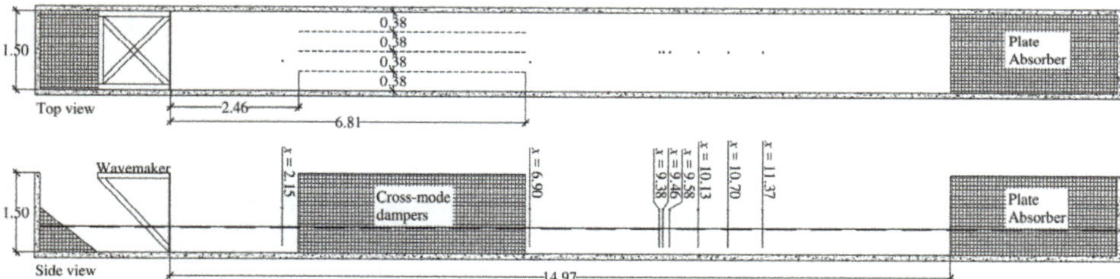

Figure 4. Experimental setup of the wave flume. Measures in meters.

The flume is equipped with a piston-type wavemaker, cross-mode dampers and a passive absorption system with perforated sheets in the end of the flume. The cross-mode dampers are a permanent installation in the flume and not installed due to specific cross-mode problems with the generated waves in the present study. The wavemaker has two sets of flush mounted resistance wave gauge on the wave board, which is used to avoid re-reflected waves from the wave board with use of the active wave absorption method by Lykke Andersen et al. [10]. The active absorption system has proven also to be effective for nonlinear irregular waves, cf. Lykke Andersen et al. [11]. The control signals are thus the wavemaker position and the surface elevation in the nearfield.

To evaluate the validity of each wavemaker theory, the measured surface elevation is compared to the theoretical surface elevation. The regular waves are compared to the predicted surface elevation by Fenton and Rienecker [8] and the irregular waves are compared to the COULWAVE Boussinesq wave model by Lynett and Liu [12]. The COULWAVE model is an accurate model when it comes to shoal waves from deep to shallow water including nonlinear interactions, see [13,14]. The results from the COULWAVE model was also used to generate the input for the ad-hoc unified irregular wave generation method by Zhang et al. [7].

The COULWAVE model was used to propagate irregular waves from deep water to the water depth used in the laboratory (0.5 m). The water depth at the wavemaker in COULWAVE is a compromise between the wave regime where the model is valid, $kh < 6$, and that regime where first-order wavemaker theory can be used. The waves were generated in the COULWAVE model on a horizontal part using a narrow banded JONSWAP spectrum ($\gamma = 10$) and truncated at 0.5 to 1.5 times the peak frequency. Thus, the primary spectrum does not significantly overlap with the bound harmonics. Then followed a 1:100 slope to the depth of 0.5 m and finally a horizontal part with a water depth of 0.5 m. The point where the 1:100 slope end, the surface elevation time series was extracted and used as a target for the different wave generation methods. The last horizontal part is where the target values are extracted to be compared with the measured surface elevation in the laboratory. Table 1 shows the numerical model parameters and the sea state conditions at the generation point and at $h = 0.5$ m.

Table 1. Wave conditions of the irregular waves at the generation point in COULWAVE.

Model Parameters			Generation Point			$h = 0.5$ m
Case	Cells per Wavelength	Courant Number	h [m]	H_{m0} [m]	T_P [s]	H_{m0} [m]
F	200	0.5	2.5	0.102	2	0.083
G	200	0.5	4.0	0.087	3	0.087
H	300	0.5	8.0	0.068	5	0.091

The number of cells pr. wavelength is based on the wavelength of the peak wave period at the generation point. Thus, the discretisation of the model is, Δx = (cells per wavelength)/wavelength. The courant number is calculated by $C = c_0 \Delta t / \Delta x$, where c_0 is the shallow water celerity at the generation point and Δt is the timestep.

The method by Zhang et al. [7] uses the averaged velocity in the wave direction, but since only the surface elevation is provided from the COULWAVE model, a conversion to depth-averaged velocity is performed by assuming shallow water wave theory being valid. That is not entirely correct in intermediate water, but a fair approximation.

Because the generated primary spectrum in COULWAVE is narrow banded and truncated (0.5 to 1.5 times the peak frequency), the primary components can be separated from the bound harmonics by bandpass filtering at $0.5 f_P < f < 1.5 f_P$. This truncated spectrum was used as primary spectrum for the second-order wave generation method. The signal without bandpass was used as input for the method by Zhang et al. [7]. The measured waves in the laboratory were then compared with the unfiltered signals from the numerical model in the wave gauge locations.

The theoretical surface elevations calculated with COULWAVE and Fenton and Rienecker [8] might differ from the measured surface elevation due to the cross-mode dampers. The cross-mode dampers might slightly dissipate some of the energy in the physical flume. This is likely to be seen as a reduction in wave height when comparing the measured surface elevation at x = 2.15 m and x = 6.90 m. Furthermore, the theoretical surface elevation is without reflected waves which is difficult to entirely avoid in the physical model. Therefore, reflection in the experimental tests should be reduced to a minimum. The amplitude reflection coefficient in the physical tests was in the range of 11% to 16%, which was calculated with the nonlinear irregular wave separation method by Eldrup and Lykke Andersen [15]. The amount of reflection in the physical experiments is found acceptable for comparing the total measured surface elevation with the theoretical surface elevation.

5. Regular Wave Results

For regular waves, the different wavemaker theories are evaluated against the theoretical stream function wave theory. This evaluation is performed at different distances from the wavemaker. Linear to highly nonlinear waves were generated and compared to the theoretical wave profiles. The measured surface elevations are shown after the ramp-up of the wavemaker is completed and if possible before reflection is present in the signals.

Figure 5 shows the results for Sea State A. The measured profiles are almost identical for the different wave generation methods. However, a reduction in the amplitude is observed for $x \geq 6.90$ m. The reduction in amplitude might be due to the cross-mode dampers. The shape of the measured surface elevations for all the wavemaker theories are similar to the theoretical. Therefore, it can be concluded that all the tested wavemaker methods are valid for Sea State A, and thus the first and second-order wave generation methods lead to acceptable waves, in a more extensive area than given by the Le Méhauté diagram. However, Sea State A is within the validity range given by Schäffer [3] ($S < 1$).

The results for Sea State B are shown in Figure 6 from which it appears that the first-order wavemaker method lead to some minor deviations in the wave shape, indicating some free higher harmonics energy exist. The surface profile for the second-order methods and the ad-hoc unified generation method are similar, and these are close to the theoretical profile. The wave height of the generated waves is slightly lower than the theoretical, but the wave shape is a close match to the target. From the section with the theoretical analysis of an optimum of S_{max}, a difference was expected to be seen between the original second-order and the modified second-order method when $S > 1.0$, but this is not observed from Figure 6 for a case with $S = 1.46$.

Results for Sea State C shows that the wave profile generated with first and second-order wavemaker theory is not constant in the various gauge positions, cf. Figure 7. This is due to free unwanted waves being generated by these theories when the nonlinearity is too high. The second-order wavemaker theory is slightly better than first-order theory, but the ad-hoc unified wave generation is a close match to the theoretical profile. For $x = 2.15$ m, the wave crest of the generated waves by the second-order wavemaker theory by Schäffer [3] is much larger than the theoretical wave crest, and the proposed correction to the second-order method leads to a wave profile closer to the theoretical profile. These results shows that Schäffer [3] overestimates the second-order amplitude when $S > 1$.

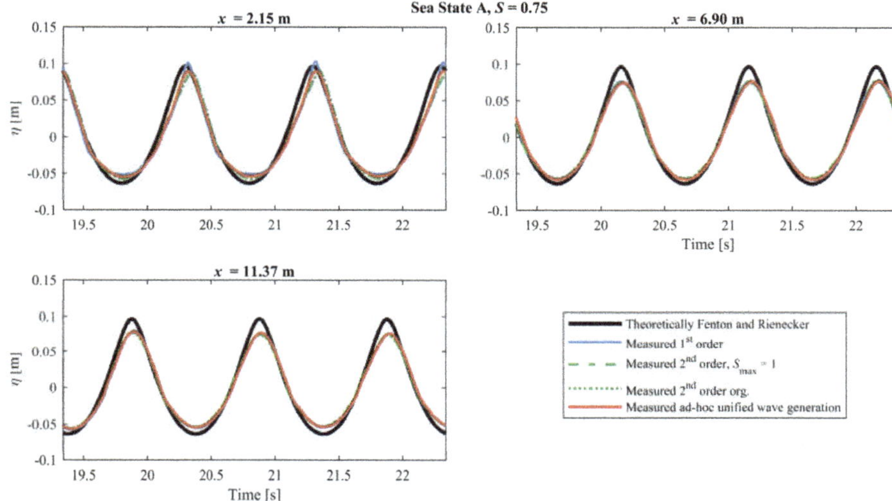

Figure 5. Theoretically surface elevation compared with the first-order, second-order, modified second-order and the ad-hoc unified wavemaker methods for Sea State A.

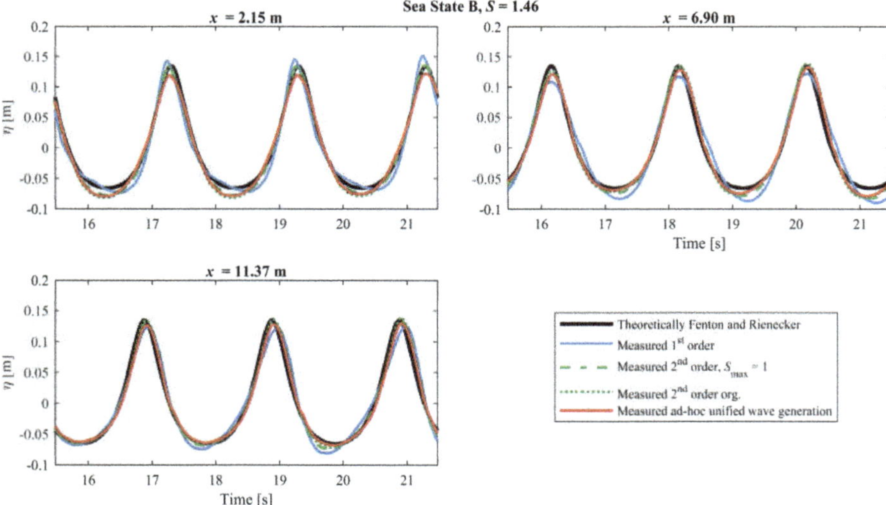

Figure 6. Theoretically surface elevation compared with the first-order, second-order, modified second-order and the ad-hoc unified wavemaker methods for Sea State B.

Sea State D is shown in Figure 8. Again, large waves are observed in $x = 2.15$ m for the second-order wavemaker theory, and a small wave crest is seen in the wave trough. The first-order wavemaker theory is actually better than the second-order wavemaker theory for $x = 2.15$ m. This indicates the second-order method actually increases the amplitude of the free waves compared to first-order wavemaker theory. The large waves generated by second-order wavemaker theory were causing breaking waves, which did not occur for the other generation methods. The modified second-order wavemaker theory is performing better than first-order and second-order wavemaker theory, but free waves are also observed for that method. The ad-hoc unified wave generation is a close match to the theoretical wave profile except for the observed reduction in amplitude for $x \geq 6.90$ m, which is likely due to the cross-mode dampers.

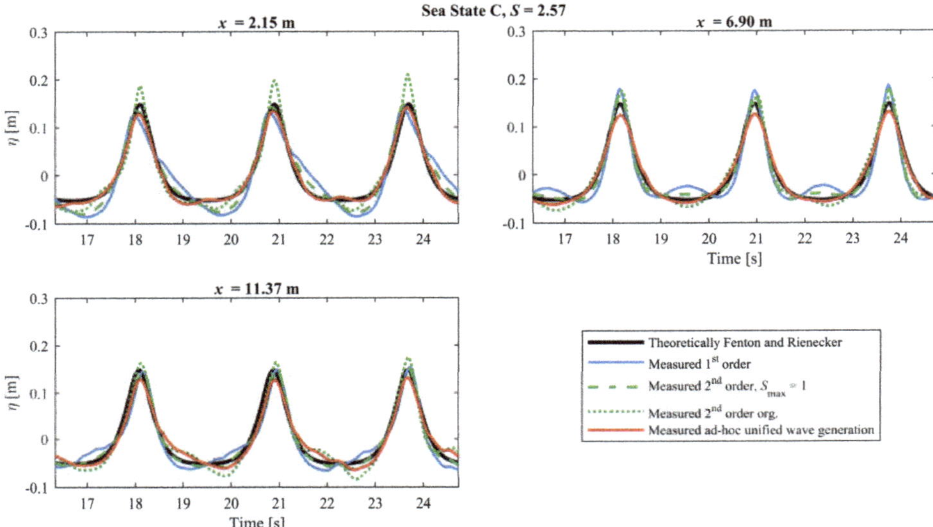

Figure 7. Theoretically surface elevation compared with the first-order, second-order, modified second-order and the ad-hoc unified wavemaker methods for Sea State C.

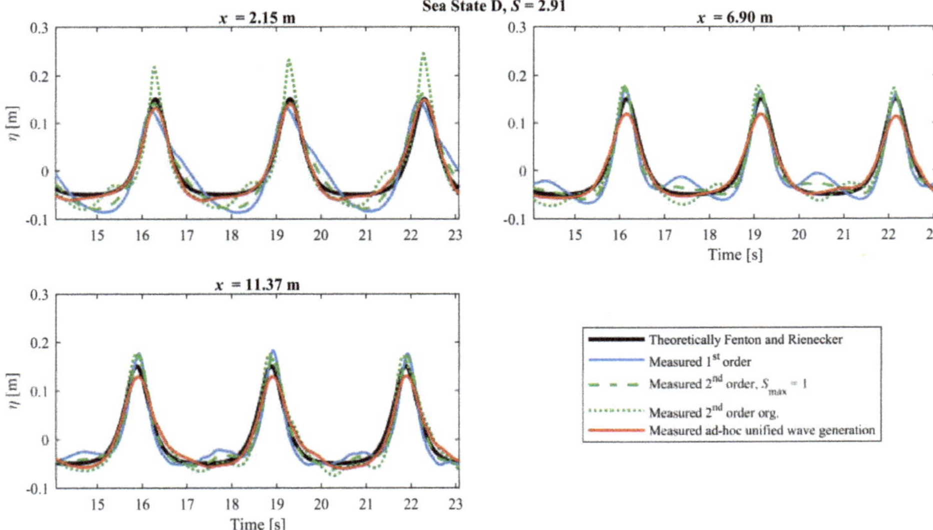

Figure 8. Theoretically surface elevation compared with the first-order, second-order, modified second-order and the ad-hoc unified wavemaker methods for Sea State D.

Results for Sea State E is shown in Figure 9. For that case second-order wavemaker theory generates too large waves in x = 2.15 m and a secondary wave is also seen in the wave trough of the primary wave. The first-order wavemaker theory is significantly better than the second-order wavemaker theory as no secondary waves are seen in x = 2.15 m. As in Figure 8, the waves generated by second-order wavemaker theory are breaking during the tests, which is the reason for the large wave crest are not observed for the two other locations in the figure. The modified second-order wavemaker theory is slightly better than the first-order wavemaker theory, but they are both far from the theoretical profile. The ad-hoc unified wave generation is a close match to the theoretical profile except for small deviations in the trough, but this is most likely due to reflections from the passive absorber in the wave flume.

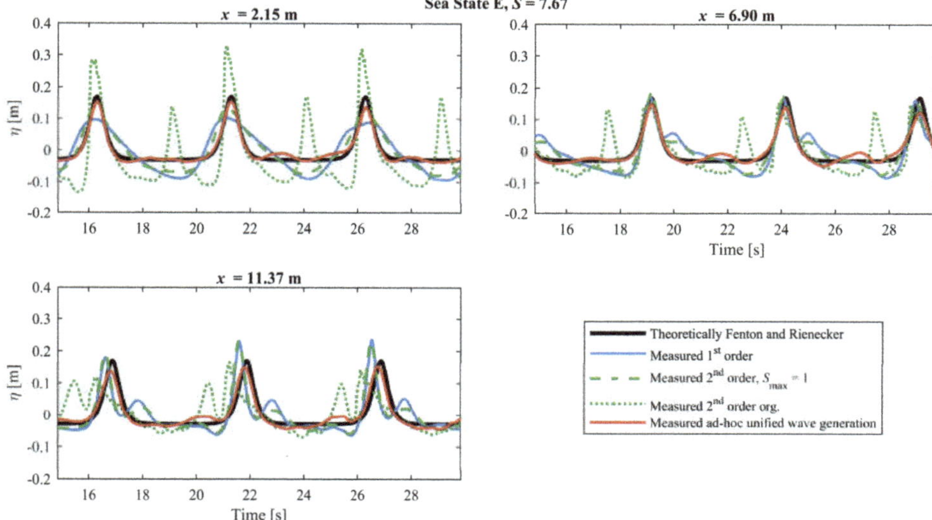

Figure 9. Theoretically surface elevation compared with the first-order, second-order, modified second-order and the ad-hoc unified wavemaker method for Sea State E.

A more detailed analysis of the measured waves for Sea State E is shown in Figure 10 for x = 2.15 m. The time series of the primary and the first three superharmonics is calculated by Fourier transformation on the discrete frequencies $n\omega$. Thus, the measured components can be compared with the theoretical stream function wave components. The figure shows that the original second-order wavemaker theory generates higher harmonics that are significantly higher than the theoretical stream function wave. It also shows that only the ad-hoc method generates the correct amplitudes of the different contributions. Furthermore, a phase shift between the superharmonics and the first-order component is observed especially for the first-order generation method and the original second-order method. This phase shift demonstrates significant free energy to be present. Only the ad-hoc method has correct amplitudes and phases for all components.

It has been shown that first-order wavemaker theory can be used for $S \leq 0.8$ and that the modified second-order wavemaker theory gives identical results as the second-order theory by Schäffer [3] when the second-order theory is valid. However, the modified second-order method performs significantly better when used outside the validity area. This is due to a more realistic value of the amplitude for the second-order harmonic, which for the method by Schäffer [3] has no upper limit and for the present tests is calculated to approximately two times the first-order amplitude in the worst case (Sea State E). In that case, the higher order harmonics have a significant influence, and therefore second-order theory is not valid for this case. This is shown in Figure 10 by comparing the contributions of the first four harmonics with the theoretical amplitudes. The second-order wavemaker theory can be used for $S \leq 1.5$ with reasonable results. The ad-hoc unified generation can be used for all the tested conditions (tested up to $S = 7.7$).

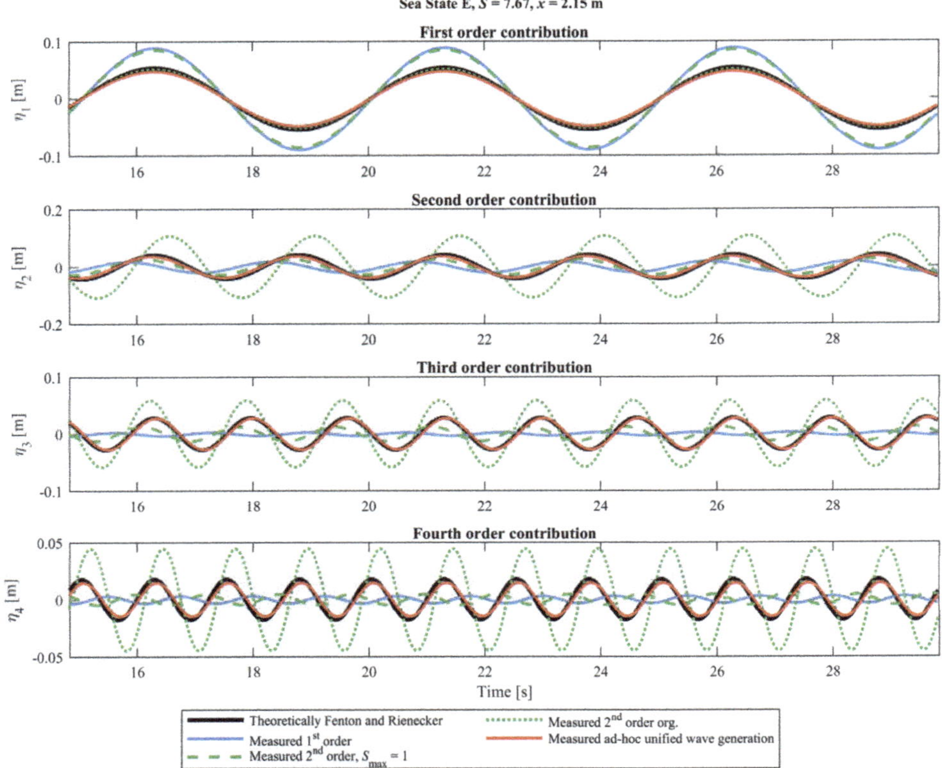

Figure 10. Contribution of the first four harmonics to the surface elevation for Sea State E at x = 2.15 m. First-order, second-order, modified second-order and the ad-hoc unified wavemaker method are compared with the theoretical harmonics in stream function waves.

6. Irregular Wave Results

The performance of the different wavemaker methods for the irregular waves are compared with the results from the COULWAVE model at different distances from the wavemaker. The measured surface elevations are shown for a time window including the highest wave.

Results for Sea State F are given in Figure 11. The first and second-order wavemaker theories are according to the Le Méhauté diagram not expected to be valid for this sea state. Furthermore, the test is also outside the applicability range given by Schäffer [3]. However, the figure shows that all the tested methods give similar results and that they are a close match to the COULWAVE model with only minor deviations.

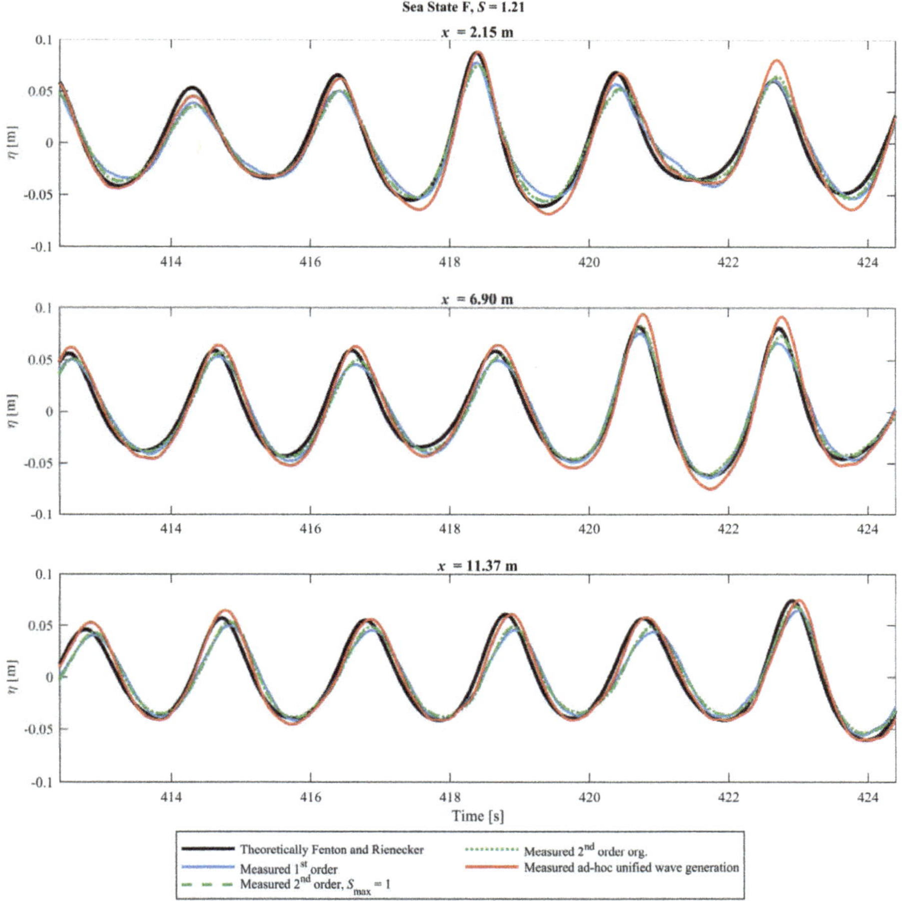

Figure 11. Numerically surface elevation compared with the first-order, second-order, modified second-order and the ad-hoc unified wavemaker methods for Sea State F.

Figure 12 presents the results for Sea State G. For this case is the wave crest of the largest wave with the first-order wavemaker theory significantly smaller than the theoretical wave profile at x = 2.15 m. The original second-order wavemaker theory is closer to the numerical profile except for some deviations in the wave trough for x = 2.15 m. For the three wave gauges, it can be seen that the wave crest is too small for the largest waves when the waves are generated with the modified second-order wavemaker method, but except for that, it is a close match. The shape of the waves generated by the ad-hoc unified generation method is a close match but with a deeper wave trough at x = 2.15 m, but this is likely due to long reflected waves as it is not seen for the two other gauges.

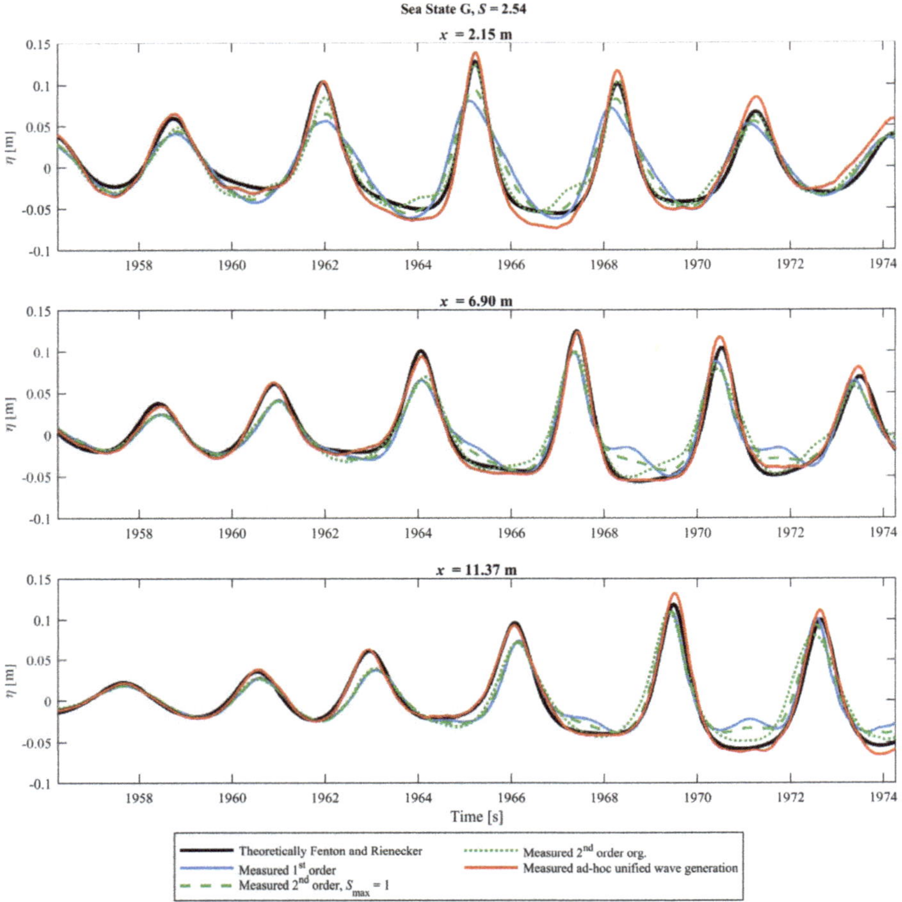

Figure 12. Numerically surface elevation compared with the first-order, second-order, modified second-order and the ad-hoc unified wavemaker methods for Sea State G.

For Sea State H, the first-order wavemaker theory is deviating significantly from the numerical profile as the wave trough is not flat and long and the wave crest is not high and narrow, cf. Figure 13. The wave profile generated by the first-order wavemaker theory has a step front followed by a gentler rear as the waves propagate away from the wavemaker—this is due to the free and bound waves having different celerity. For $x = 11.37$, the free and bound waves are phase shifted to such a degree that the crest of both the free and the bound waves are visible. The second-order wavemaker method has a secondary crest in the wave trough, which is reduced significantly with the modified second-order method. The wave crest with the modified second-order wavemaker method is though significantly smaller than the theoretical but is still closer to the theoretical compared to the first-order wavemaker method. The modified method is better than the original second-order method as less free energy is generated, but for this case the amount of secondary energy is also much smaller than the target. The ad-hoc unified wavemaker method provides a close match to the numerical profile with only minor deviations, and for this sea state, only this method leads to acceptable results.

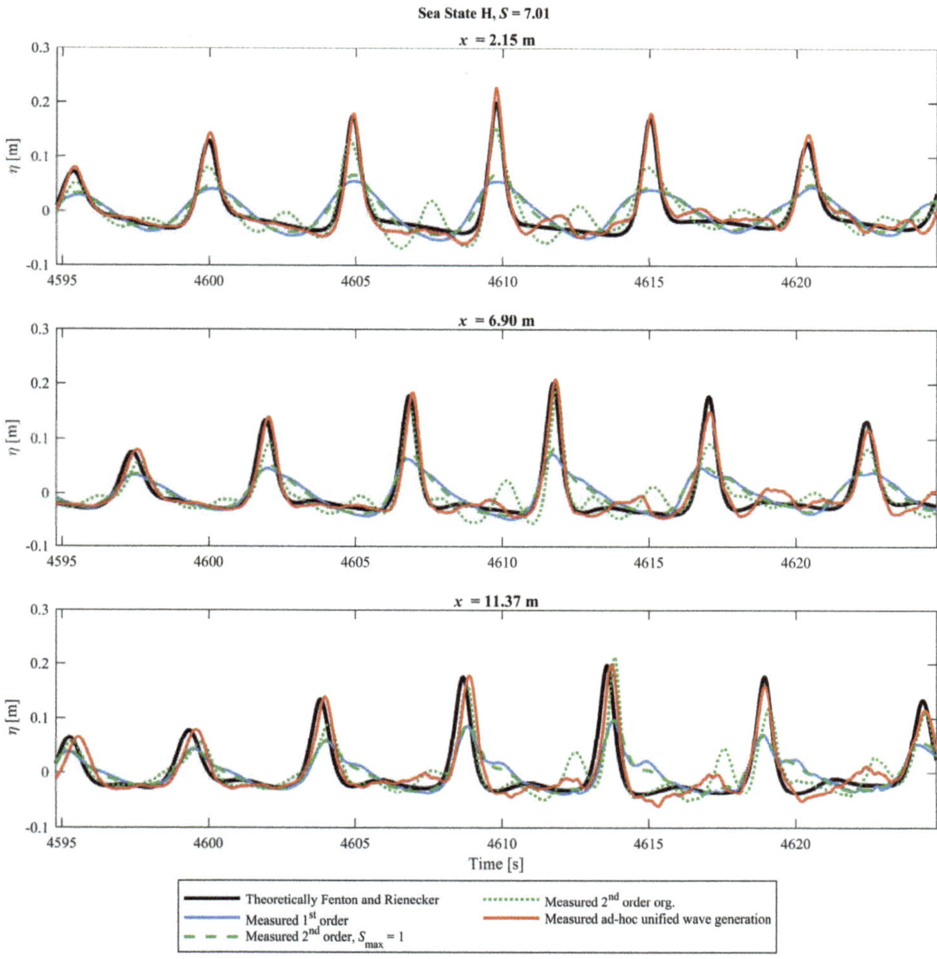

Figure 13. Numerically surface elevation compared with the first-order, second-order, modified second-order and the ad-hoc unified wavemaker methods for Sea State H.

For irregular waves, the present tests showed that first-order wavemaker theory can be used for $S \leq 1.2$, while the second-order wavemaker theory can be used for $S \leq 2$ with reasonable results. The ad-hoc unified generation showed good performance and can be used for all the tested conditions (tested up to S = 7.0).

7. Conclusions

Model tests have been performed with the purpose to generate waves of various nonlinearity and evaluate different wavemaker methods. First-order, second-order and ad-hoc unified generation methods have been tested, and the validity range of each method has been found by physical experiments. Furthermore, a modification to the second-order wavemaker method was proposed to increase the validity to more nonlinear waves. The ad-hoc unified generation method for the regular waves was the most accurate wavemaker method, and if possible, it should be used, otherwise the modified second-order is preferable. For the irregular waves, the ad-hoc unified generation method is also the most accurate one, but it is also much more time-consuming in synthesize the signals. This is because a time series has to be prepared by a numerical model before waves can be generated in the physical model. Therefore, it is recommended to use the modified second-order wavemaker theory when it is valid.

Table 2 summarizes the applicability ranges of each wavemaker method. For the regular waves, acceptable results are found with first-order wavemaker when $S < 0.8$, while second-order methods extends the validity to $S < 1.5$. The ad-hoc method is applicable for all the tested conditions ($S < 7.7$), but is expected to be reliable for all S values. For the irregular waves, acceptable results are found with first-order theory when $S < 1.2$. For both second-order methods acceptable results for irregular waves were found when $S < 2.0$. The ad-hoc method showed also for irregular waves good results for all the tested conditions. As it has not been possible to test all possible sea states these results should be taken as preliminary, but can be used until more tests have been performed.

Table 2. Applicability of each tested wavemaker method.

	Regular Waves				Irregular Waves			
	First-order	Second-order	Modified second-order	Ad-hoc unified	First-order	Second-order	Modified second-order	Ad-hoc unified
S	0.8	1.5	1.5	∞	1.2	2.0	2.0	∞

Author Contributions: M.R.E. wrote the original draft and performed all the simulations, experiments and analysis. Furthermore, he established the modified second-order wavemaker theory; T.L.A. reviewed the paper and contributed to the performed analysis and in establishing the modified second-order wavemaker theory.

Funding: This research received no external funding.

Conflicts of Interest: The authors declare no conflict of interest.

References

1. Havelock, T.H. LIX. Forced surface waves on water. *Philos. Mag.* **1929**, *8*, 569–576. [CrossRef]
2. Schäffer, H.A. Laboratory Wave Generation Correct to Second Order. In *Wave Kinematics and Environmental Forces: Papers Presented at a Conference Organized by the Society for Underwater Technology and Held in London, U.K., 24–25 March 1993*; Springer: Dordrecht, The Netherlands, 1993; pp. 115–139.
3. Schäffer, H.A. Second-order wavemaker theory for irregular waves. *Ocean Eng.* **1996**, *23*, 47–88. [CrossRef]
4. Orszaghova, J.; Taylor, P.H.; Borthwick, A.G.L.; Raby, A.C. Importance of second-order wave generation for focused wave group run-up and overtopping. *Coast. Eng.* **2014**, *94*, 63–79. [CrossRef]
5. Sriram, V.; Schlurmann, T.; Schimmels, S. Focused wave evolution using linear and second order wavemaker theory. *Appl. Ocean Res.* **2015**, *53*, 279–296. [CrossRef]
6. Zhang, H.; Schäffer, H.A. Approximate Stream Function wavemaker theory for highly non-linear waves in wave flumes. *Ocean Eng.* **2007**, *34*, 1290–1302. [CrossRef]
7. Zhang, H.; Schäffer, H.A.; Jakobsen, K.P. Deterministic combination of numerical and physical coastal wave models. *Coast. Eng.* **2007**, *54*, 171–186. [CrossRef]
8. Fenton, J.D.; Rienecker, M.M. Accurate Numerical Solutions for Nonlinear Waves. In *Coastal Engineering 1980*; American Society of Civil Engineers: New York, NY, USA, 1980; pp. 50–69.
9. Méhauté, B. *An Introduction to Hydrodynamics and Water Waves*; Springer: Berlin/Heidelberg, Germany, 1976.
10. Lykke Andersen, T.; Clavero, M.; Frigaard, P.; Losada, M.; Puyol, J.I. A new active absorption system and its performance to linear and non-linear waves. *Coast. Eng.* **2016**, *114*, 47–60. [CrossRef]
11. Lykke Andersen, T.; Clavero, M.; Eldrup, M.R.; Frigaard, P.; Losada, M. Active Absorption of Nonlinear Irregular Wwaves. In Proceedings of the Coastal Engineering Conference, Baltimore, MD, USA, 30 July–3 August 2018.
12. Lynett, P.; Liu, P.L.-F. A two-layer approach to wave modelling. *Proc. R. Soc. Lond. Ser. A Math. Phys. Eng. Sci.* **2004**, *460*. [CrossRef]
13. Hsiao, S.-C.; Lynett, P.; Hwung, H.-H.; Liu, P.L.-F. Numerical simulations of nonlinear short waves using a multilayer model. *J. Eng. Mech.* **2005**, *131*, 231–243. [CrossRef]

14. Teixeira, P.R.F.; Pinheiro, L.; Fortes, C.M.J.; Carreiros, C.; Grande, R. Comparison of Three Nonlinear Models to Analyze Wave Propagation Over Submerged Trapezoidal Breakwaters. In Proceedings of the V European Conference on Computational Fluid Dynamics, Lisbon, Portugal, 14–17 June 2010; pp. 14–17.
15. Eldrup, M.R.; Lykke Andersen, T. Estimation of Incident and Reflected Wave Trains in Highly Nonlinear Two-Dimensional Irregular Waves. *J. Waterw. Port Coast. Ocean Eng.* **2019**, *145*.

Journal of
Marine Science and Engineering

Article

Cubipod® Armor Design in Depth-Limited Regular Wave-Breaking Conditions

M. Esther Gómez-Martín [1,*], María P. Herrera [2], Jose A. Gonzalez-Escriva [1] and Josep R. Medina [1]

1 Laboratory of Ports and Coasts, Institute of Transport and Territory, Universitat Politècnica de València, 46022 Valencia, Spain; jgonzale@upv.es (J.A.G.-E.); jrmedina@upv.es (J.R.M.)
2 PROES Consultores S.A., 28020 Madrid, Spain; mherrera@proes.engineering
* Correspondence: mgomar00@upv.es; Tel.: +34-963-877-375

Received: 17 October 2018; Accepted: 4 December 2018; Published: 6 December 2018

Abstract: Armor stability formulas for mound breakwaters are commonly based on 2D small-scale physical tests conducted in non-overtopping and non-breaking conditions. However, most of the breakwaters built around the world are located in breaking or partially-breaking wave conditions, where they must withstand design storms having some percentage of large waves breaking before they reach the structure. In these cases, the design formulas for non-breaking wave conditions are not fully valid. This paper describes the specific 2D physical model tests carried out to analyze the trunk hydraulic stability of single- and double-layer Cubipod® armors in depth-limited regular wave breaking and non-overtopping conditions with horizontal foreshore ($m = 0$) and armor slope (α) with $\cot\alpha = 1.5$. An experimental methodology was established to ensure that 100 waves attacked the armor layer with the most damaging combination of wave height (H) and wave period (T) for the given water depth (h_s). Finally, for a given water depth, empirical formulas were obtained to estimate the Cubipod® size which made the armor stable regardless of the deep-water wave storm.

Keywords: mound breakwater; armor stability; Cubipod®; breaking waves; non-overtopping; horizontal foreshore; regular waves

---

1. Introduction

Breakwaters are usually tested using small-scale models in non-breaking wave conditions. Since the failure of large breakwaters in the late 1970s and early 1980s (e.g., Sines West Breakwater, Portugal), most physical tests have been carried out using irregular waves to better reproduce the highest waves in prototype conditions. However, most mound breakwaters are placed in depth-limited wave breaking conditions, where the highest waves break on the foreshore before reaching the structure.

Since the pioneering work of Iribarren (1938) [1], several formulas have been developed to characterize the hydraulic stability of rock armors, such as those provided by Hudson (1959) [2], Van der Meer (1988) [3], Van Gent et al. (2003) [4] and other authors. An in-depth literature review was undertaken by Herrera et al. (2017) [5]. Most of these formulas are based on 2D physical tests with models in non-breaking wave conditions. Spectral significant wave height, $H_s = H_{m0} = 4(m_0)^{0.5}$, at the toe of the structure and wave height with a 2% exceedance probability, $H_{2\%}$, are usually considered to describe the incident wave characteristics in non-breaking wave conditions. $H_{2\%}$ is strongly correlated to H_{m0} in deep water when wave heights are Rayleigh-distributed, but this is not the case in depth-limited breaking wave conditions (see Battjes and Groenendijk 2000) [6].

For shallow water and horizontal foreshore, the water depth (h_s) is the key variable for designing mound breakwaters because the maximum wave height (H_{max}) attacking the structure mostly depends on h_s. In these conditions, irregular waves may cause less armor damage than regular waves. In a given wave storm, the highest waves may break on the foreshore while smaller waves cause irrelevant armor damage; only a very low proportion of waves will have a wave height (H) close to the maximum wave height ($H \approx H_{max}$) to be relevant when analyzing the hydraulic stability of the armor layer. By contrast, runs of regular waves with the appropriate wave height (H) and wave period (T) can be generated in 2D physical experiments to attack the structure with many waves, say 100 waves, close to the maximum wave height, which would never be obtained with runs of a few thousand irregular waves propagating on a horizontal foreshore.

Given a water depth (h_s), a test matrix of regular wave runs covering the full range of wave height (H) and wave period (T), guarantees that the structure will be attacked by many of the most damaging waves. In these conditions, armor damage under the full test matrix of regular waves is higher than damage caused by a conventional test matrix of irregular wave runs (1,000 waves). Design rules based on the results using regular waves are on the safe side because the structure in shallow water and horizontal foreshore will never be attacked by such a large number of high waves, regardless of the design storm.

This paper describes the specific 2D physical model tests carried out to analyze the hydraulic stability of single- and double-layer Cubipod® armored breakwaters on a horizontal foreshore, in depth-limited regular wave breaking and non-overtopping conditions. New empirical formulas depending on the design water depth (h) are proposed to estimate the size of Cubipod® units needed to design stable armors with cotα = 1.5 on a horizontal bottom (m = 0), for any deep-water wave climate. The paper is structured as follows: Section 2 describes the background of the research. Then, Sections 3 and 4 describe the experimental methodology and present the experimental results, respectively. Section 5 summarizes and details the proposed design criterion.

2. Background

The armor layer of mound breakwaters is commonly designed using empirical formulas, such as those proposed by Hudson (1959) [2] or Van der Meer (1988) [3], based on 2D small-scale physical tests conducted in non-breaking and non-overtopping conditions. However, most of the breakwaters built around the world are located in breaking wave conditions, where they must withstand design storms having some percentage of large waves which break before reaching the structure. In these cases, it is not clear whether the design formulas for non-breaking wave conditions should be applied. The Hudson formula given below (Equation (1)), based on the pioneering work of Iribarren (1938) [1], was popularized by USACE (1975 and 1984) [7,8]:

$$N_s = \frac{H_s}{\Delta D_n} = (K_D \cot\alpha)^{1/3} \tag{1}$$

where D_n is the nominal diameter of the units, $\Delta = (\rho_r - \rho_w)/\rho_w$ is the relative submerged mass density, ρ_r is the mass density of the units, ρ_w is the mass density of the sea water, K_D is the stability coefficient, α is the armor slope, H_s is the significant wave height ($H_s = H_{1/3}$ = average of one-third highest waves or $H_s = H_{m0}$ = spectral significant wave height), and N_s is the stability number.

Equation (1) was based on regular wave tests in non-breaking and non-overtopping conditions. USACE (1975 and 1984) [7,8] recommended a change in the stability coefficient (K_D) to use Equation (1) in breaking wave conditions, reducing K_D values for breaking waves. K_D takes into account the geometry of the armor unit, number of layers, breakwater section (trunk or head), armor slope (α), and also an implicit safety factor for design (see Medina and Gómez-Martín 2012 [9]). Equation (1) does not explicitly consider the duration of the wave storm, the wave period or wave steepness, the permeability of core and filter layers, or other relevant factors affecting the hydraulic stability of the armor layer. Nevertheless, Equation (1) is still widely used by practitioners for preliminary designs and feasibility studies to compare construction costs of different breakwater designs using different armor units.

In addition, the generalized Hudson formula is extensively used in practice for single- and double-layer armors; however, this formula is not reasonable for interlocking units because interlocking increases with the armor slope. Furthermore, armor design to Initiation of Damage (IDa) popularized by USACE (1975) [7] for double-layer randomly-placed armors must be adapted to single-layer armors with brittle failure functions when IDa and Initiation of Destruction (IDe) are very close. In order to maintain a reasonable safety factor to IDe, the recommended K_D for single-layer armors corresponds to an armor damage level much lower than IDa. The values of K_D for single- and double-layer armors are based on small-scale physical tests and may change slightly over time, depending on experience and implicit or explicit safety factors.

For double-layer rock armors in breaking wave conditions, Van der Meer (1988) [3] proposed replacing H_s with $H_{2\%}/1.4$ in Equations (2) and (3), after conducting several physical tests in breaking wave conditions with an m = 1/30 bottom slope and a permeable structure. The relationship $H_{2\%}/1.4 = H_s$ is valid if wave heights are Rayleigh distributed (non-breaking wave conditions) but it is a conservative criterion if the highest waves break before reaching the structure (breaking wave conditions).

$$N_s = \frac{H_s}{\Delta D_{n50}} = 6.2 S^{0.2} P^{0.18} N_Z^{-0.1} \xi_m^{-0.5} \text{for } \xi_m < \xi_{mc} \text{ (plunging waves)} \quad (2)$$

$$N_s = \frac{H_s}{\Delta D_{n50}} = 1.0 S^{0.2} P^{-0.13} N_Z^{-0.1} (\cot\alpha)^{0.5} \xi_m^P \text{for } \xi_m > \xi_{mc} \text{ (surging waves)} \quad (3)$$

in which N_s is the stability number, D_{n50} is the equivalent cube size or nominal diameter of the units, S is the dimensionless armor damage, $\xi_{mc} = 6.2P^{0.31}(\tan\alpha)^{0.5})^{1/(P+0.5)}$ is the critical breaker parameter, $0.1 \leq P \leq 0.6$ is a parameter which considers the permeability of the structure, N_z is the number of waves, and $\xi_m = \tan\alpha/(2\pi H_s/(gT_m{}^2))^{0.5}$ is the surf similarity parameter based on the mean period, T_m.

Although several hydraulic stability formulas for rock armors are used in breaking wave conditions, considering different characteristic wave heights, most experimental validations are carried out in non-breaking wave conditions.

Melby and Kobayashi (1998) [10] and Van Gent et al. (2003) [4] also proposed hydraulic stability formulas for double-layer randomly-placed rock armors based on specific laboratory tests in breaking wave conditions ($0.64 < H_s/h_s < 1.11$ and $0.15 < H_s/h_s < 0.78$, respectively) with different bottom and armor slopes. Furthermore, available formulas require knowing H_s or $H_{2\%}$ at the toe of the structure; however, these formulas are calibrated with waves measured at a certain distance from the structure.

Van Gent (2013) [11] established that breakwaters with a berm can significantly reduce overtopping and reduce the required rock size, especially for the slope above the berm, compared to straight slopes without a berm. Van Gent (2013) [11] provided empirical relationships to quantify the required armor rock size for non-overtopped rock slopes with a horizontal berm, with a 1:2 slope above and below the berm, or a 1:4 slope above and below the berm; this formula is only valid with a horizontal foreshore on which no wave breaking occurs. Nevertheless, the experimental range investigated by Van Gent (2013) [11] is limited to rather high berms, i.e., characterized by small values of water depth over the berm compared with the water depth at the toe.

Herrera et al. (2017) [5] proposed Equation (4) for double-layer rock armors in breaking wave conditions ($0.20 < H_s/h_s < 0.90$), considering the observed potential 6-power relationship between the equivalent dimensionless armor damage (S_e) and the spectral significant wave height (H_{m0}) at a seaward distance of $3h_s$ from the breakwater toe, where h_s is the water depth at the toe.

$$S_e = 0.066 \left(\frac{H_{m0}}{\Delta D_{n50}} \right)^{6.0} \quad (4)$$

Celli et al. (2018) [12] provided a design criterion for the armor layer elements of conventional breakwaters with submerged berm marked by a minimal thickness compared to water depth, extending the validity range of the method suggested by Van Gent (2013) [11].

CLI (2018) [13] specified stability coefficients for single-layer armors with Accropode^{TM} (K_D = 15) and other interlocking armor units (K_D = 16), valid for armor slopes with cotα = 1.33 to 1.5 in non-breaking conditions. CLI (2018) [13] as well as USACE (1975, 1984) [7,8] recommended using lower values of K_D for wave breaking conditions. On the contrary, Xbloc® (2014) [14] recommended K_D = 16 and 14.2 for armor slopes with cotα = 1.33 and 1.5, respectively, in breaking or partially breaking wave conditions ($H_s/h_s > 0.4$), and K_D = 10.7 and 8.0 for $0.29 < H_s/h_s < 0.4$ and $H_s/h_s < 0.29$, respectively, when cotα = 1.33.

To rationalize the use of K_D when characterizing the hydraulic stability of the armor layers, Medina and Gómez-Martín (2012) [9] proposed including explicit safety factors (SF) for IDa and IDe associated to each recommended K_D. Based on the experimental results for each armor unit, the safety factors SF(IDa) and SF(IDe) are the ratio between the stability number to IDa and IDe and the design stability number (N_{sd}), SF(IDa) = N_s(IDa)/N_{sd} and SF(IDe) = N_s(IDe)/N_{sd}, where N_{sd} is given by Equation (1) when $H_s = H_{sd}$ and H_{sd} is the design significant wave height. Medina and Gómez-Martin (2016) [15] recommended K_D values for single- and double-layer Cubipod® armors in non-overtopping and non-breaking conditions, with the corresponding explicit safety factors.

3. Experimental Methodology

2D physical model tests using regular waves in breaking wave conditions were conducted in the wind and wave test facility (30 × 1.2 × 1.2 m) of the Laboratory of Ports and Coasts at the Universitat Politècnica de València (LPC-UPV). The piston-type wavemaker was equipped with an active wave absorption system (AWACS), which generates regular and irregular waves. Water surface elevation was measured using capacitance wave gauges at ten points along the wave flume. One group of four wave gauges (S1, S2, S3 and S4) was placed near the wavemaker and other group of four wave gauges (S5, S6, S7 and S8) near the model. Wave gauges of each group were distanced according to the criterion given by Mansard and Funke (1980) [16], and incident and reflected waves where separated sing the LASA-V method developed by Figueres and Medina (2004) [17]. The water depth ranged between h_s(cm) = 30 and h_s(cm) = 42 at the model area, and between h(cm) = 55 and h(cm) = 67 at the wave generation zone, with a 4% slope bottom transition between the two horizontal wave flume bottoms at different levels. Figure 1 shows a longitudinal cross-section of the LPC-UPV wave flume with the location of the wave gauges used in these tests.

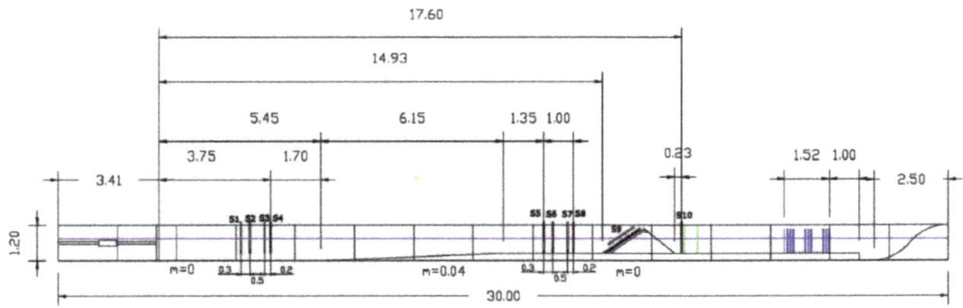

Figure 1. Longitudinal cross-section of the Laboratory of Ports and Coasts at the Universitat Politècnica de València (LPC-UPV) wave flume (dimensions in meters).

Three models with different armoring protection (single- and double-layer Cubipod® armors with toe berm and double-layer Cubipod® armor without toe berm) were tested in breaking wave conditions, using the same core and filter layers (see Vanhoutte and De Rouck 2009) [18]. The breakwater crest elevation was high enough for non-overtopping conditions, and the armor slope was α with cotα = 1.5. The armor porosities were 39% and 40% for single-and double-layer armors. The characteristics of the core, filter and armor layer are specified in Table 1, where D_{n50} is the equivalent cube size or nominal diameter of the units, D_{85} and D_{15} is the equivalent cube size corresponding to the 85% and 15% percentile, respectively, ρ_r is the mass density, and M is the unit mass.

Table 1. Characteristics of the tested models.

Materials	D_{n50} (cm)	D_{85}/D_{15}	ρ_r (g/cm^3)	M (g)
Core (G2)	0.70	2.0	2.70	0.9
Toe berm and filter (G1)	1.80	1.5	2.70	16
Cubipod®	3.82	-	2.30	128

The reference scale was 1/50, and the objective was to determine, for a given water depth, a unit size which made the armor safe regardless of the deep-water wave storm conditions. In wave breaking conditions, the maximum wave height attacking the breakwater depends on the water depth at the toe (h_s), the wave period (T) and the bottom slope (m = 0 in this study). The worst test conditions were those generating runs of regular waves with a variety of wave periods and with series of increasing wave heights (H), for each water depth (h_s). For a given water depth (h_s) and a wave period (T), the energy of the wave attack on the breakwater model increased with increasing H until waves broke in the horizontal foreshore before reaching the structure.

For each water level, h_s(cm) = 30, 35, 38, 40 and 42 (h_s(m) = 15.0, 17.5, 19.0, etc. at prototype scale), trains of 100 regular waves with different wave periods T(s) = 0.85, 1.28, 1.70, 2.13 and 2.55 (6 < T(s) < 18 at prototype scale) were generated increasing wave height (H) from no damage up to the armor being severely damaged (IDe) or wave breaking on the foreshore. H was increased in steps of 1 cm within the range $9 \leq H(\text{cm}) \leq 26$ ($4.5 < H(\text{m}) < 13$ at prototype scale). For each water depth, around 3000 waves attacked the structure, which covers the range of periods for Spanish coasts in deep water. For a given water depth (h_s) and wave period (T), if H is too high, the waves break before reaching the structure and there is no damage; if H is too small, the waves will not cause the maximum damage to the structure. The methodology described above ensured that the most damaging waves attacked the structure for a given water level.

Figure 2 shows the double-layer Cubipod® armored breakwater model without toe berm, and Figure 3 shows the cross-sections of single- and double-layer Cubipod® armors with toe berm. A detail of the toe berm used in these experiments is shown in Figure 4. The toe berm was built with filter material (G1). Equivalent dimensionless armor damage (S_e) was measured with the virtual net method described by Gómez-Martín and Medina (2014) [19], considering armor unit extractions, armor layer sliding as a whole, and heterogeneous packing (HeP) failure modes simultaneously. The Cubipods in the bottom armor layer were painted white to enhance color contrast, while those in the upper armor layer were painted with stripes of different colors to facilitate the visual counting. The armor was photographed before and after each run of 100 regular waves, allowing armor damage to be measured with the virtual net method. The virtual net method was applied to the photographs taken perpendicular to the armor to calculate the corresponding equivalent dimensionless damage parameter, S_e. After each series of tests with constant water depth, the armor damage was obtained, considering the previous armor damage obtained for each water depth. The accumulated armor damage was measured to obtain the maximum armor damage associated to each water depth, increasing water levels from h_s(cm) = 30 to 42, for each model. Different armor damage was observed for each water depth (h_s); the higher the water depth, the greater armor damage.

Some long irregular wave trains were generated in this study, but they caused little damage compared to the armor damage caused by the runs of regular waves corresponding to the experimental methodology described above. Most of the higher waves of the irregular trains broke on the horizontal foreshore and only a few relevant waves attacked the structure. The methodology used in this study was defined to ensure that 100 waves attacked the armor with the most damaging combination of wave height and period for the given water depth. Therefore, the design based on the results of these experiments with regular waves in breaking conditions is on the safe side if the breakwater is placed on a horizontal sea bottom.

Figure 2. View of a double-layer Cubipod® armored breakwater model without toe berm tested at the LPC-UPV.

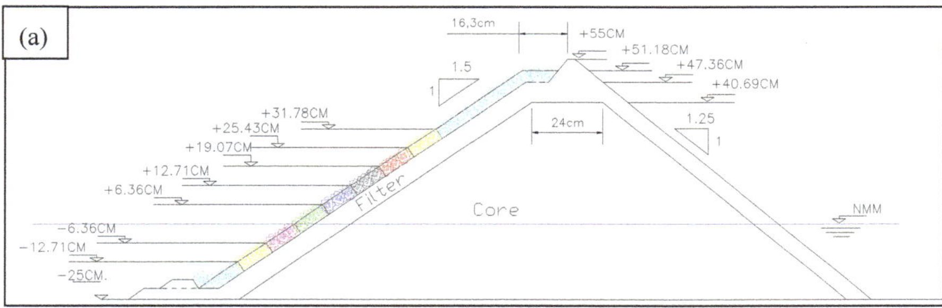

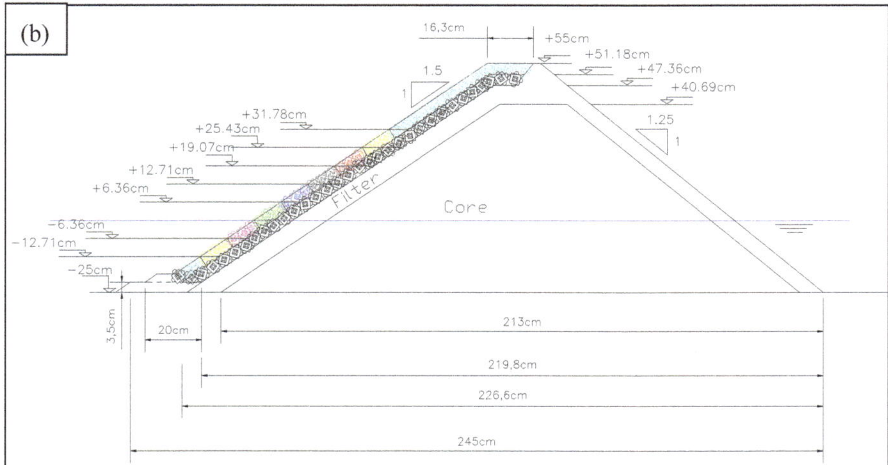

Figure 3. Cross-sections of (**a**) single- and (**b**) double-layer Cubipod® models with the toe berm.

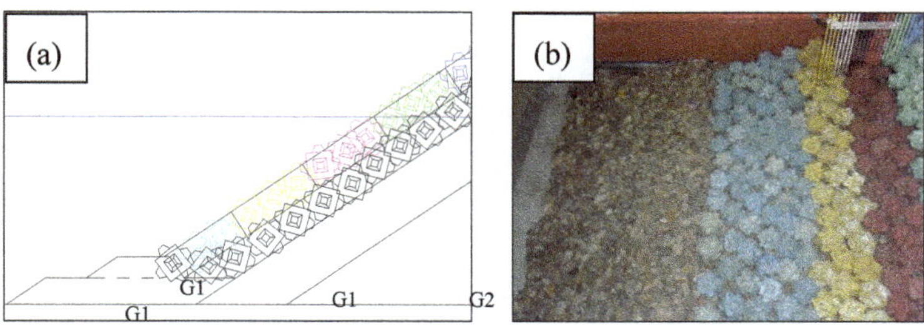

Figure 4. Toe berm used in the experiments: (**a**) Cross-section and (**b**) zenithal view.

4. Experimental Results

According to Medina et al. (1994) [20], rough quarrystone armor damage and wave height given by USACE (1984) [8] and Van der Meer (1988) [3] follow the one-fifth power relationship, and according to Gómez-Martín and Medina (2014) [19], cube and Cubipod® armor damage observations in non-breaking conditions also follow this relationship. Therefore, the linearized equivalent dimensionless armor damage $S_e^{1/5}$ is used in this study to show the stability of Cubipod® armors in depth-limited wave breaking conditions.

Herrera et al. (2017) [5] showed that the rough quarrystone armor damage data presented in the USACE (1975) [7] fit with the empirical Equation (5), considering K_D = 3.5. Where H is the wave height corresponding to damage S, and $H_{D=0}$ is the design wave height corresponding to dimensionless armor damage parameter, S = 0.7. Based on Equation (5), a general damage function (Equation (6)) for any given armor unit can be calculated taking into account the corresponding K_D value.

$$\frac{H}{H_{D=0}} = \left(\frac{S}{1.4}\right)^{1/5} \tag{5}$$

$$S^{\frac{1}{5}} = 1.4^{0.2}\left(\frac{3.5}{K_D}\right)^{1/3} \frac{H}{H_{D=0[K_D=3.5]}} \tag{6}$$

The accumulated linearized equivalent dimensionless armor damage, $S_e^{1/5}$, measured in single- and double-layer Cubipod® armor experiments with and without toe berm is represented in Figure 5, as a function of the dimensionless wave height (considering the wave height, $H_{D=0}$, that produces IDa in rough quarrystone (K_D = 3.5) with the Hudson formula). It was observed that the K_D for single- and double-layer Cubipod® armors is approximately 10 times higher than that for rough quarrystone armors in breaking-wave conditions (USACE 1975 [7]). For damage levels below IDa, the toe berm slightly increases the hydraulic stability of the double-layer Cubipod® armor; however, for armor damage above IDa, the toe berm has no significant influence on armor damage. The stability coefficient for single- and double-layer Cubipod® armors with breaking waves is $K_D \approx 35$ (10 times higher than $K_D \approx 3.5$ for rough quarrystone). The double-layer Cubipod® armors with a toe berm were tested up to h_s(cm) = 42 (corresponding to 21 m at prototype scale). The other two armors were tested up to h_s(cm) = 40 (20 m at prototype scale).

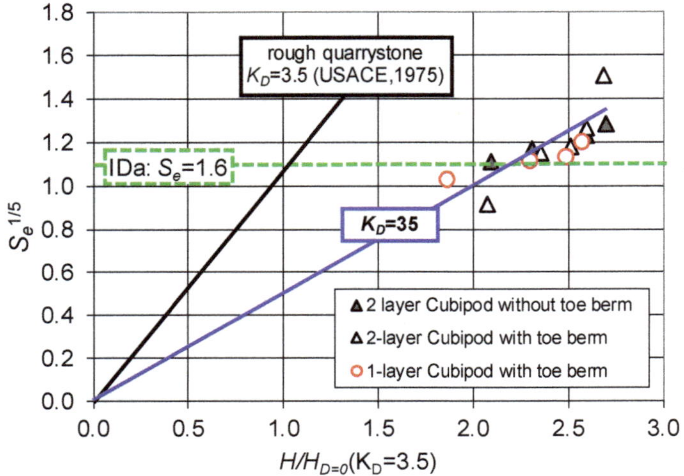

Figure 5. Accumulated linearized equivalent dimensionless armor damage ($S_e{}^{1/5}$) as a function of dimensionless wave height in wave breaking conditions for single- and double-layer Cubipod® armors with and without toe berm.

Figure 6 shows the accumulated dimensionless armor damage as a function of the measured water depths at the toe of the structure. Given the observed relationship (Figure 6) between the dimensionless water depth ($h_s/\Delta D_n$) and the linearized equivalent dimensionless armor damage ($S_e{}^{1/5}$), Equation (7) is proposed as the dimensionless failure function to describe the hydraulic stability of Cubipod® armored breakwaters on horizontal bottoms in depth-limited wave breaking conditions.

$$S_e = 10^{-4}\left[\frac{h_s}{\Delta D_n}\right]^5 \tag{7}$$

where S_e is the equivalent dimensionless armor damage, h_s is the water depth at the toe of the structure, Δ is the relative submerged mass density, and D_n is the nominal diameter or equivalent cube side length.

Considering that IDa corresponds to an equivalent dimensionless damage $1.0 < S_e < 2.0$, say $S_e = 1.6$, then $h_s < 7.0(\Delta D_n)$ is the water depth which guarantees that both single- and double-layer armors will not exceed IDa under any deep water wave climate conditions. Furthermore, the tests carried out indicate a relevant safety margin to IDe. Taking a safety factor SF(IDe) = 1.15 for double–layer Cubipod® armors and SF(IDe) = 1.30 for single–layer Cubipod® armors, the following design values (Equations (8) and (9)) are obtained for safe Cubipod® armored breakwaters placed on horizontal seafloors ($m = 0$):

$$h_s < 7.0(\Delta D_n) \quad \text{for double-layer Cubipod® armors} \tag{8}$$

$$h_s < 6.2(\Delta D_n) \quad \text{for single-layer Cubipod® armors} \tag{9}$$

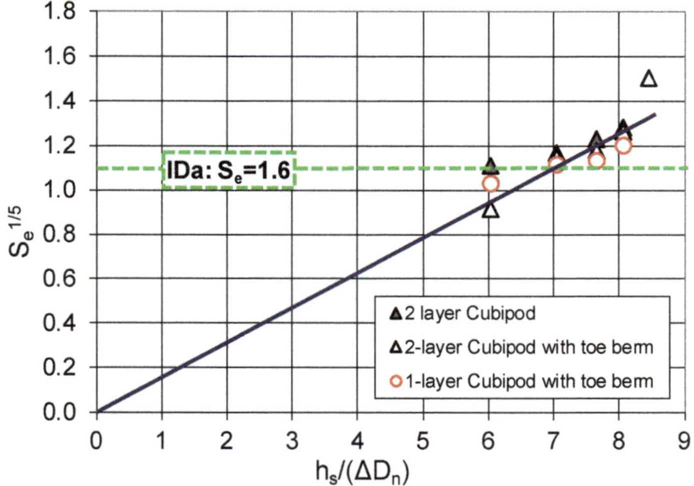

Figure 6. Linearized equivalent dimensionless armor damage $Se^{1/5}$ as a function of the dimensionless water depth.

Figure 7 shows an example of the incident wave height (H) measured at the model area as a function of the incident wave height measured at the wave generation zone, corresponding to a water depth h_s(cm) = 30 and double-layer Cubipod® armored model without toe berm.

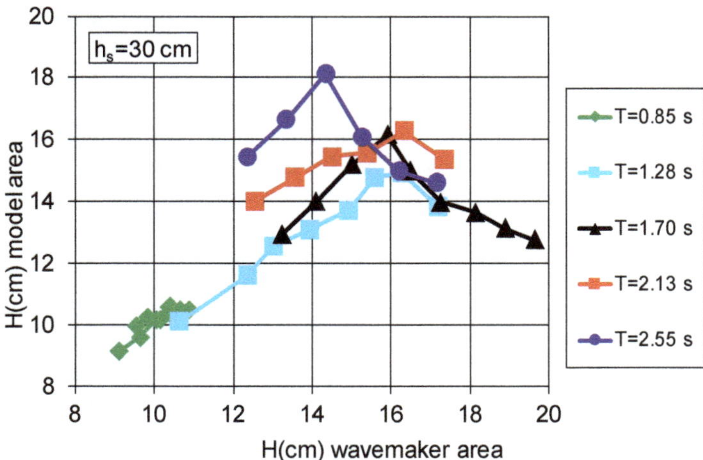

Figure 7. H measured at the model area as a function of H measured at the wave generation zone (m = 0).

The highest wave heights measured in the model area were approximately 60% of the water depth, $H_{max}/h_s \approx 0.60$ with a horizontal seafloor (m = 0). For gentle bottom slopes ($m > 0$), a criterion similar to that proposed by Goda (2010) [21] and Herrera et al. (2017) [5] is reasonable. The design water depth (h) is calculated at a distance three times the water depth at the toe of the structure (h_s), $h = h_s(1 + 3m)$. This approximation is reasonable for gentle bottom slopes ($m \leq 0.02$); for steep bottom slopes (e.g., m = 0.10), the toe berm design is critical, and armor layer and toe berm must be studied simultaneously (see Herrera and Medina 2015 [22] or Herrera et al. 2016 [23]). Small-scale models are highly recommended when breakwaters are placed on steep sea bottoms in depth-limited breaking wave conditions.

Once the design water depth is calculated, $h = h_s(1 + 3\text{m})$, the size of Cubipod® units can be estimated to design safe Cubipod® armored breakwaters (cotα = 1.5 and perpendicular wave attack) for any deep water wave climate, regardless of the wave height at the toe of the structure, considering the following equations:

$$D_n > h/7.0\Delta = h_s(1 + 3\text{m})/(7.0\Delta) \quad \text{for double-layer Cubipod® armors} \tag{10}$$

$$D_n > h/6.2\Delta = h_s(1 + 3\text{m})/(6.2\Delta) \quad \text{for single-layer Cubipod® armors} \tag{11}$$

5. Summary and Conclusions

The armors of conventional mound breakwaters are usually pre-designed using simple empirical equations such as the Hudson formula, based on results from small-scale physical tests with Froude similarity. Most of the tests used to characterize the hydraulic stability of different armor units are conducted in non-overtopping and non-breaking conditions. However, most of the world's breakwaters are built to withstand design storms having some percentage of large waves breaking before reaching the structure and in depth-limited breaking wave conditions. Nevertheless, there are few studies in the literature, and often with contradictory recommendations, to design breakwaters in breaking wave conditions based on formulas validated with experiments in non-breaking wave conditions.

A specific experimental methodology is described to analyze the hydraulic stability of Cubipod® armored breakwaters on horizontal seafloors in depth-limited breaking wave and non-overtopping conditions. This methodology ensures that the main armor is attacked by runs of 100 regular waves with the most damaging combination of wave height (H) and period (T) for the given water depth (h_s). In wave breaking conditions, the maximum wave height attacking the breakwater (H_{max}) depends on the water depth at the toe (h_s), the wave period (T) and the bottom slope (m = 0 in this study). This research verifies that, for a given water depth, any irregular wave train caused less armor damage compared to that caused by the series of regular waves corresponding to the experimental methodology used in this study. The worst test conditions were obtained generating regular waves for a range of wave height and periods, which can be observed in nature; for each water depth (h_s) and wave period (T), a series of runs of 100 waves with increasing wave heights (H) was generated until waves broke on the foreshore before reaching the structure. The accumulated equivalent dimensionless armor damage (S_e) was measured to obtain the maximum armor damage associated with each water depth (h_s) and multiple combinations of wave height (H) and wave period (T).

Considering that IDa corresponds to an equivalent dimensionless damage S_e = 1.6 and a safety factor SF(IDe) = 1.15 and 1.30 for double- and single-layer Cubipod® armors, respectively, the water depths at the toe of the structure which guarantee that both double- and single-layer armors will not exceed IDa are $h_s < 7.0(\Delta D_n)$ and $h_s < 6.2(\Delta D_n)$, respectively. Armor unit size can be estimated with Equations (10) and (11) to design safe Cubipod® armored breakwaters on horizontal seafloors with cotα = 1.5 for any deep-water wave climate. Taking a design water depth (h) at a distance three times the water depth at the toe of the structure, $h = h_s(1 + 3\text{m})$, Equations (10) and (11) can be safely be used to design Cubipod® armored breakwaters on gentle bottom slopes ($0 \leq m \leq 0.02$) regardless of the wave height at the toe of the structure and deep-water wave climate.

Author Contributions: M.E.G.-M. wrote the original draft and was responsible for the conceptualization, experimental methodology, and analysis. M.P.H., J.A.G.-E. and J.R.M. supervise the investigation, review and approved the manuscript.

Funding: This research was funded by Conselleria d'Educació, Investigació, Cultura i Esport (Generalitat Valenciana) under grant GV/2017/031.

Acknowledgments: The authors thank data provided by SATO-OHL Group (Cubipod Project 2007–2009). The authors thank Debra Westall for revising the manuscript.

Conflicts of Interest: The authors declare no conflict of interest.

References

1. Iribarren, R. *Una Fórmula Para el Cálculo de los Diques de Escollera*; M. Bermejillo-Pasajes: Madrid, Spain, 1938.
2. Hudson, R.Y. Laboratory investigation of rubble-mound breakwaters. *J. Waterw. Harbors Div.* **1959**, *85*, 93–121.
3. Van der Meer, J.W. Rock Slopes and Gravel Beaches under Wave Attack. Ph.D. Thesis, Technical University of Delft, Delft, The Netherlands, 1988.
4. Van Gent, M.R.A.; Smale, A.J.; Kuiper, C. Stability of rock slopes with shallow foreshores. In Proceedings of the Coastal Structures 2003, Portland, OR, USA, 26–30 August 2003; pp. 100–112.
5. Herrera, M.P.; Gómez-Martín, M.E.; Medina, J.R. Hydraulic stability of rock armors in breaking wave conditions. *Coast. Eng.* **2017**, *127*, 55–67. [CrossRef]
6. Battjes, J.A.; Groenendijk, H.W. Wave height distributions on shallow foreshores. *Coast. Eng.* **2000**, *40*, 161–182. [CrossRef]
7. United States Army Corps of Engineers (USACE). *Shore Protection Manual*; U.S. Army Coastal Engineering Research Center, U.S. Army Engineer Waterways Experiment Station: Vicksburg, MI, USA, 1975.
8. United States Army Corps of Engineers (USACE). *Shore Protection Manual*; U.S. Army Coastal Engineering Research Center, U.S. Army Engineer Waterways Experiment Station: Vicksburg, MI, USA, 1984.
9. Medina, J.R.; Gómez-Martín, M.E. K_D and safety factors of concrete armor units. *Coast. Eng. Proc.* **2012**, *1*, 29. [CrossRef]
10. Melby, J.A.; Kobayashi, N. Progression and variability of damage on rubble mound breakwaters. *J. Waterw. Port Coast. Ocean Eng.* **1998**, *124*, 286–294. [CrossRef]
11. Van Gent, M.R. Rock stability of rubble mound breakwaters with a berm. *Coast. Eng.* **2013**, *78*, 35–45. [CrossRef]
12. Celli, D.; Pasquali, D.; De Girolamo, P.; Di Risio, M. Effects of submerged berms on the stability of conventional rubble mound breakwaters. *Coast. Eng.* **2018**, *136*, 16–25. [CrossRef]
13. Concrete Layer Innovation (CLI). 2018. Available online: http://www.concretelayer.com/documentation (accessed on 8 Febrary 2018).
14. Xbloc. Guidelines for Xbloc Concept Designs. 2014. Available online: https://www.xbloc.com/sites/default/files/domain-671/documents/xbloc-design-guidelines-2014-671-15039173271578936988.pdf (accessed on 8 February 2018).
15. Medina, J.R.; Gómez-Martín, M.E. *Cubipod® Manual 2016*; Editorial Universitat Politècnica de València: Valencia, Spain, 2016; Available online: http://hdl.handle.net/10251/72310 (accessed on 8 February 2018).
16. Mansard, E.P.D.; Funke, E.R. The measurement of incident and reflected spectra using a least squares method. In Proceedings of the 17th International Conference on Coastal Engineering, Sydney, Australia, 23–28 March 1980; pp. 154–172.
17. Figueres, M.; Medina, J.R. Estimation of incident and reflected waves using a fully non-linear wave model. In Proceedings of the 29th International Conference on Coastal Engineering; ASCE: Reston, VA, USA, 2004; Volume 1, pp. 594–603.
18. Vanhoutte, L.; De Rouck, J. Hydraulic Stability of Cubipod Armour Units in Breaking Conditions. Master's Thesis, Ghent University, Ghent, Belgium, June 2009. Available online: https://lib.ugent.be/catalog/rug01:001418208 (accessed on 19 February 2018).
19. Gómez-Martín, M.E.; Medina, J.R. Heterogeneous packing and hydraulic stability of cube and Cubipod armor units. *J. Waterw. Port Coast. Ocean Eng.* **2014**, *140*, 100–108. [CrossRef]
20. Medina, J.R.; Hudspeth, R.T.; Fassardi, C. Breakwater armor damage due to wave groups. *J. Waterw. Port Coast. Ocean Eng.* **1994**, *120*, 179–198. [CrossRef]
21. Goda, Y. *Random Seas and Design of Maritime Structures*, 3rd ed.; World Scientific: Singapore, 2010; p. 708.
22. Herrera, M.P.; Medina, J.R. Toe berm design for very shallow waters on steep sea bottoms. *Coast. Eng.* **2015**, *103*, 67–77. [CrossRef]
23. Herrera, M.P.; Molines, J.; Medina, J.R. Hydraulic stability of nominal and sacrificial toe berms for mound breakwaters on steep sea bottoms. *Coast. Eng.* **2016**, *114*, 361–368. [CrossRef]

Journal of
Marine Science and Engineering

MDPI

Article

Wave Impact Pressures on Stepped Revetments

Nils B. Kerpen *, Talia Schoonees and Torsten Schlurmann

Ludwig-Franzius-Institute for Hydraulic, Estuarine and Coastal Engineering, Leibniz University Hannover, 30167 Hannover, Germany; schoonees@lufi.uni-hannover.de (Ta.S.); schlurmann@lufi.uni-hannover.de (To.S.)

* Correspondence: kerpen@lufi.uni-hannover.de; Tel.: +49-511-762-3740

Received: 19 October 2018; Accepted: 5 December 2018; Published: 13 December 2018

Abstract: The wave impacts on horizontal and vertical step fronts of stepped revetments is investigated by means of hydraulic model tests conducted with wave spectra in a wave flume. Wave impacts on revetments with relative step heights of $0.3 < H_{m0}/S_h < 3.5$ and a constant slope of 1:2 are analyzed with respect to (1) the probability distribution of the impacts, (2) the time evolution of impacts including a classification of load cases, and (3) a special distribution of the position of the maximum impact. The validity of the approved log-normal probability distribution for the largest wave impacts is experimentally verified for stepped revetments. The wave impact properties for stepped revetments are compared with those of vertical seawalls, showing that their impact rising times are within the same range. The impact duration for stepped revetments is shorter and decreases with increasing step height. Maximum horizontal wave impact loads are about two times larger than the corresponding maximum vertical wave impact loads. Horizontal and vertical impact loads increase with a decreasing step height. Data are compared with findings from literature for stepped revetments and vertical walls. A prediction formula is provided to calculate the maximum horizontal wave impact at stepped revetments along its vertical axis.

Keywords: Stepped revetment; wave impact; physical model test

1. Introduction

The increasing population living in coastal areas poses new demands in terms of the environmental and touristic compatibility of coastal protection structures. As aesthetically pleasing coastal protection structures gain increasing importance, and accessible revetments, e.g., stepped revetments become more attractive. Recent installations can be found in Margate, UK (finished 2013), Blackpool, UK (finished 2017), or Chicago shoreline project, US (finished 2018). Furthermore, the stepped surface of a stepped revetment induces additional turbulence in the flow, which leads to increased energy dissipation compared to smooth impermeable structures, however, traditional permeable structures provide more dissipation due to filtration and percolation effects [1,2]. Consequently, wave energy available as kinetic energy for the wave run-up process is reduced. Presently however, practical design guidance is limited for stepped revetments. A comprehensive overview of existing research focusing on the general wave-interaction with stepped revetments is presented in Reference [3]. References [2,4] highlight the reduction capabilities of stepped revetments for wave run-up and overtopping compared to smooth impermeable structures for regular and irregular waves. Reference [5] discusses the energy dissipation within the wave run-up at stepped revetments and detected similarities to steady flow conditions over stepped surfaces like spillways. Thus far, wave impacts on stepped revetments have not been analyzed systematically and are discussed in this study. The paper is structured as follows: First, the current understanding of wave impacts on stepped revetments is presented in Section 2 and key knowledge gaps outlined. Section 3 describes the geometrical and hydraulic boundary conditions of the conducted model tests. The analysis and discussion of the results for wave impact pressures with special focus on (1) the probability distribution,

(2) the time evolution including load case definitions, and (3) the specific evolution of wave-induced pressure is given in Section 4. Section 5 focuses on the underlying laboratory and scale effects in the physical model. Finally, the results are summed-up and contrasted in context of the present state-of-the-art and the advancement of knowledge outlined through this new study.

2. Previous Studies on Wave Impacts on Stepped Revetments

Knowledge of loads on a coastal protection system is essential to determine its final design and performance over its lifetime. Hydraulic loads can be classified as either hydrostatic or hydrodynamic. Only the hydrodynamic load cases are addressed in this paper.

A general classification of breaking wave loads on vertical structures for different breaker types is given by Reference [6]. Dynamic impact loads most often occur due to plunging waves impacts on a marine structure. The progress of a single wave impact is dependent on the hydraulic- and geometry-related boundary conditions (described in this section). A stepped revetment has horizontally and vertically aligned fronts in relation to the direction of gravity. The loads on these two fronts are different due to the asymmetry and phase-shift of orbital velocities in shallow waters and the asymmetric structure of the wave-form represented by the wave steepness (wave height $H \ll$ wave length L). Furthermore, the effect of wave loads is dependent on the overall slope (n) of a structure. A unique loading case is represented by a plain vertical wall without horizontal planes where an overturning, i.e., breaking wave crest violently slams against the wall without any damping or dissipation due to preceding waves. The other extreme case is a very gentle slope (nearly horizontal) where plunging and spilling breaking waves run-up on an inclined, often impermeable plane (dependent on the Iribarren number $\xi = \tan\alpha/\sqrt{H/L}$). In the case of spilling breaking waves the exposed load on the structure is regularly damped by previous waves. On a structure with horizontal and vertical fronts, tongues, and jets of water can be created from the breaking wave and be directed upwards like an up-rushing jet of water [1]. This jet of water instantaneously collapses and falls downwards, and in turn, induces an additional impact on the horizontal fronts. It is evident that this loading on a horizontally aligned front (tread of the step) is smaller than on a vertical front (riser of the step).

Recent literature provides only little insight on wave impacts on stepped revetments. According to Reference [7], the forces on stepped structures should be calculated for design purposes with the same method as for vertical walls since the dynamic pressures are in the same range. Reference [8] conducted hydraulic model tests with irregular waves for a stepped seawall in a 76 m long wave flume following a Froude scale of 1:19. These tests focused on measuring the reduction of wave overtopping and the wave-induced impacts on stepped seawalls. Two sloping structures (n = 1.5, n = 2) with step heights of S_h = 0.026 m were analyzed. A recurved seawall was incorporated at the crest of both sloping structures. Reference [8] analyzed the wave loads on stepped structures in a specific range of Iribarren numbers of $2.8 < \xi < 6.3$ and step ratios in a range of $9.0 < H_{m0}/S_h < 11.0$ with H_{m0} as zeroth moment wave height. Similarly, Reference [8] remarks the importance of the short duration shock pressures (impacts) resulting from the rapid compression of an air pocket trapped between the front of a breaking wave and the wall. For vertical walls the authors in Reference [6] note that 'the shock pressure exerted by a breaking wave is due to the violent simultaneous retardation of a certain limited mass of water that is brought to rest by the action of a thin cushion of air, which in the process becomes compressed by the advancing wave front' [9]. The position of the highest measured impacts was dependent on the initial still water level (SWL). According to the analysis of impact distributions, the maximum impacts at different wall elevations rarely occur simultaneously. This finding is particularly valid in the case of a non-vertical wall, such as the stepped wall studied here, since some wave energy is dissipated through turbulence. In some cases, a negative impact duration was measured, which is interpreted as a characteristic of turbulence and air entrainment occurring at the base of each seawall step. Finally, Reference [8] summarizes a discussion about the importance of shock pressures for the actual design of a stepped seawall. According to the discussion, pressures of such short duration should not be used for establishing the design load case. Rather, it is recommended to consider the smaller surge pressures with a longer duration to determine the critical dynamic load.

In addition, Reference [10] conducted hydraulic model tests in a Froude scale of 1:20 for stepped revetments (S_h = 0.015 m), focusing on wave run-up and wave overtopping. The vertical wave impact was measured on a single step with a sampling rate of only 100 Hz, which is considered as rather inappropriate for measuring rapid, i.e., almost instantaneous, wave impact pressures. As such, analyzed data represent merely averaged maximum impact pressures P_{max} of six test repetitions (with a standard deviation of STD~0.1 P_{max}) to depict the inherent loading bias within each experiment.

Both studies lack a comprehensive discussion on the instantaneous pressure impact events with corresponding wave conditions. Hence, the systematic analysis of the wave impacts on stepped revetments conducted in the experiments in the present study includes a comparison with the data and evaluations provided by References [8,10].

3. Experimental Set-Up, Test Conditions and Procedures

Hydraulic model tests focusing on the wave interaction with stepped revetments were conducted in a wave flume, which has a length of 110 m, a width of 2.2 m and an overall depth of 2.0 m. For these tests, wave spectra were calculated with second order wave theory routines. The irregular wave profiles ($H_{1/3,max}$ = 0.42 m with T_{max} = 2.0 s) were generated with a piston-type wave maker. Two model set-ups, constructed from plywood, with varying step heights (large steps: S_h = 0.3 m, small steps: S_h = 0.05 m) are placed in a 0.7 m wide sub section over a horizontal flume bottom at a distance of L_F = 81.6 m from the wave paddle (Figure 1). The relative flume length with respect to the tested wave length L_p is $10 < L_F/L_p < 36$.

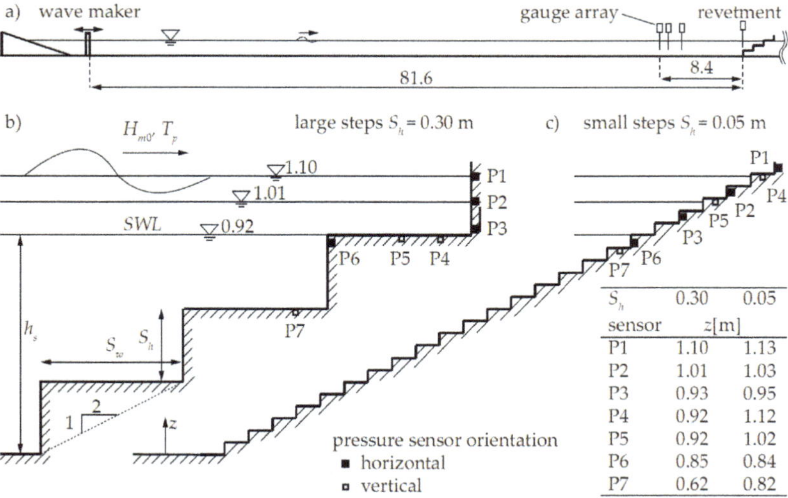

Figure 1. Model set-up of pressure sensors and position of the *SWL* for two 1:2 inclined stepped revetments. (**a**) Cross-section of the flume set-up, (**b**) detail of the large step configuration (S_h = 0.3 m), and (**c**) detail of the small step configuration (S_h = 0.05 m).

The surface elevation is measured by five ultrasonic sensors with a measuring range of 200 to 1200 mm, a superior resolution of 0.36 mm and a sampling rate $f_{sample.}$ = 50 Hz. Three of the sensors are positioned at a distance larger than two wave lengths L from the toe of the revetment, to determine incident wave conditions, as calculated by a reflection analysis. One sensor is placed at the toe of the stepped revetment and another in the shallow water region of the still water level. Pressure impacts on the stepped revetment are recorded by seven pressure transducers, which are placed along the horizontally ($f_{sample.}$ = 2.4 kHz) and vertically ($f_{sample.}$ = 19.2 kHz) orientated step fronts (Figure 1). An impression of the set-ups is given in Figure 2 for the analyzed step heights of 0.05 m (a) and 0.3 m (b). In order to capture a profile of wave induced loads on an 1:2 inclined stepped revetment, sensor locations are varied in relative water depths —6.0 < z/H_{m0} < 2.0, relative to the still water level. The pressure sensors (ATM.1ST/N fabricated by sts-sensors) have a range from zero to 150 mbar and a non-conformity of ±0.1% from full scale. The sensors are connected by a serial interface connection (RS232) to the data acquisition and provide an output signal from zero to 10 V. The configuration of the probes allows a local (over a single step) and a more global interpretation (for the whole revetment).

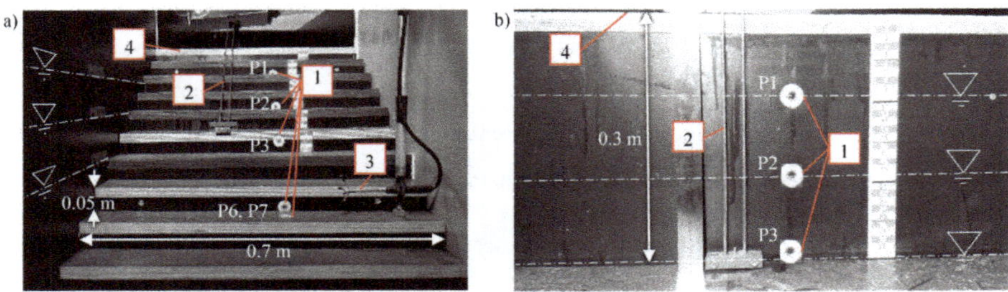

Figure 2. Model set-up with instrumentation for (**a**) step height 0.05 m and (**b**) step height 0.3 m.

The hydraulic boundary conditions of the wave impact experiments are listed in Table 1. The parameter choice covers a wide range of dimensionless variables with a wave steepness of 0.015 < H_{m0}/L_p < 0.04 (with H_{m0}: spectral wave height, L_p: wave length calculated from the peak period T_p measured at the gauge array in a distance of x = 8.6 m (1 L_p < x < 3.8 L_p) from the toe of the revetment) and Iribarren numbers of 2.5 < ξ < 4.9. Three different water levels h_s with intermediate water depths (0.13 < h_s/L_p < 0.49) are tested. Additionally, the corresponding freeboard height R_c and the number of waves N in each test are given. A total of 13 tests are conducted for steps with a height of S_h = 0.05 m (1.1 < H_{m0}/S_h < 2.8) and 10 tests with a step height of S_h = 0.3 m (0.2 < H_{m0}/S_h < 0.6).

Table 1. Hydraulic boundary conditions for the wave impact tests.

Test	S_h	R_c	H_{m0}	T_p	h_s	N	ξ	H_{m0}/L_p	H_{m0}/h_s	H_{m0}/S_h
#	(m)	(m)	(m)	(s)	(m)	(-)	(-)	(-)	(-)	(-)
101	0.05	0.121	0.056	1.43	1.100	1298	3.7	0.018	0.051	1.12
102			0.063	1.20	1.100	427	3.0	0.028	0.057	1.26
103			0.084	1.38	1.100	1261	2.9	0.029	0.076	1.68
104			0.082	1.38	1.100	1256	3.0	0.028	0.074	1.63
105			0.082	1.38	1.100	1261	3.0	0.028	0.075	1.64
106			0.084	1.37	1.100	167	2.9	0.029	0.076	1.67
107			0.088	2.11	1.100	1422	4.2	0.015	0.080	1.76
108			0.114	2.20	1.100	171	3.8	0.019	0.104	2.28
109			0.119	2.81	1.100	179	4.6	0.016	0.108	2.38
110			0.143	2.26	1.100	162	3.5	0.023	0.130	2.86
111		0.211	0.085	2.09	1.010	1410	4.2	0.016	0.084	1.71
112			0.085	2.07	1.010	1434	4.1	0.016	0.084	1.70
113			0.085	2.08	1.010	1371	4.1	0.016	0.084	1.70
201	0.30	0.121	0.111	1.37	1.100	1413	2.6	0.038	0.101	0.37
202			0.129	3.18	1.100	307	4.9	0.016	0.117	0.43
203			0.116	1.38	1.100	1428	2.5	0.040	0.106	0.39
204			0.167	2.08	1.100	278	3.0	0.030	0.151	0.56
205		0.211	0.064	1.36	1.010	1586	3.3	0.023	0.064	0.21
206			0.091	1.34	1.010	1390	2.8	0.033	0.090	0.30
207			0.166	2.01	1.010	1515	2.9	0.032	0.164	0.55
208		0.300	0.064	1.41	0.921	1339	3.4	0.021	0.069	0.21
209			0.089	1.38	0.921	1294	2.8	0.031	0.097	0.30
210			0.170	2.11	0.921	1468	2.9	0.032	0.184	0.57

Raw data from the pressure sensors are offset-corrected by means of the first five seconds of the data. The time and amplitude of the peaks are calculated for minimum peak heights of 10 mbar and a minimum distance between two peaks of 0.8 times the wave period T_p.

Every single wave in a wave spectrum causes an individual impact p on the structure. The magnitude of the single impact depends on its individual wave kinematics and the influence of the previous wave (remaining water layer over a pressure sensor and amount of aeration in the wave). The analysis of wave loads includes a number of parameters such as the maximum impact pressure P_{max}. The quantity of a finite number N of waves that cause N individual impacts p within a single test is defined as P. If the quantity P is sorted in a descending order, the maximum recorded impact P_{max} is defined as max$\{P\}$ or $P_{(i=1)}$, as given in Equations (1) and (2).

$$P = \{p_1, p_2, p_3, \cdots, p_N\} \quad with: \; p_1 > p_2 >, p_3 > \cdots > p_N \tag{1}$$

$$P_{max} = \max\{P\} = P_{(i=1)} = p_1 \tag{2}$$

While the maximum induced impact P_{max} is very important for the design of a structure, it presents significant scatter when formulating predictions for practical design purposes. Therefore, the collected pressure magnitudes will rather be described with a probability of exceedance (e.g., 2% of all incident waves reveals the probability of exceedance impact pressure $P_{98\%}$, Equation (3)), which offers more representative and reliable predictions.

$$P_{98\%} = \frac{\sum_{i=1}^{n} p_i}{n} \quad with: \; n = N(1 - 0.98) \in \mathbb{N} \tag{3}$$

In order to compare the measured data to data from previous investigations (e.g., Reference [11]), the impact magnitudes exceeded by the four highest waves out of a 1000-wave test should be calculated. Hence, $P_{99.6\%}$ is calculated on the basis of the number of waves (N) given in Table 1. The impact with a certain probability of exceedance ($P_{99.6\%}$ in this case) is calculated finally by the mean of the n highest impact events according to Equation (4).

$$P_{99.6\%} = \frac{\sum_{i=1}^{n} p_i}{n} \quad with: \; n = N(1 - 0.996) \in \mathbb{N} \tag{4}$$

4. Results and Discussion

The general differences in the wave impact and wave run-up for a plain slope and stepped revetments with two different step ratios are given in Figure 3 for a better comparison and understanding of the results presented in this section. The color-scheme of the initial black and white images recorded with a frame rate of 50 Hz is inverted for the sake of visual clarity. Small tracers (tiny black dots in the images) that follow the main flow field induced by the up-rushing wave are visible on each image. Intense dark areas represent a high amount of aeration due to wave breaking and wave-structure interaction, as such, a corresponding high flow velocity. Figure 3 shows the wave-induced processes on a 1:2 sloped revetment with smooth impermeable surface in chronological order, i.e., the wave impact (a), the wave run-up (b,c), and finally, the wave run-down (d) of a breaking wave. The time step t of each frame is given in relation to the wave period T. In analogy, Figure 3e–h shows the wave impact and wave run-up of a wave with the same boundary conditions over a revetment with small steps (H/S_h = 1.7) and Figure 3i–l over a revetment with large steps (H/S_h = 0.3). For the two stepped configurations, the still water level SLW is at the level of a step edge.

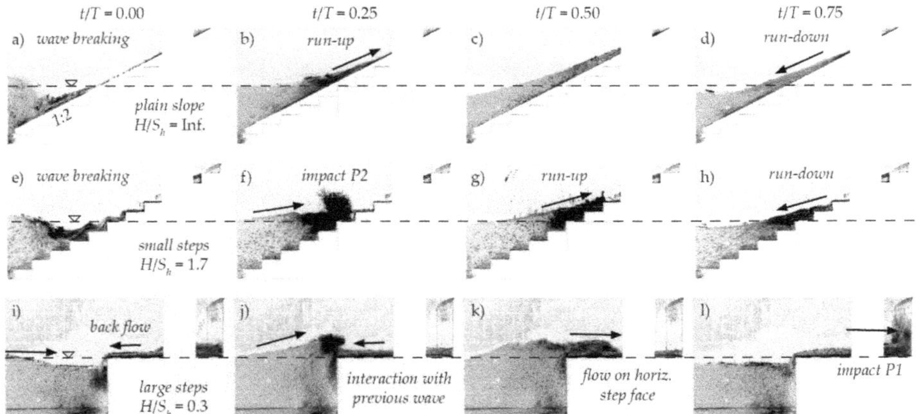

Figure 3. Demonstration of flow processes for relative time-steps in the run-up process on revetments with smooth surface (**a–d**), small step heights (**e–h**), and large step heights (**i–l**).

On a smooth plain slope, a breaking wave induces air in the water body. The wave run-up (b) contains water and air. At the maximum point of wave run-up excursion (c) the entire fraction of air is leaked from the water body, and in consequence, the wave run-down (d) contains only water. During the wave run-down the preceding wave then interacts with the next incoming wave and the described run-up and aeration process is repeated. For revetments with small steps (H/S_h = 1.7), the general flow processes of wave breaking, wave run-up and wave run down sequence is almost in analogy with the processes on a smooth slope. The main differences are: (1) The impact of the breaking wave causes a splash-up at the vertical step front (f) during the wave run-up process, (2) the aeration during the wave run-up is more intense during the run-up process caused by larger friction induced by the steps while interacting with the incoming waves, and (3) higher turbulence and continuous air intrusion (g). Yet, in contrast, the wave run-down induces air into the water body and is retarded compared to the smooth revetment (h). For large step heights (H/S_h = 0.3), the flow processes differ significantly from a normal wave run-up as shown in (a–h). The incident breaking wave interacts with the reflected backflow of a previous wave at the first step edge (i). The interaction leads to an impact on this first step edge, an upwards-directed flow and a wave breaking in very shallow waters over the step edge (j). Then, the wave is propagating as spilling breaker over the horizontal step front (k) until it impacts and resonates with the second vertical step front (l). The impact induces a second splash up.

The previous described processes during the wave run-up over a plain and stepped revetment resemble the major dissimilarities, namely (1) the higher volume of aeration, (2) the stronger interaction with and the influence of previous waves, and (3) the higher breaking wave impact characteristic (impact direct on the slope/step front, impact on a water cushion induced by the previous wave). In summary, these previously outlined effects feature the structure and functioning of stepped revetment by means of reflecting and dissipating incoming wave energy on a sloping beach leading to less wave run-up and, in consequence, to lower wave overtopping rates. In order to quantify the influence of the formerly described processes, the induced wave impact pressures on stepped revetments are analyzed. This examination focuses on (1) the probability distribution of impact pressures, (2) the temporal evolution and typical load cases of impact pressures, and (3) the spatial distribution of impact pressures. Attention is drawn to these three foci in reference to the step ratio. Recent literature on wave impacts on impermeable plain slopes, stepped revetments and vertical seawalls are presented to complement the findings of this research and ensure a comprehensive analysis of processes.

4.1. Probability Distribution of Impact Pressures

According to References [12,13], the probability distributions of wave impact pressures can be described in agreement with mathematical fitting approaches based log-normal functions. Additionally, it has been argued that these log-normal functions are valid in model scale and prototype.

Hence, Figure 4 gives the log-normal probability distribution over the maximum impact pressure for two exemplary tests (test number 103 and 209 according to Table 1). The tests were conducted with a spectral wave height of H_{m0} = 0.084 m and H_{m0} = 0.089 m (for test 103 and 209, respectively) and a corresponding Iribarren number of ξ = 2.9 (2.8 for test 209). Figure 4a provides measurements of wave impacts on slopes with a small step height (H_{m0}/S_h = 1.68, h_s = 1.1 m, sensor *P1*). A total of 876 individual wave impacts have been recorded with pressure sensor *P1* with a maximum recorded pressure impact of P_{max} = 1.26 kPa. The wave impacts are presented with a certain probability of exceedance. The figure presents an idealized normal-distribution, shown as a dashed line, as reference. As is evident from the presented figure, the wave impacts on this stepped revetment follow a log-normal distribution. Over and under predictions from the idealized normal-distribution given with the dashed line appear for impacts larger than $P_{95\%}$ and smaller than $P_{10\%}$.

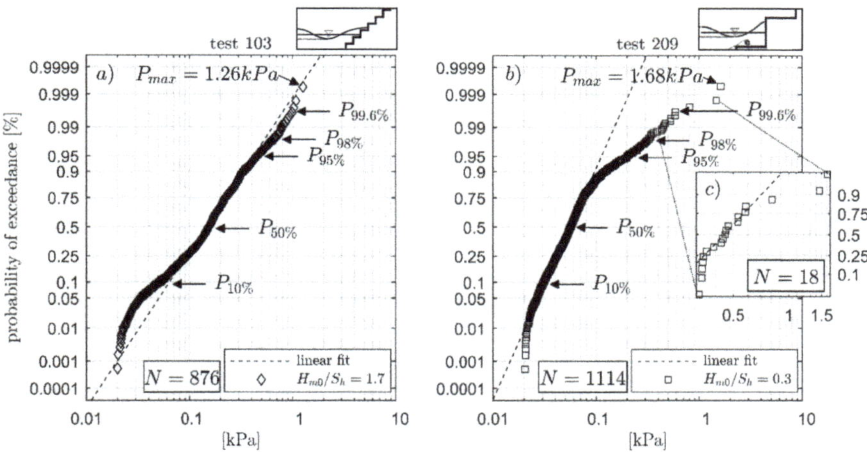

Figure 4. Recorded pressure impacts and log-normal probability distributions of the maximum impact pressures for (**a**) small step heights (test 103, pressure sensor *P1*, H_{m0}/S_h = 1.7) and (**b**) large step heights (test 209, pressure sensor *P2*, H_{m0}/S_h = 0.3). (**c**) Gives a detail of the log-normal probability distribution of the 18 largest impacts in test 209.

Figure 4b presents the log-normal distribution of wave impact pressures on a revetment with large steps (H_{m0}/S_h = 0.3, h_s = 0.921 m, sensor *P2*) for the same hydraulic boundary conditions, as in Figure 4a. The maximum pressure impact P_{max} of 1114 individual impacts is 1.68 kPa, which is about 35% larger when compared to the small steps. As for Figure 3c, slight deviations from the ideal log-normal distribution are detected for impacts smaller than $P_{10\%}$, while impacts larger than $P_{90\%}$ deviate significantly from this trend. These deviations indicate that the influence of the wave-interaction with large steps differs for high and low impact scenarios. The apparent difference in the wave breaking and wave run-up at these large steps compared to slopes with and without smaller steps (Figure 3) is a significant transformation of the incident wave which breaks over the very shallow horizontal step front. Hence, the log-normal distribution of the induced pressures of incident wave heights is disturbed and transformed over large steps directly at the structure, and in analogy, the wave impact distribution is affected. The increase in the wave impacts for large steps can be explained by the fact that the vertical step fronts fall completely dry during the run-down process, whereas for small step heights, there is always a thin water flow remaining from the previous run-down. This thin water layer (often aerated) reduces the wave impact of the next breaking wave.

In parallel, the geometry of a wave front has a significant influence on its resulting wave impact. Generally, the wave height plays a dominant role. It is observed, that small wave heights in the spectrum lead to pulsating extreme loads, whereas the largest waves lead to impacting loads. Distribution of wave impacts deviate strongly from the log-normal distribution for less frequent waves indicating occasional extreme loads on the structure, which is most relevant for practical design attempts. Reference [14] indicated that for vertical walls impacts smaller than 0.4 P_{max} are based on pulsating loads and larger impacts on impacting loads. Reference [15] found that often only a part of all impacts show a good agreement with the log-normal distribution according to the stochastic pattern of the pressures in the impact area. This observation is confirmed for the present experimental configuration. In contrast, when re-considering only the highest impacts during one wave impact event on the area, it turns out that measured data points follow an almost linear relation in a log-normal plot. Hence, Figure 4c gives a detailed log-normal probability distribution of only the 18 largest impacts (>$P_{98\%}$) of test 209, confirming evaluations stemming from Reference [15] to be also valid for stepped revetments.

It can be seen from the left-hand side of this diagram that only a part of them show good agreement with the log-normal function according to the stochastic pattern of the pressures in the impact area. However, if the highest pressure on the area during one wave impact event is selected and plotted in log-normal scale the measured points nearly fall on a straight line which verifies the log-normal distribution.

4.2. Temporal Evolution and Load Cases of Impact Pressures

The temporal evolution of wave impact pressures on vertical structures was initially categorized by Reference [6] and analyzed in depth by Reference [16]. According to Reference [16], impacting loads on vertical structures are defined as impact peaks that are minimum 2.5 times larger than the subsequent quasi-static peak P_q. Partially breaking waves induce peaks in a range of $1.0 < P_{max}/P_q < 2.5$. Pulsating impacts due to standing waves occur when the impact peak and its subsequent quasi-static peak P_q are in the same order of magnitude. Figure 5 sketches a parameter definition describing both a pressure impact event (a), and an exemplary time series of the maximum impact event during test 204 for horizontal oriented pressure sensors *P1*, *P2*, and *P3* (b). A typical time evolution with an impact and subsequent quasi-static peak is observed. The highest impact is recorded by sensor *P2*, which is located near the still water level *SWL*. The oscillation of the signal and negative pressures, as also observed by Reference [8], are interpreted as characteristically mimicking physical processes of turbulence and air entrainment occurring at the base of each revetment step. With increasing distance to the *SWL*, the impact amplitude decreases (*P1* and *P3*). While sensor *P2* recorded a violently impacting load, *P1* was impacted by a slightly breaking wave and *P3* experienced a pulsating impact.

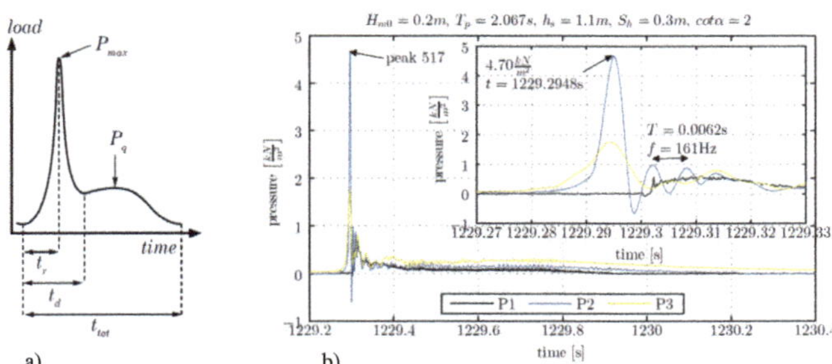

Figure 5. (**a**) Parameter definition describing a pressure impact event and (**b**) exemplary time series of the maximum impact event during test 204 for pressure sensors *P1*, *P2* and *P3*.

Figure 6 presents the normalized time evolution of impact events with a certain probability of exceedance for a 1:2 stepped slope with large (test 103) and small step heights (test 209). These tests were conducted with a similar spectral wave height and wave steepness (Table 1). Raw data (black line) are given to indicate the largest impact. A filtered time series (red line, 6th order low pass filter with 38.4 Hz cut-off frequency) is superimposed in order to highlight a clearer temporal progress. Subfigures (a–f) correspond to step ratios of H_{m0}/S_h = 1.7 (small steps) and subfigures (g–l) to step ratios of H_{m0}/S_h = 0.3 (large steps). Each subfigure displays the measured impact event of a certain probability of exceedance (P_{max} to $P_{10\%}$) over the relative time. The time t is normalized by the wave period $T_{m-1,0}$. Time step t/T = 0 is set at the time of maximum impact. Each peak impact (based on the raw data) is given proportionally to the maximum measured peak P_{max} of the entire test duration.

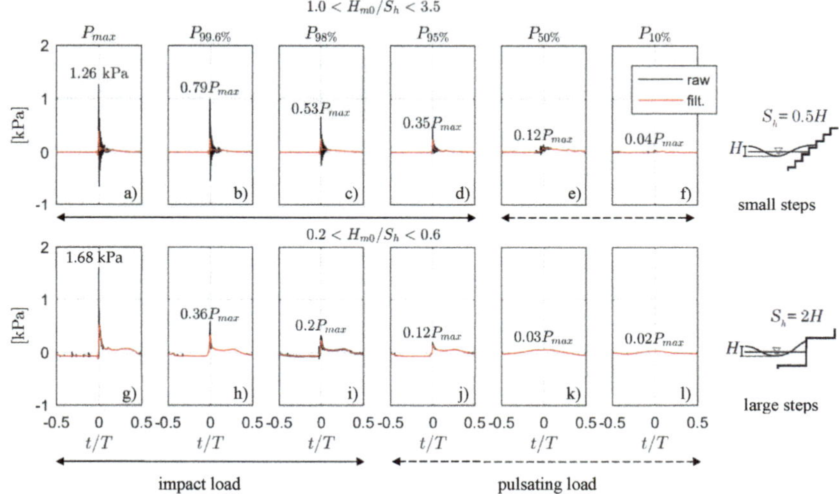

Figure 6. Comparison of pressure events with a certain probability of exceedance between small steps ((**a–f**), H_{m0}/S_h = 1.7, test 103) and large steps ((**g–l**), H_{m0}/S_h = 0.3, test 209).

The maximum measured impact for the small steps (a) was 1.26 kPa, which is about 30% lower than the maximum impact for large steps (g). For the large steps the maximum peak amplitude decays more rapidly with increasing probability of exceedance than for the small steps. In the case with smaller steps, a peak amplitude of about 35% of the maximum impact P_{max} is reached by 5% of all impact events ($P_{95\%}$) (d), whereas in the case of the large steps (h), it is only 0.4% ($P_{99.6\%}$) of all impact events.

A comparison of the occurring load cases defined by Reference [16] elucidates that large steps ($H_{m0} < S_h$) demonstrate a behavioral function like vertical walls in reflecting incoming wave energy. Functional processes are mimicked in two distinct phases: A clear initial impact followed by a quasi-static peak P_q. For small steps ($H_{m0} > S_h$) the impact peak is also clearly visible, but the subsequent quasi-static peak P_q is not as prominent as for large steps or vertical walls. These differences are caused due to different flow principles. At large steps, the water level in front of the step front is constantly rising after the initial impact, thus inducing the quasi-static load. At small steps, which dissipate energy from the up-rushing wave tongue, a highly aerated flow emerges after the initial impact (Figure 3f,g) and the absolute depth under the wave crest is evidently lower. This leads in general to smaller quasi-static peaks. For the specific case presented in Figure 6, impacting loads are observed in the range of about 5% of all impacts at the small steps and 2% of all impacts at the large steps. The initial violent impact dissipates more energy than a pulsating load induced by wave run-up. As a result, it can thus be deduced that the overall energy dissipation at small step heights is larger, i.e., more effective. Furthermore, the impact-mitigating effect due to the practically permanent existence of a thin water layer on top of the individual small steps of the revetment during the wave run-down phase stemming from the preceding wave (Figure 3) plays a dominant role in impact reduction.

The characteristics of an impact pressure on stepped revetments can be described by the rising time t_r and impact pressure duration t_d, as defined in Figure 7 and depicts the correlation between the dimensionless impact duration t_d/t_r and the dimensionless impact rising time $t_r/T_{m-1,0}$. The data are grouped in terms of the step ratios H_{m0}/S_h, the direction of the measured wave impact (horizontal or vertical with respect to gravity) and the position of the impact (above or below the still water level *SWL*). As a meaningful reference, Figure 7 presents the empirical projections for rising times and durations of impacts at vertical walls provided by Reference [16]. Impacting load cases ($P_{max} > 2.5$ P_q) are characterized by a short relative rising time ($0.02 < t_r/T_{m-10} < 0.2$), while pulsating loads are distinguished by their longer peak durations ($t_r/T_{m-10} > 0.2$). The dominance of a peak is characterized by the relative impact duration t_d/t_r. The highest impact loads (e.g., $P_{\max}$ or $P_{99.6\%}$) correspond to short relative impact durations t_d/t_r in combination with a short relative rising time t_r/T_{m-10}. The larger the relative impact duration t_d/t_r, the more critical the impact gets in terms of causing damage or instability of the revetment.

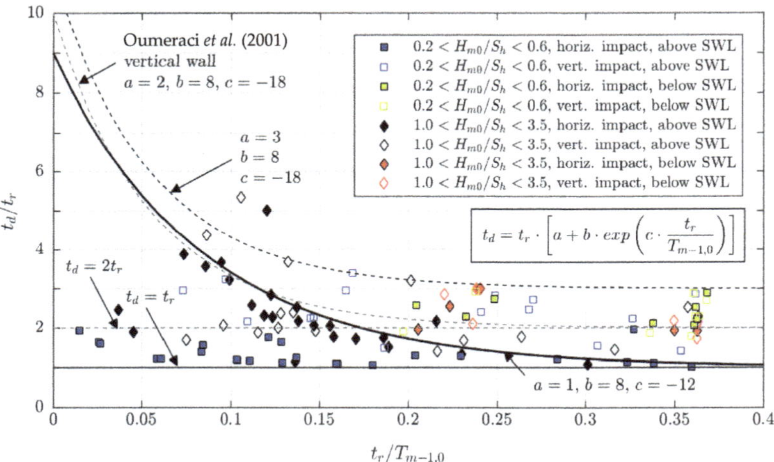

Figure 7. Correlation of the dimensionless impact duration t_d/t_r and the dimensionless impact rising time $t_r/T_{m-1,0}$ at stepped revetments compared to findings for vertical walls. Given trend lines following Equation (5) represent upper envelopes of the findings. Straight horizontal lines represent the corresponding lower envelope line.

Impacts measured on stepped revetments below the *SWL* (green and red markers) show relative rising times of $t_r/T_{m-10} > 0.2$. Horizontal impacts (filled marker) or vertical impacts (empty marker) have comparable impact rising times and impact durations. Below the *SWL*, no impacting load case was detected for either small or large steps. The recorded loads have a pure hydrostatic nature induced due to the water level changes induced by wave run-up and run-down over the pressure sensors. Stronger impacts are mitigated by means of a sufficient water layer ("cushion") effectively sheltering the submerged steps' fronts from more violent impacts. The stepped shape of the slope retards the wave run-down leading to a permanent water cover of the revetment below the *SWL*.

The relative impact rising time of horizontal impacts above the *SWL* (filled blue squares for large steps $0.2 < H_{m0}/S_h < 0.6$ and filled black diamonds for small steps $1.0 < H_{m0}/S_h < 3.5$) ranges from very short impacts ($t_r/T_{m-10} < 0.05$) up to long load cases ($t_r/T_{m-10} > 0.3$). The latter ones represent a full run-up and run-down phase inducing pulsating loads. While the impacts at small steps show an increase in the peak duration for decreasing relative peak rising times up to $t_d = 4t_r$, the maximum duration of $t_d = 2t_r$ is observed for large steps. This finding can be explained with the fact that impacts at large steps are often caused by breaking waves that directly hit the step front, whereas for small steps, the step fronts are covered by thin water layers. Rising times and peak durations of vertical impacts above SWL (empty blue squares for large steps $0.2 < H_{m0}/S_h < 0.6$) and empty black diamonds for small steps $1.0 < H_{m0}/S_h < 3.5$) scatter significantly and do not follow a clear, yet visual trend. The formation and progression of an aerated water layer over the horizontal step front influences the recorded impact significantly and explains the scattering in the data. Pulsating loads can be explained by the water layer thickness during the wave run-up and run-down. Impacting conditions occur only very close to the SWL. The impact rising times for vertical walls and stepped revetments are in the same range. Reference [16] found a minimum peak duration of $t_d = 2t_r$ for vertical walls, whereas the minima for stepped revetments is $t_d = t_r$. These differences may be due to the differences in the sampling frequency of the impact pressure sensors (0.6 to 1.0 kHz at [16] and 19.2 kHz in the present study). Lines of best fit for the different load cases are calculated according to Equation (5) with corresponding regression coefficients a, b and c given in Table 2.

$$t_{d,max} = t_r \cdot \left[a + b \cdot exp\left(c \cdot \frac{t_r}{T_{m-1,0}} \right) \right] \tag{5}$$

Table 2. Coefficients a, b and c for Equation (5) and corresponding minimal impact duration $t_{d,min}$.

Geometry	Impact Direction	a	b	c	$t_{d,min}$
Vertical walls [16]	horizontal	2	8	−18	2.0 t_r
Stepped revetments	horizontal	1	8	−12	1.0 t_r
	vertical	3	8	−18	1.5 t_r

4.3. Spatial Distribution of Impact Pressures

For sloping structures the maximum wave impact resulting from depth-induced wave breaking occurs slightly below the SWL [12]. This section presents the maximum wave impacts on stepped revetments at different locations relative to the SWL. Figure 8 gives the horizontal (a) and vertical (b) relative maximum pressure impact distribution over large (dashed, Test 209) and small (solid, Test 103) stepped revetments. For each pressure sensor P_j, the maximum impact $P_{j,max}$ is normalized by the maximum pressure impact of all compared sensors ($P_{max} = \max\{P_{j,max}\}$). For this case, the absolute maximum was recorded by sensor $P2$ of the revetment with the small steps (compare Figure 1).

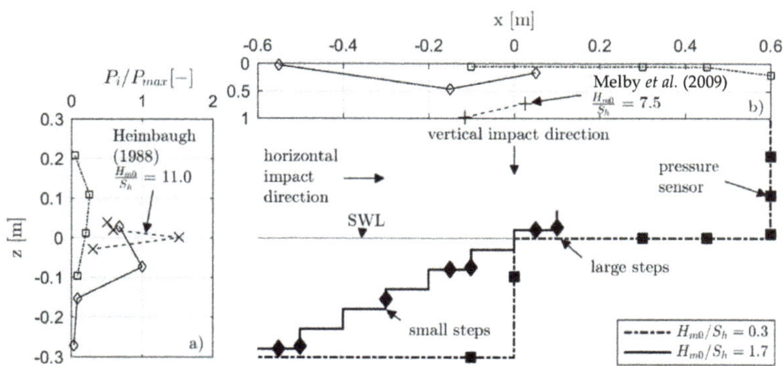

Figure 8. Horizontal (**a**) and vertical (**b**) relative maximum pressure impact distribution over large (dashed, Test 209, H_{m0} = 0.089 m, and ξ = 2.8) and small (solid, Test 103, H_{m0} = 0.084 m, and ξ = 2.9) stepped revetments.

The vertical distribution of horizontal wave impacts (Figure 8a) shows that the maximum wave impact for all configurations occurs close to the *SWL*. The impact loads decrease with increasing distance to the *SWL* ($|z| > 0$). Below the *SWL*, the pressures are reduced by the presence of the water body, which mitigate the impact of a breaking wave. Above the *SWL* the energy of the breaking wave is less due to the decrease of the potential energy (reduced drop height). Impacts above the SLW are often caused by the secondary impact of the breaking wave during the wave run-up or due to splashes induced by the primary wave impact close the *SWL*. For Test 209 the *SWL* is directly located at the edge of the large step (H_{m0}/S_h = 0.3), while the maximum horizontal impact is about four times lower than at the small steps (Test 103, H_{m0}/S_h = 1.7). Additionally, data from Reference [8] are shown. These data represent very small step heights (H_{m0}/S_h = 11) and indicate a maximum horizontal wave impact of about 1.5 times larger than for the steps with a step ratio of H_{m0}/S_h = 1.7. The smaller the step height in relation to the wave height, the higher the recorded horizontal loads become. The reduced impact for an increasing step height can be ascribed to the delayed run-down of the previous wave impact (compare Figure 3a,e). This run-down causes a constant water layer on the revetment, which buffers the wave impact. Moreover, the plunging wave breaking at a 1:2 slope with small steps ($H_{m0}/S_h \geq 1.7$) triggers impacting wave loads whereas the transformation of the breaking wave by large dominant step edges (H_{m0}/S_h = 0.3) leads to the formation of spilling wave breaking over the horizontal step front and triggers, thereby pulsating wave impacts.

The horizontal distribution of vertical wave impacts (Figure 8b) shows that the maximum vertical wave impact for the small steps is located slightly below the *SWL*. The maximum vertical impact has an amplitude of about 50% of the maximum horizontal impact ($P_{max,vertical} = 0.5\ P_{max,horizontal}$). Note that, in the context of the onshore-orientated wave propagation and wave breaking process, this finding is reasonable for design and constructional purposes in any practical applications of stepped revetments. The vertical impact for the large steps is negligible as it represents only the hydrostatic pressure induced by the overflow of the incident wave. Additionally, data from Reference [10] are given. The relative step height of H_{m0}/S_h = 7.5 is comparable to the data set provided by Reference [8] for horizontal wave impacts. The maximum presented impact is about two times larger than for the steps with a step ratio of H_{m0}/S_h = 1.7. An increase in the vertical impact loads is observed for decreasing step heights and is explained analogous to the horizontal wave impact.

Figure 9 provides a non-dimensional relation between the pressure impact normalized by water density ρ, gravity g, and spectral wave height to $P_{99.6\%}/(\rho g H_{m0})$ on the abscissa and the relative position to the *SWL* (z/H_{m0}) on the ordinate. The probability wave impact $P_{99.6\%}$ is selected following the approach of Reference [11] to allow a comparison with the underlying data for wave impacts on vertical walls. For both examined step heights, the maximum pressure impact is close to the *SWL*. The maximum pressure decreases significantly within a range of $\pm\, z/H_{m0}$ around the SWL. Within a range of $\pm 2\ z/H_{m0}$, the normalized pressure impact becomes low. The $P_{99.6\%}$ wave loads measured over the stepped revetment tend to be about 50% smaller than those measured at a vertical wall. On the contrary, the impacts on the steps around the *SWL* exhibit pressures peak impacts which are comparable to those measured on vertical walls by Reference [11] (data points of single impacts not given in this figure scatter significantly along the abscissa). A dual-sided envelope curve (Equation (6)) with coefficients a, b and c according to Table 3 representing the best fit of data (coefficient of determination: R^2 = 0.62, STD = 0.189) is given to describe the correlation of the vertical distribution z/H_{m0} of horizontal impacts $0.01 < P_{99.6\%}/(\rho g H_{m0}) \leq 3.6$.

$$\frac{P}{\rho g H_{m0}} = min\left\{ tan\left[\frac{z/H_{m0} + a}{b}\right]/c, 3.6 \right\} \tag{6}$$

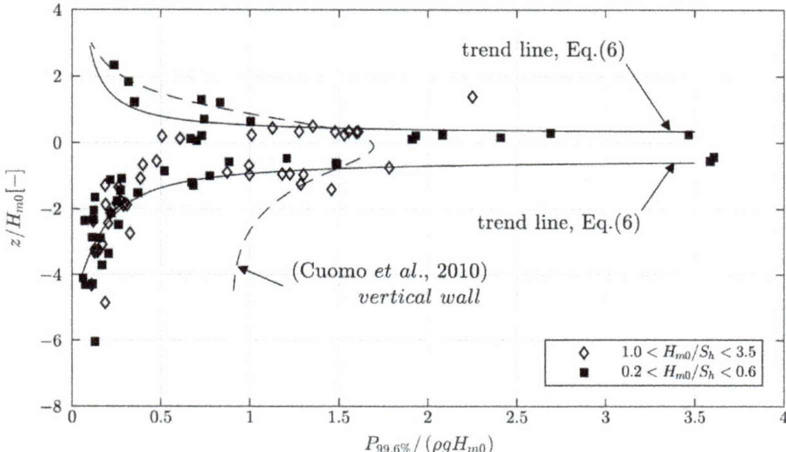

Figure 9. Normalized pressure impact relative to the *SWL* (z = 0) for a 1:2 inclined stepped revetment.

Table 3. Coefficients *a*, *b*, and *c* for Equation (6).

z (*SWL* at z = 0)	***a***	***b***	***c***
$z \geq 0$	−1171.64	745.72	−2831.66
$z < 0$	4.97	−2.87	−6.15

Normalized horizontal pressure impacts with varying probability of exceedance and corresponding relative water depth z/H_{m0} for revetments with large ($0.2 < H_{m0}/S_h < 0.6$, Figure 10a–d) and small ($1.0 < H_{m0}/S_h < 3.5$, Figure 10e–h) steps are shown. Each row represents data with a certain probability of exceedance, as described in Section 4.1. For comparison, the reference line for the horizontal impact forces [11] and the trend line according to Equation (6) are given. The pressure distribution predicted for $P_{99.6\%}$ values by Equation (6) also shows a good agreement for P_{max} values, if the step height is larger than the wave height (a). P_{max} values for revetments with step heights smaller than the wave height (e) show higher impacts, which scatter in a range of $\pm 2\ z/H_{m0}$. The higher and more variable distributed impacts for small steps compared to large steps can again be ascribed to the influence of higher aeration [8,9]. With increasing probability of exceedance ((c,d) for large steps and (g–h) for small steps) the peaks around the *SWL* become less prominent, although the peaks are still visible. This trend is reasonable as the statistical distribution of the individual wave heights in a wave spectrum follows a Rayleigh distribution and the mean wave height of all waves, exceeding a certain threshold, decreases significantly with increasing probability of exceedance. Additionally, the shape of the front of a breaking wave governs the subsequent wave impact as the wave's asymmetry has a secondary importance on the resulting impact. The overall distribution over the water depth becomes more homogeneous with the decreasing probability of exceedance for reasons discussed in Section 4.2.

It has to be clarified that Equation (6) is based on experimental data by means of mathematical fitting routines. Each data-point represents the average of the four maximum impact values recorded by all implemented pressure sensors during a test. Hence, these four impacts do not necessarily correspond to or stem from the same wave or wave train. Furthermore, Equation (6) is based on a fitting through all data points. Individual under-estimations larger than the given mean standard deviation (up to factor 3) are present for single measurements. Similar observations are stated by Reference [11]. Despite these constrains, the approach is applicable to reliably estimate the location, magnitude and trend of impact pressures on stepped revetments. Further, the formula enables an estimation of the wave-induced local impacts and loads on a stepped revetment, and thus provides the first robust estimate in the practical design and dimensioning of adequate foundations and anchoring systems.

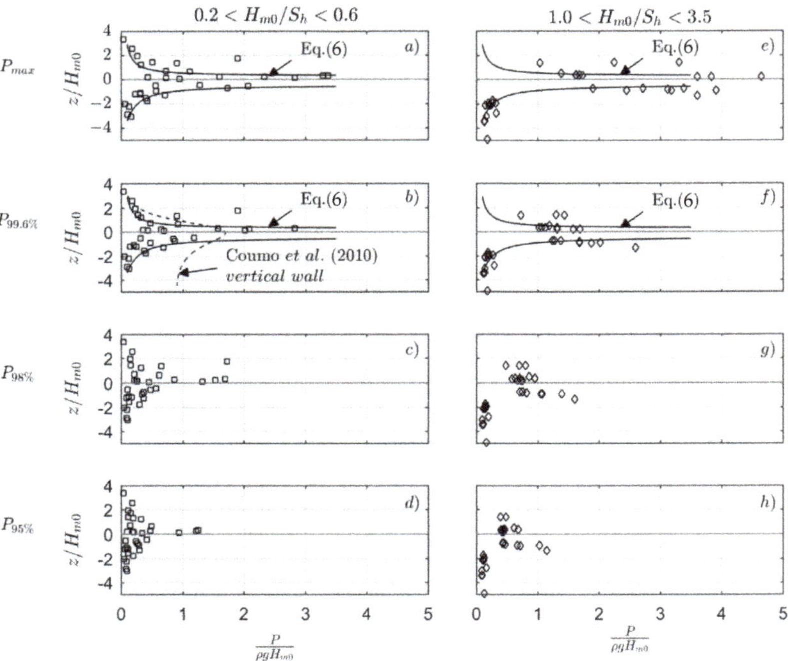

Figure 10. Normalized horizontal pressure impacts with varying probability of exceedance and corresponding relative water depth z/H_{m0} for large ($0.2 < H_{m0}/S_h < 0.6$) and small ($1.0 < H_{m0}/S_h < 3.5$) stepped revetments including theoretical laws for stepped revetments according to Equation (6) and for vertical walls according to Reference [11].

5. Laboratory and Scale Effects

The tests are conducted in a 2D wave flume. The natural multi-directionality of prototype wave conditions is simplified. This affects the wave breaking and accordingly the measured impact forces. Waves are generated in intermediate water depth over a horizontal flume bottom. Hence, waves transform non-linearly during propagation onto and in interaction with the structure. Furthermore, reflected waves from the structure are re-reflected at the piston wave board as experiments were conducted without active reflection compensation. The influence of both boundary conditions on the calculated wave parameters (H_{m0}, T_p), minimized by an analysis of the incident waves very close to the structure.

Assuming gravity as dominant physical force, the model tests are Froude scaled. Therefore, inertia forces are balanced to prototype conditions whereas other physical forces as viscosity, elasticity, surface tension are incorrectly reproduced [17]. As the latter forces have an influence on the process of aeration after wave breaking and during wave run-up flow on the structure, the aeration in the Froude scaled model is reproduced incorrectly. In accordance with Reference [18], the phenomenon and process of aeration in hydraulic models, e.g., in hydraulic jumps is significant lower in smaller scales and cannot be achieved under Froude similitude. This outcome is somehow evident in any hydraulic modeling attempts and holds true for stepped revetments likewise. Hence, the presented impact pressures (in small scaled experiments) are probably over predicted as already confirmed for vertical walls by Reference [19]. As of this, an attempt of upscaling the measured impact pressures is not taken in this paper as a detailed comparison with data from large scale model tests on stepped revetments still have to be done in due time. Accordingly, the presented data enable a discussion of the physical phenomena and triggered processes as well as relative magnitudes of impact pressures on stepped revetments and differences in trends for varying step ratios without providing exact, i.e., real-world estimates for structural design or dimensioning.

6. Conclusions

Wave-induced impacts on stepped revetments have been investigated by means of physical model tests conducted in a wave flume. The tests focused on the wave impact on stepped revetments with relative step heights in a range of $0.3 < H_{m0}/S_h < 3.5$.

It was found that the probability distributions of wave impact pressures on stepped revetments follow a log-normal distribution, which confirm initial findings by Reference [12,13,15]. The maximum peak amplitude in a test series decreases with increasing probability of exceedance. The decay is more intensive for step heights larger than the wave height due to significant wave transformation processes over the dominant horizontal step fronts. Large steps ($H_{m0} < S_h$) show similar load cases compared to vertical walls. For small steps ($H_{m0} > S_h$) the impact peak is also clearly visible but the subsequent quasi-static peak P_q is not as prominent as for large steps or vertical walls. Therefore the recommendation of Reference [7] to calculate the forces on stepped structures with the same method as for vertical walls, which only holds true for large steps.

It was found that at small steps the highly aerated flow that emerges after the initial wave impact generally leads to smaller quasi-static peaks. The initial violent impact dissipates more energy than a pulsating load induced by wave run-up. As a result, it can be deduced that the overall energy dissipation at small step heights is larger. Impacts measured at stepped revetments below the *SWL* show relative rising times of $t_r/T_{m-10} > 0.2$. No impacting load case was detected at stepped revetments with large or small step height below the *SWL*. Real impacts are buffered by a water layer protecting the steps from violent impacts. Impacting conditions occur only very close to the *SWL*. Compared to impacts on vertical walls, the impact rising times are in the same range. The minima for stepped revetments is $t_d = t_r$. As the importance of aeration in the run-up to the wave impact is identified as a future study, it should focus on this effect in analogy to Reference [15].

The spatial distribution of the impact loads generally decrease with increasing distance to the *SWL*. The smaller the step height compared to the wave height, the higher the horizontal loads become. The reduced impact for increasing step height can be ascribed to the delayed run-down of the previous wave impact. The maximum vertical wave impact for the small steps ($H_{m0} > S_h$) is located slightly below the *SWL* with an amplitude of about 50% of the maximum horizontal impact. The vertical impact for the large steps ($H_{m0} < S_h$) is negligible as it represents only the hydrostatic pressure induced by the overflow of the incident wave. An increase in the vertical impact loads is seen for decreasing step heights analogous to the horizontal wave impact. The maximum impact decreases significantly within a range of $\pm\, z/H_{m0}$, mainly in the range of $\pm 2\, z/H_{m0}$. The highest ($P_{99.6\%}$) wave loads measured over the stepped revetment tend to be about 50% smaller than those measured at a vertical wall. On the contrary, the impacts on the steps around the *SWL* showed impacts comparable to those measured on a vertical wall. Higher and more variable distributed impacts for small steps ($H_{m0} > S_h$) compared to large steps can be explained by the influence of higher aeration. With increasing probability of exceedance, the peaks around the *SWL* are less significant.

The analysis of the data including insights from the literature [17–19] showed the significant influence of scale effects on the results, especially in terms of the aeration. The future work will address these topics in order to enable the upscaling of the presented data. Furthermore, additional slopes have to be analyzed to determine this influence on the magnitudes of impact pressures in context of the step ratios and fully align the research in the existing findings for vertical walls and plain slopes.

Author Contributions: N.B.K. and To.S. conceived of the presented idea. N.B.K. developed the theory and performed the study. Ta.S. and To.S. verified the analytical methods. To.S. encouraged N.B.K. to investigate a comparison with wave impacts on vertical walls and supervised the findings of this work. All authors discussed the results and contributed to the final manuscript.

Funding: The presented findings were developed within the framework of the research project 'waveSTEPS – Wave run-up and overtopping at stepped revetments' (03KIS118) funded by the Federal Ministry of Education and Research (BMBF) through the German Coastal Engineering Research Council (KFKI).

Conflicts of Interest: The authors declare no conflict of interest.

References

1. Van der Meer, J.W.; Allsop, N.W.H.; Bruce, T.; De Rouck, J.; Kortenhaus, A.; Pullen, T.; Schüttrumpf, H.; Troch, P.; Zanuttigh, B. EurOtop: Manual on Wave Overtopping of Sea Defences and Related Structures. An Overtopping Manual Largely Based on European Research, But For Worldwide Application. 2016. Available online: www.overtopping-manual.com (accessed on 11 July 2018).
2. Kerpen, N.B. Wave-Induced Responses of Stepped Revetments. PH.D. Thesis, Gottfried Wilhelm Leibniz Universität Hannover, Hannover, Germany, 2017.
3. Kerpen, N.B.; Schlurmann, T. Stepped revetments–Revisited. In Proceedings of the 6th International Conference on the Application of Physical Modelling in Coastal and Port Engineering and Science (Coastlab16), Ottowa, ON, Canada, 10–13 May 2016; p. 10.
4. Kerpen, N.B.; Goseberg, N.; Schlurmann, T. Experimental Investigation on Wave Overtopping on Stepped Embankments. In Proceedings of the 5th International Conference on Application of Physical Modelling to Port and Coastal Protection; Black Sea-Danube Coastal Research Association (BDCA): Varna, Bulgaria, 2014; pp. 262–269.
5. Kerpen, N.B.; Bung, D.B.; Valero, D.; Schlurmann, T. Energy dissipation within the wave run-up at stepped revetments. *J. Ocean Univ. China* **2017**, *16*, 649–654. [CrossRef]
6. Oumeraci, H.; Klammer, P.; Partenscky, H.W. Classification of Breaking Wave Loads on Vertical Structures. *J. Waterw. Port Coast. Ocean Eng.* **1993**, *119*, 381–397. [CrossRef]
7. *SPM Shore Protection Manual*; Dept. of the Army, Waterways Experiment Station, Corps of Engineers, Coastal Engineering Research Center: Vicksburg, MS, USA; For Sale by the Supt. of Docs., U.S. G.P.O.: Washington, DC, USA, 1984.
8. Heimbaugh, M.S. *Coastal Engineering Studies in Support of Virginia Beach, Virginia, Beach Erosion Control and Hurricane Protection Project: Report 1, Physical Model Tests of Irregular Wave Overtopping and Pressure Measurements*; National Technical Information Service: Springfield, VA, USA, 1988.
9. Bagnold, R.A. Interim report on wave-pressure research. *J. Inst. Civil Eng.* **1939**, *12*, 200–227. [CrossRef]
10. Melby, J.A.; Burg, E.C.; McVan, D.; Henderson, W. *South Florida Reservoir Embankment Study*; Engineer Research and Development Center Vicksburg MS Coastal and Hydraulic Laboratory: Vicksburg, MS, USA, 2009.
11. Cuomo, G.; Allsop, W.; Bruce, T.; Pearson, J. Breaking wave loads at vertical seawalls and breakwaters. *Coast. Eng.* **2010**, *57*, 424–439. [CrossRef]
12. Führböter, A. *Der Druckschlag durch Brecher auf Deichböschungen*; Mitt. Franzius-Institut der TU Hannover: Hannover, Germany, 1966; Volume 28.
13. Weggel, J.R. Discussion on: Shock pressures on coastal structures (by Adel M. Kamel). *J. Waterw. Harb. Coast. Eng. Division* **1971**, *WW3*, 584–588.
14. Hull, P.; Müller, G.; Allsop, N.W.H. A Vertical Distribution of Wave Impact Pressures for Design Purposes. In *MAST III, PROVERBS-Project: Probabilistic Design Tools for Vertical Breakwaters*; Belfast: Northern Ireland, 1998; p. 16.
15. Führböter, A. Model and prototype tests for wave impact and run-up on a uniform 1:4 slope. *Coast. Eng.* **1986**, *10*, 49–84. [CrossRef]
16. Oumeraci, H.; Kortenhaus, A.; Allsop, W.; de Groot, M.; Crouch, R.; Vrijling, H.; Voortman, H. *Probabilistic Design Tools for Vertical Breakwaters*; CRC Press: Boca Raton, FL, USA, 2001; ISBN 978-90-5809-249-6.
17. Hughes, S.A. *Physical Models and Laboratory Techniques in Coastal Engineering*; World Scientific: Singapore, 1993; ISBN 978-981-02-1541-5.

18. Chanson, H.; Murzyn, F. Froude Similitude and Scale Effects Affecting Air Entrainment in Hydraulic Jumps. In Proceedings of the World Environmental and Water Resources Congress 2008, Honolulu, Hawaii, 12–16 May 2008; Volume 316, p. 10.
19. Cuomo, G.; Allsop, W.; Takahashi, S. Scaling wave impact pressures on vertical walls. *Coast. Eng.* **2010**, *57*, 604–609. [CrossRef]

Journal of
Marine Science and Engineering

MDPI

Article

Physical Modelling of Blue Mussel Dropper Lines for the Development of Surrogates and Hydrodynamic Coefficients

Jannis Landmann [1,*], Thorsten Ongsiek [1], Nils Goseberg [2], Kevin Heasman [3], Bela H. Buck [4,5], Jens-André Paffenholz [6] and Arndt Hildebrandt [1]

1. Ludwig-Franzius-Institute for Hydraulic, Estuarine and Coastal Engineering, Leibniz Universität Hannover, 30167 Hannover, Germany; ongsiek@lufi.uni-hannover.de (T.O.); hildebrandt@lufi.uni-hannover.de (A.H.)
2. Department of Hydromechanics and Coastal Engineering, Leichtweiß-Institute for Hydraulic Engineering and Water Resources, Technische Universität Braunschweig, 38106 Braunschweig, Germany; n.goseberg@tu-braunschweig.de
3. Cawthron Institute, 7010 Nelson, New Zealand; kevin.heasman@cawthron.org.nz
4. Alfred Wegener Institute for Polar and Marine Research, Faculty of Biosciences, Shelf Sea Systems Ecology, Marine Aquaculture, 27570 Bremerhaven, Germany; bela.h.buck@awi.de
5. Faculty 1, Applied Marine Biology and Aquaculture, Bremerhaven University of Applied Sciences, 27568 Bremerhaven, Germany
6. Geodetic Institute, Leibniz Universität Hannover, 30167 Hannover, Germany; paffenholz@gih.uni-hannover.de

* Correspondence: landmann@lufi.uni-hannover.de; Tel.: +49-511-762-2580

Received: 30 December 2018; Accepted: 6 March 2019; Published: 12 March 2019

Abstract: In this work, laboratory tests with live bivalves as well as the conceptual design of additively manufactured surrogate models are presented. The overall task of this work is to develop a surrogate best fitting to the live mussels tested in accordance to the identified surface descriptor, i.e., the Abbott–Firestone Curve, and to the hydrodynamic behaviour by means of drag and inertia coefficients. To date, very few investigations have focused on loads from currents as well as waves. Therefore, tests with a towing carriage were carried out in a wave flume. A custom-made rack using mounting clamps was built to facilitate carriage-run tests with minimal delays. Blue mussels (*Mytilus edulis*) extracted from a site in Germany, which were kept in aerated seawater to ensure their survival for the test duration, were used. A set of preliminary results showed drag and inertia coefficients C_D and C_M ranging from 1.16–3.03 and 0.25 to 1.25. To derive geometrical models of the mussel dropper lines, 3-D point clouds were prepared by means of 3-D laser scanning to obtain a realistic surface model. Centered on the 3-D point cloud, a suitable descriptor for the mass distribution over the surface was identified and three 3-D printed surrogates of the blue mussel were developed for further testing. These were evaluated regarding their fit to the original 3-D point cloud of the live blue mussels via the chosen surface descriptor.

Keywords: aquaculture; drag; inertia; Abbott–Firestone Curve; laboratory tests

---

1. Introduction

In recent decades, aquaculture production has served as an essential source of protein for large parts of the world population. An ever-increasing demand for aquatic products suggests that aquaculture will continue to be one of the fastest-growing sectors for the production of protein-based foods [1]. An important part of its production is of bivalve origin, i.e., clams, oysters, mussels and other species. As of 2014, more than 16 million tons of bivalve produce were farmed around the world [1]. Bivalve cultivation continues to play an important role in providing food for the growing world population.

Currently, mussel farming is situated near or inshore, mostly because sheltered sites have less technological requirements. However, these sites can also pose a variety of different problems. These range from navigational issues for marine vessels [2], nutrient depletion in the wake of mussel farms and undesirable changes in the species assemblage to negative alterations to the benthic environment below the farms due to the mussels' pseudofaeces and marine litter [3]. These issues in combination with more and more contested space nearshore have stipulated a push towards offshore developments. The perspective to use larger volumes of high-quality water to decrease the stress on cultured organisms offshore and the chance to create more revenue by offsetting seafood deficits have led to an increasing effort to move mussel farming into open waters [4].

Aquaculture design is predominantly governed by environmental constraints, and depending on the in situ conditions, different aspects have to be considered in, near or offshore. Especially at offshore sites, high energy is acting on the structures as a result of wave and current conditions. Although hydrodynamic forces acting on fixed and floating structures in such conditions are commonly investigated, existing research related to shellfish aquaculture is generally limited [5]. The feasibility of moving bivalve-related aquaculture offshore is mainly evaluated in regard to farming techniques that are already in use. Suspended farm systems can be divided into intertidal, raft and long-line cultures. Intertidal farms use lines connected to stakes driven into the intertidal sea bed from which the dropper lines are suspended [6]. In raft systems, the mussels are grown on ropes which are hung from a moored raft into the water column [7,8]. Furthermore, so-called "long-line systems" are a possible candidate to be moved further offshore [9]. Long-line systems consist of floatation elements that are connected by ropes to form a mussel-bearing backbone. The backbone is kept in place at each end via an anchor warp connected to the mooring systems. The mussels are cultivated on ropes, so called "dropper lines" or "collectors", and suspended perpendicular from the backbone. The dropper lines are usually spaced evenly along the backbone, with lengths between 5–30 m depending on the water depth and available nutrition. The hydrodynamic forcing on long-line systems has been observed in near-shore conditions to identify the dominant modes of flow-structure interaction and to provide a baseline for designs of future structures [10]. On this basis, work regarding the physics of offshore bivalve-aquaculture has been presented [11]. A description of the hydrodynamic implications of large long-line farms based on observations and scaling arguments is available [12]. A study using a rigid, artificial mussel crop rope constructed from the shells of *Perna canaliculus* provides C_D-values for the towing velocities of 0.05 to 0.40 m/s.

The aim of the study was to determine the influence of the inhalants and exhalents of fluids on the drag [13]. An investigation regarding the drag coefficients of suspended canopies, i.e., mussel dropper lines, derived from a physical model with circular cylinders as a representation of the dropper lines is available by Plew, D.R. [14]. Research regarding numerical simulations of suspended bivalve farms is available to a limited extent and mostly details the flow patterns in the wake of farm systems [15]. A numerical study by Raman-Nair and Colbourne [16] to quantify the loads induced into long-line systems is the basis for a numerical model of the three-dimensional dynamics of a submerged mussel long-line system presented by the same author [17]. Further, a flow analysis around mussels is part of the current research activities, however mainly focusing on naturally bedded mussel cultures [18] or on netted structures in association with biofouling [19] in contrast to the long-line systems. While observations of long-line systems are available alongside numerical work, to date no laboratory tests are available. However, experimental data are of crucial concern for the calibration of numerical models facilitating the design process. Thus, this work depicts the first detailed examination of the response of a live-mussel long-line to hydrodynamic forcing in laboratory conditions.

The available literature shows that research gaps concerning the behaviour of suspended long-lines in current and wave conditions exists. Especially the behaviour of dropper lines under waves has received only little attention. A series of physical tests with full-scale blue mussel dropper lines was conducted to evaluate the corresponding drag and inertia characteristics in the future as well as to determine the magnitude of wave and current forces acting on the mussel dropper lines. To that end, a custom-made test rack consisting of aluminium profiles was attached to a traversing carriage located over a wave flume. Three different specimens of live blue mussel (*Mytilus edulis*) dropper lines were fitted to the test rack. These were tested at various current speeds simulated by the carriage runs to obtain the drag coefficients. Independently, wave tests with varying parameters were conducted to obtain the inertia coefficients. Additionally, surrogate models were created from a digital model based on the 3-D scanned data of the live-mussel dropper lines. The live and surrogate shellfish dropper lines were compared subsequently. The surrogate models were evaluated regarding the statistical mean values of sampled single mussels as well as a surface descriptor. By this means, a number of different surrogates were created and assessed in regard to their fit to the corresponding Abbot–Firestone curve.

2. Materials and Methods

Experiments involving live-mussel specimens were carried out at the medium wave and towing tank "Schneiderberg" (WKS) at the Ludwig-Franzius Institute for Hydraulic, Estuarine and Coastal Engineering of the Leibniz Universität Hannover, Germany. The WKS is 110 m long, is 2.2 m wide and contains up to 1.1 m water with a total depth of 2.0 m. In this facility, regular and irregular waves can be generated up to 0.5 m in wave height. A towing carriage allows for testing at constant velocities of up to 1.5 m/s. A plan and side view of the WKS including the towing carriage and wave maker are outlined in Figure 1. All data chosen for the current velocities and wave characteristics were related to potential offshore sites, e.g., off the coast of New Zealand or Canada, and scaled down to allow for future experiments with scaled surrogates. A 1:10 scale was selected, and a Froude similarity was applied.

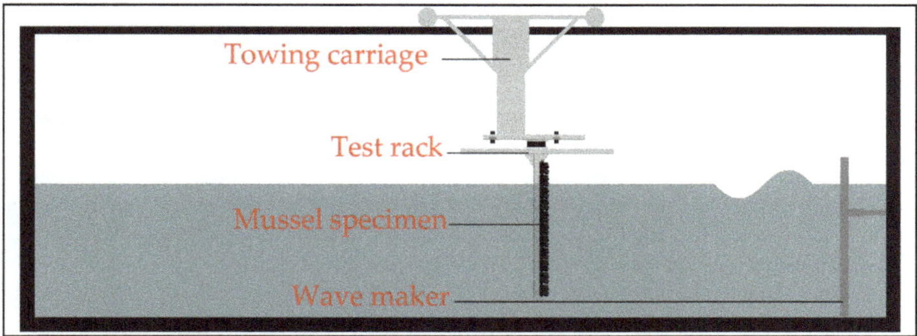

Figure 1. A side view of the "Schneiderberg" (WKS) wave tank with a towing carriage, test rack, attached mussel specimen and wave maker (right end); the sketch is not to scale.

A vertical aluminium beam of 2.52 m length, attached to the carriage, allowed the connection of additional equipment. In this work, a load frame was fixed to the main beam that consists of interconnected aluminium profiles. To ensure a sufficient rigidity, a 1.08 m × 0.72 m aluminium frame, horizontally oriented, was used as the top part of the rack. This frame was braced and reinforced by two 1.0 m-long aluminium profiles on the inside of the frame, running in line with the 1.08 m-long outer profiles. The lower part of the test rack was also assembled with an aluminium frame of 1.00 m × 0.80 m, vertically oriented, and used primarily for the attachment of the live-mussel specimens. Figure 2 presents schematic sketches of the test rack with the attached measuring equipment. The main beam, not visible in the sketches, is represented by a connecting plate. Custom-made clamps with a coarse interlocking grid were CNC-milled. These allowed for an easy connection of the mussel dropper line to the test rack while also being able to restrict unwanted movements for the test duration. The clamps were attached to the bottom of the lower frame and the interior profiles of the upper frame. This allowed for a horizontal orientation of the mussel dropper lines in a defined distance from the main beam. Four wire connections ran from the outer corners of the upper frame to the bottom of the lower frame to further increase the stability. Further, the wires were tensioned to ensure the stability of the test rack. The test rack, as a whole, provided a structure for mussel droppers with a maximal length of 1.1 m. The carriage speed for drag testing was checked via an incremental rotary encoder (SICK DBV50) with a resolution of 12.5 pulses/mm. The rotary encoder allowed for the exact determination of the associated towing velocity by means of single differentiation over the travelled distance. Testing was conducted in either direction, along-flume; hence the rotary encoder installed on the carriage track was either used as a trailing or guiding encoder, depending on the direction of the test run.

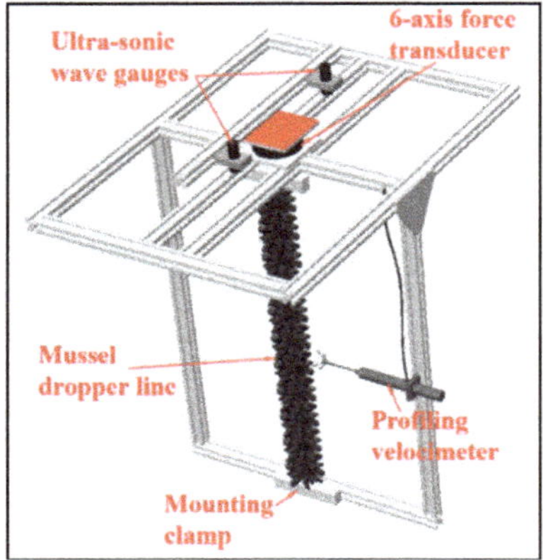

Figure 2. A conceptual sketch of the test rack with the measuring equipment and mussel specimen attached in free view: The red plate represents the connection to main beam.

Most importantly, the forces on the mussel dropper lines were of interest when subjected to waves and currents. To obtain this information on the test rack, a six-axis force-transducer (K6D110 ME-Meßsysteme GmbH, 16761 Henningsdorf, Germany) was rigidly attached to the main beam. The K6D110 has a nominal force range of up to 10 kN in x-, y- and z-directions and a nominal torque range of 750 Nm in Mx, My and Mz. Accurate measurements as low as 0.1 kN were possible, based on the specifications by the manufacturer. The x-direction corresponds to a movement from the point of origin along the flume; the y-direction represents the lateral direction, while the z-direction described the direction towards the bottom of the flume. The recorded momentum forces in Mx, My and Mz describe the torques around these axes, respectively. Furthermore, a profiling, acoustic Doppler velocimeter, ADV, (Vectrino Profiler, Nortek, 1351 Rud, Norway) with a maximum sample rate of 100 Hz was attached to the test rack. With the ability to profile three-component velocities over a vertical range of 3 cm and with a resolution of 1 mm, insights into the turbulence dynamics alongside the mussel dropper lines becomes possible. The ADV Profiler was attached in a distance of approximately 0.45 m from the lowest aluminium profile. That approximately corresponds to the midsection of the tested blue mussel dropper lines. Furthermore, measurements of the time-history of the free surface elevation in close vicinity to the dropper line were carried out by ultrasonic wave gauges to obtain information concerning the local wave field. To this end, two locations roughly 0.20 m in front of and behind the dropper line are selected.

All tests were recorded via two sets of cameras. A GoPro Hero4 with a high-definition resolution and a sample rate of 100 fps as well as a Logitech C920 webcam with high-definition resolution and a sample rate of 30 fps were added to the test setup. A linear Field-of-View (FOV) setting for the GoPro Hero4 (Firmware v5.00) was used to correct the convex distortion of the camera. The C920 recorded with a linear FOV as the default setting. The GoPro was installed underwater to the side of the test rack with a slight offset to the dropper line. The offset ensured an unobstructed view on the test specimen. The webcam was installed on the upper frame of the test rack directed at the blue mussel dropper line in the direction of the wave flume. It was fitted to monitor the wake and oscillatory behaviour of the dropper line. During inertia testing, the carriage, as a whole, was positioned in front of a window section of the flume. A tripod holding another Logitech C920 webcam was used. All video files recorded allowed hindsight into the testing conditions and also provided data regarding the motion response of the dropper-lines under different current velocities. The water depth was kept constant at 0.93 m and was recorded together with temperature periodically throughout the testing procedure. The constant water depth allowed for continuous and identical testing conditions. Figure 3 shows the test rack with the attached measuring equipment and mussel dropper line.

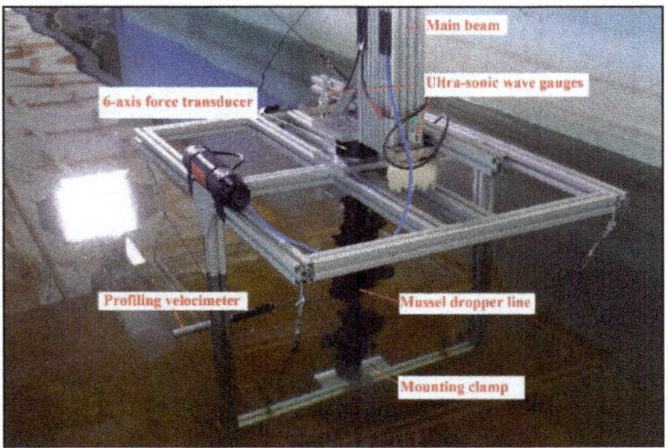

Figure 3. The test rack with the Vectrino II Profiler, ultrasonic wave gauges and visible mounting clamps with the attached blue mussel specimen viewed from inside the flume.

A 3.62 m long dropper line of marketable, adult blue mussels (Mytilus edulis) encrusted with newly seeded spat was selected for testing. The blue mussel was selected due to its wide distributional pattern. Furthermore, blue mussels are prime candidates for aquaculture [20]. This is mainly due to its ability to withstand wide fluctuations in salinity, desiccation, temperature and oxygen levels [21]. Blue mussels are found throughout European waters, occupying habitats from Russia to the Bay of Biscay off the French coast. Its zonatial range stretches from the high intertidal to subtidal regions, and its salinity adaptability extends from estuarine areas to fully oceanic seawaters. The blue mussel is euryhaline and proliferates in offshore sea- as well as in brackish-waters down to 4% salinity. Blue mussels are also eurythermal, withstanding freezing conditions for several months. The species is acclimated for a 5–20 °C temperature range, with an upper sustained thermal tolerance limit of about 29 °C for fully grown conditions [21]. These behavioural characteristics allow for a broad application of the insights gathered into drag and inertia characteristics. The blue mussels used in the drag and inertia tests were obtained from an aquaculture farm located in Kiel, Germany and were transported and stored in a controlled sea water tank with aeration and temperature regulation systems in use. Thus, the survival of the mussels for a prolonged time was possible with no loss of adhesive qualities of the byssus, the bundled filaments secreted by the bivalves. The byssus function was the attachment points to the dropper line and was weakened when subjected to adverse conditions.

The blue mussel dropper line was segmented into three specimens with lengths of 1.03 m to 1.05 m. These were labelled 1 to 3 to distinguish in between testing. The specimens were individually weighed, and their width was measured in steps of 5 cm. This was required to determine the mean diameter of the blue mussel specimens. The volumetric displacement of each specimen was determined by immersing them into a container of known dimensions. The water level was then measured before and after the immersion of the specimen. The extant blue mussel dropper line with a length below

50 cm was treated the same way. Additionally, the length, diameter, displacement and weight of individual blue mussels were also determined. Some mussel individuals were selected randomly to obtain information on the individuals of the encrusted rope.

The mean values for the individual mussels correspond to a mean single mussel length $msml$ = 4.7 cm, a mean single mussel thickness $msmt$ = 2.2 cm and a mean single mussel weight $msmw$ = 9.1 g. Figure 4 depicts a specimen prepared before testing alongside a single mussel evaluated regarding the mean values.

(a)

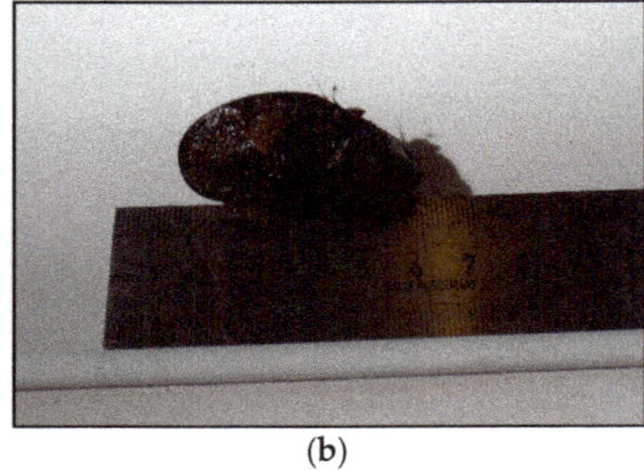

(b)

Figure 4. (**a**) A mussel specimen under preparation for drag and inertia testing and (**b**) an example of a single mussel randomly selected for data acquisition.

The specimens were inserted into the mounting clamps and tightened. For each specimen tested, several towing and wave tests with varying setups were conducted. However, before each set of experiments, an eigenfrequency test for the test rack itself was carried out with an impact hammer. The impact event was synchronously recorded by the 6-axis force transducer mounted to the load frame. By this means, the eigenfrequencies of the load frame as well as all possible combinations with dropper line specimens were recorded. Upon completion, the drag experiments in the wave tank were carried out with top- and bottom-mounted dropper lines that were towed at velocities of u_1 = 0.25 m/s, u_2 = 0.50 m/s, u_3 = 0.75 m/s and u_4 = 1.00 m/s. The mount at both ends ensures an equal flow velocity for the whole length of the dropper line. Each specimen, though differing in total length, was submerged over a length of L_{wet} = 0.80 m. This, in turn, allows the correct determination of the drag coefficient C_D for a specified length of mussel dropper line. The Reynolds numbers $Re = u_i * md/\nu$ covered during the tests were determined with the characteristic diameter md = 10.31 cm, the aforementioned towing velocities u_i, the kinematic viscosity of the water $\nu = 1.004/998{,}200$ m^2/s and the constant water temperature of 20 °C and range from 2.0×10^4 to 1.1×10^5. In between all drag tests, the flow disturbances potentially induced by previous testing settled during a waiting period in order to avoid biased influence. Testing of the first specimen was repeated three times for testing repeatability. In order to gain further information about the force and motion response of the specimen, the towing operation was divided into forward and backward motions such that a single specimen is towed twice yet with a 180° change in direction. For specimens 2 and 3, the repetitions were reduced to one to allow for quicker laboratory tests, as the live mussels lose their cohesive properties when exposed to non-autochthonous conditions over a prolonged period of time [20]. After the completion of these tests, the setup of the dropper line was changed by loosening the lower clamp, leaving a top-mounted dropper line. As the pretests had revealed that major oscillations of the blue mussel dropper lines in the lateral and longitudinal directions occur at velocities $u > 0.50$ m/s, the tested velocities were reduced for the drag tests with the top-mounted specimens. The reduced velocities tested were u_5 = 0.10 m/s, u_1 = 0.25 m/s, u_6 = 0.375 m/s and u_2 = 0.5 m/s. Figure 5 shows the top-mounted mussel specimen dragged at u_5 = 0.10 m/s, u_1 = 0.25 m/s and u_2 = 0.5 m/s. The progressive lift towards the surface due to the forces acting on the dropper line is visible throughout the three sections.

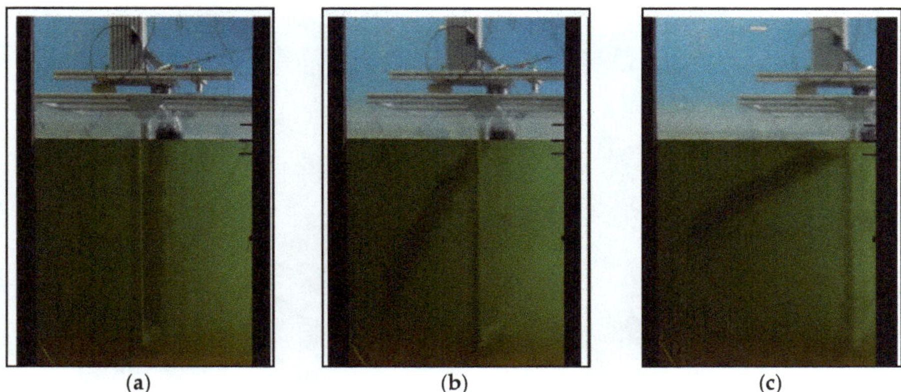

Figure 5. The drag testing of a top-mounted only specimen at velocities of 0.10 m/s (**a**), 0.25 m/s (**b**) and 0.5 m/s (**c**) with a progressive lift towards the surface visible.

After completing the towing tests, the carriage was positioned in front of a windowpane in the wall of the flume, which allows a side-view documentation of the motion response of the droppers during wave loading. The distance between the wave maker and carriage is 65 m. Three sets of waves were tested. These are described by their period T and their targeted wave height H. Thus, the wave tests were carried out with targeted wave heights H of 0.1 m, 0.12 m and 0.15 m with corresponding periods T of 1.20 s, 2.4 s and 1.65 s. Each wave test was conducted once for each of the three specimens. The chosen values regarding the wave characteristics were related to potential offshore sites, e.g., off the coast of New Zealand or Canada, and scaled down to allow for future experiments with scaled surrogates. A 1:10 scale was selected, and a Froude similarity was applied. The corresponding Keulegan–Carpenter numbers $KC = u_i \times T/md$ ranged from 4.1 to 5.9. In Table 1, an overview of the wave height, wave period, maximum horizontal and vertical velocity, maximum horizontal and vertical acceleration and wave length is provided according to Stokes 2nd order wave theory for the waves tested in the flume. The tested wave heights are displayed in model M and full-scale F according to Froude similitude and a scaling factor of 1:10. The tests were performed for the top- and bottom-mounted configurations to enable comparative studies regarding the commonly investigated cylinders under wave loads as well as the free-swinging systems. Figure 6 exemplarily shows specimen 3 in waves with a targeted wave height of 0.12 m and a period of 2.4 s in a top-and-bottom-mounted configuration in comparison to a top-mounted only configuration. The latter setup results in an additional oscillatory movement of the blue mussel dropper line. To attain the actual forces acting on the blue mussel dropper line alone, frame-only tests are necessary as a concluding step. Here, all drag and inertia tests with corresponding current velocities and wave sets were carried out without an attached specimen.

Table 1. An overview of the wave height H, wave period T, wave length L, maximum horizontal u_{max} and vertical v_{max} velocity, and maximum horizontal ax_{max} and vertical az_{max} acceleration is provided according to Stokes 2nd order wave theory in model M and full-scale F.

Parameter $_{M/F}$	**Wave Test 1**	**Wave Test 2**	**Wave Test 3**
H_M/H_F (m)	0.10/1.00	0.12/1.20	0.15/1.50
T_M/T_F (s)	1.20/3.80	2.40/7.59	1.65/5.21
L_M/L_F (m)	2.23/22.3	6.63/66.3	3.92/39.2
$u_{max,M}/u_{max,F}$ (m/s)	0.26/0.82	0.21/0.66	0.31/0.98
$v_{max,M}/v_{max,F}$ (m/s)	0.08/0.25	0.13/0.41	0.29/0.91
$ax_{max,M} = ax_{max,F}$ (m/s^2)	0.44/0.44	0.45/0.45	1.18/1.18
$az_{max,M} = az_{max,F}$ (m/s^2)	−1.37/−1.37	−0.41/−0.41	−1.08/−1.08

(a)

(b)

Figure 6. (**a**) The wave-loading on a blue mussel dropper line with a top-and-bottom-mounted clamp setup in contrast to (**b**) a top-mounted only clamp setup.

In between testing, a 3-D point cloud as a digital copy of the dropper line specimens was acquired by a terrestrial laser scanner in collaboration with the Geodetic Institute of the Leibniz Universität. The acquisition of the 3-D point cloud was carried out without immersing the specimen into the wave flume, as subaquatic scanning is less accurate and difficult to achieve with the current technology. The dropper lines were hung from the laboratory ceiling by a crane. The surrounding area is closed off for foot traffic to minimize specimen movement by unintended air flow. Around the blocked area, three tripods are pre-mounted to allow for an exact positioning of the 3-D laser scanner (Z+F IMAGER® 5010) with a measurement rate of up to 1 million points per second and a horizontal and vertical accuracy of 0.0007° (rms). Furthermore, reference points for the spatial registration were added in the vicinity of the scanning area. The 3-D point clouds from the differently positioned tripods were combined into a 3-D point cloud, with every single point containing information about x-, y-and z-directions as well as intensity.

3. Results and Discussion

3.1. Surrogate Creation

Mussels, as a structure, are subjected to natural fluctuations, e.g., crop coverage, diameter of the dropper line or mussel type. This is why the overall task of the model creation via the 3-D point cloud, acquired by the laser scanner, is to allow for the development of an artificial surrogate that features similar characteristics as compared to the live in situ specimen regarding the hydrodynamic behaviour in currents and waves. This means that the future testing of the surrogate models should lead to comparable drag and inertia coefficients C_D and C_M derived from the Morison equation [22]. The focus of this work is on the methodological aspects of the creation of the surrogates. Proceeding from the assumption that drag and inertia coefficients are significantly influenced by surface geometry, a precise description of the mussel surface is necessary for the design of a surrogate model. As a common surface descriptor, the Abbott–Firestone Curve (AFC) is selected for the 3-D point cloud of the mussel specimens as well as the surrogates. With this method, a thorough quantification of the surface geometry and porosity is possible [23]. In principle, the AFC displays the material distribution M_r (%) as a function of the fluctuation in material surface c (m). Mathematically, it is described by

$$M_r(c) = \frac{100\%}{l_n} * \sum_{i=1}^{n} l_i(c), \tag{1}$$

where l_n is the total length of the recorded section and l_i is the length of material cut at the depth c. The AFC separates its roughness profile into peak, medial and valley portions [23]. These separate planes differ in depth and main function, e.g., the peak portion has a major influence on a surface's run-in characteristics whereas the valley area defines the amount of water that may be dragged along when interacting with the flow. In case of relative movement between surfaces where friction is involved, the coefficient of friction changes over time as roughness peaks are diminished progressively. The depth of the peak, medial and valley portions is expressed by the parameters reduced peak roughness R_{PK}, medial roughness R_K and reduced valley roughness R_{VK} [24]. In order to achieve an averaging effect, unrepresentative peaks and valleys which make up less than two percent are disregarded [23].

The AFC enables the separate evaluation of a surface regarding run-in characteristics, the load-bearing capacity and the absorption of liquids. The parameters M_{r1} and M_{r2} describe the peaks' and valleys' surface portion in %. Originally designed as a descriptor for two-dimensional roughness profiles only, modern measurement techniques provide three-dimensional surface data, i.e., here, obtained from the laser scanning. Thus, a transfer of the Abbott–Firestone from 2-D to 3-D is necessary. Similar to the two-dimensional case, the 3-D AFC bases on an imaginary cutting-plane being steadily moved downwards from the profiles' highest peak to the lowest valley. This procedure is mathematically described by

$$SM_r(c) = \frac{100\%}{A} * \iint_{x,y} dx\, dy, \tag{2}$$

where A is the regarded surface within the coordinate system x, y [25]. The 2-D parameters reduced peak roughness R_{PK}, medial roughness R_K and reduced valley roughness R_{VK} may easily be transferred to 3-D by renaming them to the reduced peak roughness S_{PK}, the medial roughness S_K and the reduced valley roughness S_{VK} [26]. In 3-D, the material portions of peaks and valleys are named SM_{r1} and SM_{r2}. After having recorded a three-dimensional surface's AFC, a comparison with others is possible. In the context of this work, the AFC is used as the primary surface descriptor and basis for all surrogate models. As this paper deals with cylinder roughness featuring a large unfiltered profile depth P_t to a cylinder diameter d ratio of

$$\frac{P_t}{d} \approx 0.5, \tag{3}$$

a definition of the AFC for cylindrical geometries is introduced. Instead of filtering out the cylinder's cylindricity and determining the material portion SM_r as a function of height c of a cutting plane, the cylindricity stays untouched allowing for the profile to be cut with a cylindrical surface $A_c(r)$, providing the material portion as a function of the cutting cylinder's radius r. Accordingly, the AFC for cylindrical bodies with a large unfiltered profile depth to cylinder diameter ratio may be defined as

$$SM_{r,cylindrical}(r) = 100\% * \frac{\sum A_i(r)}{A_c(r)}, \tag{4}$$

where $\sum A_i(r)$ is the profile's cut surface and $A_c(r)$ is the nominal surface of the cutting cylinder. This is depicted in Figure 7. Due to its promising characteristics regarding the surface description, the cylindrical AFC will be applied in the analysis of mussel crop surfaces.

After the acquisition of the 3-D point cloud with 5.4 million single points via the 3-D laser scanner, the data is treated via a statistical outlier removal filter in order to remove unwanted data points. Since the surface analysis via the cylindrical AFC assumes a solid body, the 3-D point cloud needs further processing. With regards to the limited computational resources, the 3-D point cloud is, therefore, divided into ten sections analyzed separately. To allow for further processing, all generated surfaces need to be closed and must not feature any holes. Due to the 3-D point cloud's limited quality resulting from challenging scan conditions (e.g., moving live mussels, wet and black surfaces to be scanned, shift caused by moving air, etc.), the calculation of the desired surfaces requires compromises regarding shaping accuracy. The then solid bodies are analyzed with regard to the AFC. For the analysis, a total amount of 4,368,790 points within a total length of 0.915 m blue mussel dropper line is considered. The weighted arithmetic average material distribution is depicted in Figure 8. As can be seen in this figure, the mean specimen is 13.6 cm in maximum diameter and 3.3 cm in minimum diameter. Unrepresentative peaks and valleys which were not filtered by the outlier removal are not included in the results.

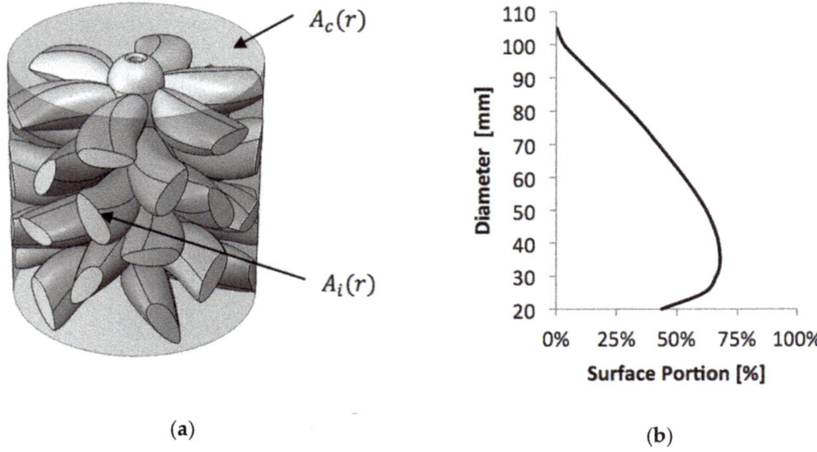

Figure 7. (**a**) A visualisation for the creation of the Abbott–Firestone Curve for cylindrical surfaces: here are a mussel dropper line with $A_i(r)$, the profile's cut surface; $A_c(r)$, the nominal surface of the cutting cylinder; and (**b**) the resulting Abbott–Firestone Curve.

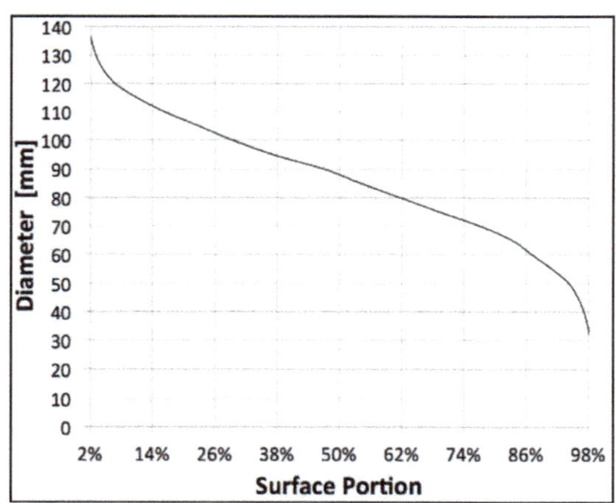

Figure 8. The weighted arithmetic average material distribution of all scanned sections.

The mean single mussel data mentioned before is the basis for the first concept of surrogate models. A uniform mussel is designed that equals the mean mussel's volume, weight, length and width. With the addition of the live specimens' statistical mean diameter of 10.32 cm for the dropper line sections, a design is possible. Copies of the uniform mussel are added to a slender cylinder at different angles of incidence until the mean weight per unit length is equal to the original live mussel data recorded. The angles of incidence are described as 0°, 90° and alternated-60°, whereby 0° is represented by mussels oriented horizontally, 90° by mussels oriented vertically and alternated-60°

by mussels changing their orientation by 60° in regard to the mussel above and below. The material distribution of the concept in comparison to the weighted arithmetic average material distribution of all sections is shown in Figure 9a alongside a side view of the alternated-60° surrogate.

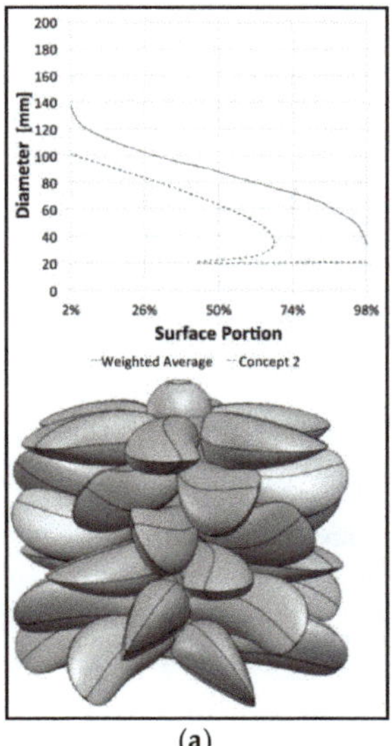

(a)

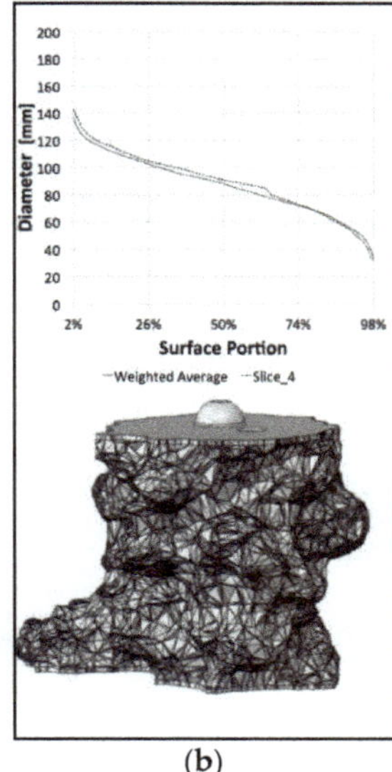

(b)

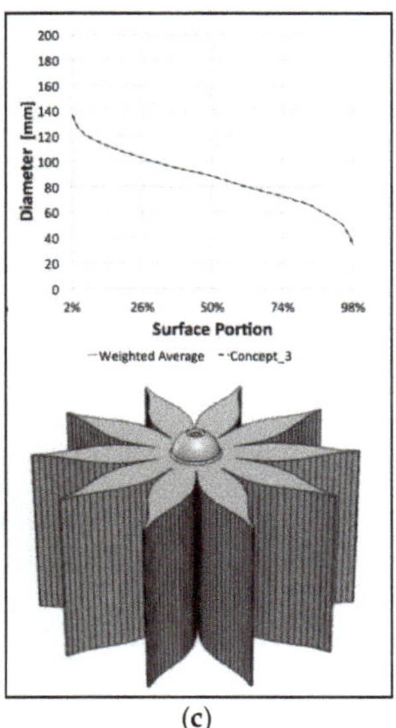

(c)

Figure 9. (**a**) Concept 1, (**b**) Concept 2 and (**c**) Concept 3 material distributions compared to the weighted arithmetic average material distribution of all sections.

The second concept for a surrogate model is based on the closest fitting 3-D point cloud section to the weighted arithmetic average material distribution. This is selected because its mass distribution features the most similarity to the original. A comparison of both AFCs as well as an exemplary side view of the surrogate model can be seen in Figure 9b.

The last concept is a reproduction of the weighted arithmetic average material distribution through a much simpler geometry that allows for easier scaling while maintaining the same AFC. The material distribution is closely fitted to the original, while the resemblance to the original live mussels is not obvious. Figure 9c, again, depicts the comparison between the two AFCs and a side view of the artificial surrogate model based on the direct AFC fit.

The resulting curves of Figure 9 display that the actual material distribution as a function of the fluctuation in material surface c (m) can be evaluated and that an evaluation by subdividing the profile into peak, medial and valley portions is possible. The AFC is applicable in 2-D and 3-D and additionally provides information on material-free and material-filled volumes. Even though the AFC is appropriate for a nonambiguous characterization of an existing surface, the reproduction of a certain surface based solely on the AFC is ambiguous. This can be seen in a comparison of the second and third surrogate concepts. The second, 3-D point cloud concept represents a close fit to the AFC, while the third concept represents the most exact results and a direct fit to the AFC. Thus, similar curves can be used to generate a variety of surrogate structures. For this reason, additional characteristics of a surface that relate to a certain Abbott–Firestone Curves should be considered. This could be the amount of break-off points along the structure or the solidity of the profile in flow. Eventually, energy dissipation and wake generation along the structure need to be better understood to find adequate additional surface descriptors. As a first step though, the above described surrogates are considered a promising starting point towards the creation of a mussel equivalent model that can be used for research regarding the behavior of suspended long-lines in current and wave conditions. In a last step, the different surrogate models are manufactured via selective laser sintering, an additive manufacturing technique. This combines a cost-effective as well as rapid prototyping possibility for the production of the surrogates. The surrogates with a paint finish for better distinguishability are depicted in Figure 10.

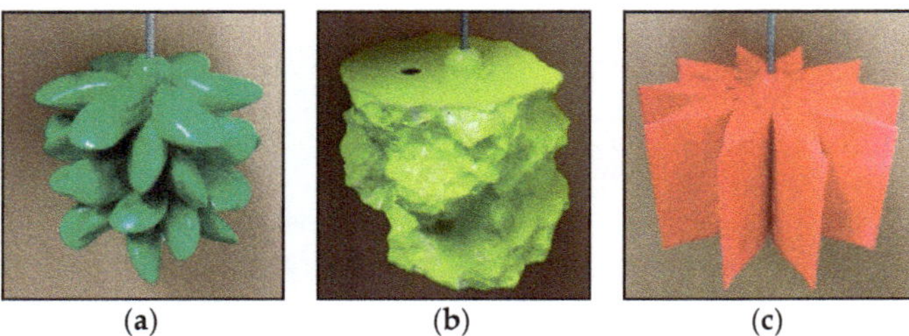

Figure 10. (**a**) Concept 1 with alternating-60° angle of incidence, (**b**) concept 2 with fit to weighted arithmetic average material distribution and (**c**) Concept 3, the direct Abbott–Firestone fit.

3.2. Drag and Inertia Coefficients

For a comparison of the created surrogates with the live specimen and an evaluation regarding the goodness-of-fit, an analysis of the hydrodynamic coefficients is needed in the future. The main focus of this work was the creation of the mussel surrogates; however, preliminary results for the live mussels were made available. The drag and inertia coefficients C_D and C_M for the live mussels under wave and current forcing were determined according to the Morison equation [22]. To this end, the time series of the forces in x-, y- and z-directions and the corresponding velocity of the towing carriage u_i and profiling velocimeter were substituted into the Morison equation to determine the drag and inertia coefficients [22]:

$$F = \frac{1}{2}\rho C_D u^2 A + \frac{\pi}{4}\rho C_M V\dot{u} \tag{5}$$

where F is the total horizontal force acting on the front face of a structure; ρ = 1000 kg/m^3 is the density of fresh water; C_D is the drag coefficient; u is the horizontal particle velocity; $A = L_{wet} * D_i$ is the projected area with reference to the flow composed of the wetted length of the tested structure L_{wet} and the characteristic, or mean, diameter D_i; C_M is the inertia coefficient; V is the volume of the structure; and $\dot{u}$ depicts the flow particle acceleration. The first term of Equation (5) corresponds to the drag forces, while the second term describes the inertia forces attributed to the flow acceleration. These terms can be isolated and the drag and inertia coefficients be determined via

$$C_D = \frac{2F}{\rho A u^2} \tag{6}$$

$$C_M = \frac{4F}{\pi \rho A^2 \dot{u}} \tag{7}$$

The tests are conducted with fresh water which has a negligible impact on the measurements as the values for CD and CM are independent of the density of the water. Tests with the same parameters, methodology and the surrogate models will be made available in upcoming research to determine the performance of the surrogates compared to the live mussels.

The preliminary results for the mean drag and inertia coefficients C_D and C_M with standard deviation and corresponding KC number as well as Reynolds number Re are shown in Table 2; Table 3. For the live mussel dropper lines tested in between Reynolds numbers of 2.0 × 104 to 1.1 × 105, resulting drag coefficients of C_D = 1.16–3.03 were recorded. It can be seen that the coefficients for smaller Re-numbers are increasing. In earlier works and the absence of experimentally derived values, the drag coefficients of mussel dropper lines were assumed to be similar to ultra-rough cylinders where C_D-values of 1.7 and lower were put forward [11,12]. A more precise study into the drag characteristics of mussel dropper lines was carries out by Plew et al. (2009) [13]. In the tests by Plew et al., a rigid, artificial mussel crop rope constructed from the shells of *Perna canaliculus* was used and a drag coefficient of 1.3 was determined within a Re-number range of 10,000 to 70,000. These values were used in a study by Dewhurst (2016) to determine the loading on a mussel raft [8]. It was concluded that these drag values may be too low for smaller Re-numbers. This is perpetuated by the results of this work where higher drag coefficients are recorded for the tests at lower velocities. The scope of this study focuses on the creation of a mussel surrogate; this is why further results and conclusions drawn regarding the C_D-coefficients and the drag tests can be seen in a work by Hildebrandt et al. [23]. The obtained CM-coefficients are still part of ongoing research, and the high standard deviation indicates that further investigations are necessary to accurately determine the inertia coefficient under waves. To the authors' knowledge, no comparable data for the inertia coefficient of mussel dropper lines is available as of now.

Table 2. The preliminary results regarding the mean C_M-coefficients with standard deviations with Re-numbers for drag tests.

$Re = \frac{u*d}{v}$	**mean**$C_D \pm SD$
2.0×10^4–3.5×10^4	2.18 ± 0.65
3.5×10^4–6.2×10^4	1.68 ± 0.23
6.2×10^4–8.6×10^4	1.60 ± 0.24
8.6×10^4–1.1×10^5	1.46 ± 0.23

Table 3. The preliminary results regarding the mean C_M ranges with standard deviations with *KC*—numbers for wave tests.

$KC = \frac{V*T}{L}$	**mean**$C_M \pm SD$
4.10	1.05 ± 0.81
5.50	1.68Y $\pm$ 0.89
5.90	1.41 ± 1.08

4. Conclusions

This work aims at a better design and the facilitation of more efficient testing for new aquaculture concepts for usage in offshore environments. The overarching aim of the live-mussel testing is the identification of drag, inertia and turbulence characteristics of suspended mussel dropper lines. To this end, current and wave tests were conducted with three specimens of blue mussels. Drag testing was carried out at four different velocities of $u_1 = 0.25$ m/s, $u_2 = 0.50$ m/s, $u_3 = 0.75$ m/s and $u_4 = 1.00$ m/s, and the wave tests regarding the inertia characteristics were conducted with three different wave setups with targeted wave heights $H_1 = 0.10$ m, $H_2 = 0.12$ m and $H_3 = 0.15$ m and corresponding periods $T_1 = 1.20$ s, $T_2 = 2.40$ s and $T_3 = 1.65$ s. From this data, the following results regarding C_D- and C_M- coefficients were attained:

- First, results regarding drag coefficients showed that the mussel dropper lines tested in between Reynolds numbers of 2.0×10^4 to 1.1×10^5 showed C_D-coefficients of 1.16–3.03.
- The results regarding inertia coefficients showed that the mussel dropper lines tested in between KC-numbers of 4.1 to 5.9 showed C_M-coefficients of 0.25–1.25.

Furthermore, a 3-D laser scan of the mussels was conducted, which resulted in the generation of a 3-D point cloud of 5.4 million data points. A systematic approach employing the Abbott–Firestone Curve was developed that allows an analysis regarding the material distribution as a function of the fluctuation in the material surface. With this approach, three concepts for surrogate models were developed and will be subsequently tested:

- Concept 1 was based on single, uniform mussels. These were added to a slender cylinder at different angles of incidence until the mean weight per unit length was equal to the original live mussel data recorded.
- Concept 2 was based on the closest fitting 3-D point cloud section of the live mussels in regard to the weighted arithmetic average material distribution.
- Concept 3 was a reproduction of the weighted arithmetic average material distribution through a much simpler geometry.

The developed surrogates featured similar characteristics to the live mussels in regard to the chosen surface descriptor. In the future, a validation of the hydrodynamic characteristics is necessary to provide a scalable surrogate dropper line that can be used in a variety of applications. The three created surrogates will allow for further testing without constraints regarding the longevity of the mussels and will be evaluated regarding their comparability to the live-mussel data. The use of a large-scale flume facility to obtain data regarding full-scale offshore conditions is worth investigating. Tests of the single surrogates under current influence as well as full-scale testing under current and wave influence have been carried out. During the resulting evaluation, a best-fit surrogate will be created and used for further testing. This allows for a better understanding of the forces acting on suspended mussel dropper lines. The results will help optimization existing and emerging aquaculture systems by finding the best orientation of the farm layout to the prevailing hydrographic conditions while also finding modern, innovative system designs adapted to high energy environments.

Author Contributions: Conceptualization, N.G. and A.H.; Data curation, J.L., T.O. and J.-A.P.; formal analysis, J.L. and T.O.; funding acquisition, N.G. and A.H.; investigation, J.L. and T.O.; methodology, N.G. and A.H.; project administration, N.G. and A.H.; supervision, N.G. and A.H.; visualization, J.L. and T.O.; writing—original draft, J.L. and T.O.; writing—review and editing, J.L., T.O., N.G., K.H., B.H.B., J.-A.P. and A.H.

Funding: This Research has been supported with funding from the New Zealand Ministry of Business, Innovation and Employment through Cawthron Institute project CAWX1607.

Acknowledgments: The authors gratefully acknowledge the support of Tim Staufenberger for providing us with the mussel specimens.

Conflicts of Interest: The authors declare no conflict of interest.

References

1. FAO. *The State of World Fisheries and Aquaculture 2016—Contributing to Food Security and Nutrition for All*; FAO: Rome, Italy, 2016.
2. Cole, R. Impacts of marine farming on wild fish populations. In *Final Research Report for Ministry of Fisheries Research Project ENV2000/08 Object. One*; OECD: Paris, France, 2002; p. 24.
3. Lloyd, B.D. *Potential Effects of Mussel Farming on New Zealand's Marine Mammals and Seabirds*; A Discussion Paper; Department of Conservation: Wellington, New Zealand, 2003.
4. McVey, J.P. Overview of Offshore Aquaculture. In Proceedings of the Open Ocean Aquaculture—Proceedings of an International Conference, Portland, ME, USA, 8–10 May 1996; Volume 1996, pp. 13–19.
5. Goseberg, N.; Chambers, M.D.; Heasman, K.; Fredriksson, D.; Fredheim, A.; Schlurmann, T. Technological Approaches to Longline- and Cage-Based Aquaculture in Open Ocean Environments. In *Aquaculture Perspective of Multi-Use Sites in the Open Ocean: The Untapped Potential for Marine Resources in the Anthropocene*; Buck, B.H., Langan, R., Eds.; SpringerOpen: Berlin/Heidelberg, Germany, 2017; pp. 71–95.
6. Garen, P.; Robert, S.; Bougrier, S. Comparison of growth of mussel, Mytilus edulis, on longline, pole and bottom culture sites in the Pertuis Breton, France. *Aquaculture* **2004**, *232*, 511–524. [CrossRef]
7. Wang, X.; Swift, M.R.; Dewhurst, T.; Tsukrov, I.; Celikkol, B.; Newell, C. Dynamics of submersible mussel rafts in waves and current. *China Ocean Eng.* **2015**, *29*, 431–444. [CrossRef]
8. Dewhurst, T. *Dynamics of a Submersible Mussel Raft*; University of New Hampshire: Durham, NH, USA, 2016.
9. Buck, B.H. Experimental trials on the feasibility of offshore seed production of the mussel Mytilus edulis in the German Bight: Installation, technical requirements and environmental conditions. *Helgol. Mar. Res.* **2007**, *61*, 87–101. [CrossRef]

10. Stevens, C.L.; Plew, D.R.; Smith, M.J.; Fredriksson, D.W. Hydrodynamic forcing of long-line mussel farms: Observations. *J. Waterw. Port Coast. Ocean Eng.* **2007**, *133*, 192–199. [CrossRef]
11. Stevens, C.; Plew, D.; Hartstein, N.; Fredriksson, D. The physics of open-water shellfish aquaculture. *Aquac. Eng.* **2008**, *38*, 145–160. [CrossRef]
12. Plew, D.R.; Stevens, C.L.; Spigel, R.H.; Hartstein, N.D. Hydrodynamic implications of large offshore mussel farms. *IEEE J. Ocean. Eng.* **2005**, *30*, 95–108. [CrossRef]
13. Plew, D.R.; Enright, M.P.; Nokes, R.I.; Dumas, J.K. Effect of mussel bio-pumping on the drag on and flow around a mussel crop rope. *Aquac. Eng.* **2009**, *40*, 55–61. [CrossRef]
14. Plew, D.R. Depth-Averaged Drag Coefficient for Modeling Flow through Suspended Canopies. *J. Hydraul. Eng.* **2010**, *137*, 234–247. [CrossRef]
15. Delaux, S.; Stevens, C.L.; Popinet, S. High-resolution computational fluid dynamics modelling of suspended shellfish structures. *Environ. Fluid Mech.* **2011**, *11*, 405–425. [CrossRef]
16. Raman-Nair, W.; Colbourne, B. Dynamics of a mussel longline system. *Aquac. Eng.* **2003**, *27*, 191–212. [CrossRef]
17. Raman-Nair, W.; Colbourne, B.; Gagnon, M.; Bergeron, P. Numerical model of a mussel longline system: Coupled dynamics. *Ocean Eng.* **2008**, *35*, 1372–1380. [CrossRef]
18. Constantinescu, G.; Miyawaki, S.; Liao, Q. Flow and Turbulence Structure Past a Cluster of Freshwater Mussels. *J. Hydraul. Eng.* **2012**, *139*, 347–358. [CrossRef]
19. Gansel, L.; Endresen, P.; Steinhovden, K.; Dahle, S.; Svendsen, E.; Forbord, S.; Jensen, Ø. OMAE2017-62030 Drag On Nets Fouled with Blue Mussel (Mytilus Edulis) and Sugar Kelp (Saccharina Latissima) and Parameterization of Fouling. In Proceedings of the ASME 2017 36th International Conference on Ocean, Offshore and Arctic Engineering OMAE, Trondheim, Norway, 25–30 June 2017; pp. 1–16.
20. Gosling, E. *Bivalve Molluscs: Biology, Ecology and Culture*; Wiley: Hoboken, NJ, USA, 2015; Volume 1542.
21. Favrel, P.; Mathieu, M. *Cultured Aquatic Species Information Programme Mytilus edulis (Linnaeus, 1758) Fisheries*; Food and Agriculture Organization of the United Nationsfor a World without Hunger—Fisheries and Aquaculture Department: Rome, Italy, 1996; Volume 205, pp. 210–214.
22. Morison, J.R.; O'Brien, M.P.; Johnson, J.W.; Schaaf, S.A. The forces exerted by surface waves on piles. *Pet. Trans. AIME* **1950**, *2*, 149–154. [CrossRef]
23. Abbott, E.J.; Firestone, F.A. Specifying surface quality—A method based on accurate measurement and comparison. *Mech. Eng.* **1933**, *55*, 569–577.
24. DIN-EN-ISO-13565-2: 1996-12. *Geometrical Product Specifications (GPS)—Surface Texture: Profile Method—Surfaces Having Stratified Functional Properties—Part 2: Height Characterization Using the Linear Material Ratio Curve*; International Organization Standardization (ISO): Geneva, Switzerland, 1996.
25. Lemke, H.W.; Seewig, J.; Bodschwinna, H.; Brinkmann, S. Kenngrößen der Abbott-Kurve zur integralen Beurteilung dreidimensional gemessener Zylinderlaufbahn-Oberflächen. *MTZ-Motortech. Z.* **2003**, *64*, 438–444. [CrossRef]
26. DIN-EN-ISO-25178-3: 2012-07. *Geometrical Product Specifications (GPS)—Surface Texture: Areal—Part 3: Specification Operators*; International Organization Standardization (ISO): Geneva, Switzerland, 2012.

Journal of
Marine Science and Engineering

Article

Contribution of Infragravity Waves to Run-up and Overwash in the Pertuis Breton Embayment (France)

Christopher H. Lashley [1,*], Xavier Bertin [2], Dano Roelvink [3] and Gaël Arnaud [2,4]

1 Hydraulic Engineering, Delft University of Technology, 2628 CN Delft, The Netherlands
2 UMR 7266 Littoral, Environment and Societies, CNRS-University of La Rochelle, 17000 La Rochelle, France
3 Water Science and Engineering, IHE-Delft Institute for Water Education, 2611 AX Delft, The Netherlands
4 Laboratory of Research in Geosciences and Energies, University of the French West Indies, 97157 Pointe-à-Pitre, Guadeloupe, France
* Correspondence: c.h.lashley@tudelft.nl; Tel.: +31-684097811

Received: 20 May 2019; Accepted: 28 June 2019; Published: 2 July 2019

Abstract: Wave run-up and dune overwash are typically assessed using empirical models developed for a specific range of often-simplistic conditions. Field experiments are essential in extending these formulae; yet obtaining comprehensive field data under extreme conditions is often challenging. Here, we use XBeach Surfbeat (XB-SB)—a shortwave-averaged but wave-group resolving numerical model—to complement a field campaign, with two main objectives: i) to assess the contribution of infragravity (IG) waves to washover development in a partially-sheltered area, with a highly complex bathymetry; and ii) to evaluate the unconventional nested-modeling approach that was applied. The analysis shows that gravity waves rapidly decrease across the embayment while IG waves are enhanced. Despite its exclusion of gravity-band swash, XB-SB is able to accurately reproduce both the large-scale hydrodynamics—wave heights and mean water levels across the 30 × 10 km embayment; and the local morphodynamics—steep post-storm dune profile and washover deposit. These findings show that the contribution of IG waves to dune overwash along the bay is significant and highlight the need for any method or model to consider IG waves when applied to similar environments. As many phase-averaged numerical models that are typically used for large-scale coastal applications exclude IG waves, XB-SB may prove to be a suitable alternative.

Keywords: combined field experiment and numerical modeling; overwash; wave run-up; infragravity waves; XBeach; coastal flooding; dune erosion

1. Introduction

1.1. Background

Extreme wave run-up is the maximum landward point that storm waves can reach as they break and move up a natural beach profile or structure. It is the combined result of the time-averaged water surface elevation at the shoreline (setup) and the time-varying fluctuations about that mean (swash). This swash may be further classified by frequency into infragravity (low-frequency, see Bertin et al. [1] for a recent review) and gravity (high-frequency) bands. The relative contribution of these frequency bands to the total wave run-up is dependent on offshore wave conditions and local bathymetry.

As a proxy for the potential impact related to a given storm, the accurate prediction of wave run-up and the identification of its individual components are essential for effective coastal engineering and coastline management. Considering the widely-cited storm impact scale of Sallenger [2], the extent of the morphological impact to a beach-dune system during a storm may be directly related to the magnitude of wave run-up. This is described by four regimes:

i. Swash regime, where run-up is limited to the foreshore and any storm related erosion is recovered post-storm (net effect is zero);
ii. Collision regime, where the run-up exceeds the threshold of the base of the foredune ridge and is no longer confined to the foreshore area. The swash is then able to erode the dune (net erosion effect);
iii. Overwash regime, where the run-up overtops the berm or foredune ridge and a net landward transport of sediment occurs. This regime often results in washover deposits of sediment; and
iv. Inundation regime, where the combined effect of surge, tide and wave run-up is sufficient to completely and continuously submerge the beach or dune area.

In this paper, we focus on the overwash regime which can be seen as a precursor to dune breaching and coastal inundation. The ability to estimate the location and extent of dune overwash is of significant importance to many coastal communities in low-lying areas who rely on dunes as natural flood defences. In light of this, several empirical formulations to estimate wave run-up on coastlines have been developed [3–9]. These formulae—often developed using laboratory experiments or with data specific to a particular beach, under mild to moderate conditions—may perform well on relatively straight, open coasts where the hydrodynamics are somewhat predictable [10]. However, their applicability to other coastline types—such as, embayments or those fronted by reefs—is highly uncertain, as each section of the beach is not necessarily exposed to the same incident wave conditions [7]. Likewise, the individual contributions of the gravity and IG components of wave run-up may also vary considerably.

These limitations, acknowledged in several studies [10–13], are associated with the difficulty in obtaining comprehensive field measurements under extreme conditions. Energetic waves and significant beach change can shift or damage observation equipment and introduce uncertainties in the data collected. In light of this, numerical models—such as XBeach [14]—are now widely used in complex coastal environments (e.g., references [15–21]) and have shown sufficient capability in handling not only the hydrodynamics, but also the morphological response to extreme events. Hence, these models may be used to complement field campaigns with limited data, towards a more comprehensive analysis of a particular phenomenon [13].

In the present study, we combine field measurements and the XBeach Surfbeat (XB-SB) numerical model to assess the contribution of IG waves to washover development in the Pertuis Breton Embayment (France) under storm conditions during the winter of 2013–2014. Additionally, as XB-SB is applied using an unconventional nesting approach, we also assess the general performance of the approach. In Section 1.2 of this paper, a brief description of the study area is provided. Section 2 describes the field experiment and numerical modeling approach applied; and describes the data post-processing and the error metrics used to assess model accuracy. In Section 3, the combined results of the field experiment and numerical modeling are presented with a discussion on the hydro-morphodynamics of the area and the importance of low-frequency motions. Section 4 concludes the paper by noting the strengths, limitations and potential application of the method to future research.

1.2. Study Area

The coastal village of La Faute-sur-Mer, situated in the Bay of Biscay (France) is particularly vulnerable to dune erosion, overwash and coastal flooding because it is an urbanized low-lying coastal community. The village is established on a sandy spit between the ocean and the Lay River in the inner part of the so-called Pertuis Breton embayment. The area is further characterized by a highly complex bathymetry that is somewhat sheltered from the sea by the island, Île de Ré (Figure 1). The 30 km × 10 km embayment has depths of approximately 25–30 m at its seaward boundary, to a maximum depth of 58 m which then shoals eastwards to a typical depth of 5 m. With over 60% of the foreshore being less than 10 m deep, the area may be considered to be shallow [22]. To further add to its complexity, the sediment characteristics of the area include a mixture of bedrock, gravel sands and mud. Moving seaward (westerly) from La Faute-sur-Mer, one first encounters fine sand and mud, followed by gravels and finally bedrock at the entrance of the area [23].

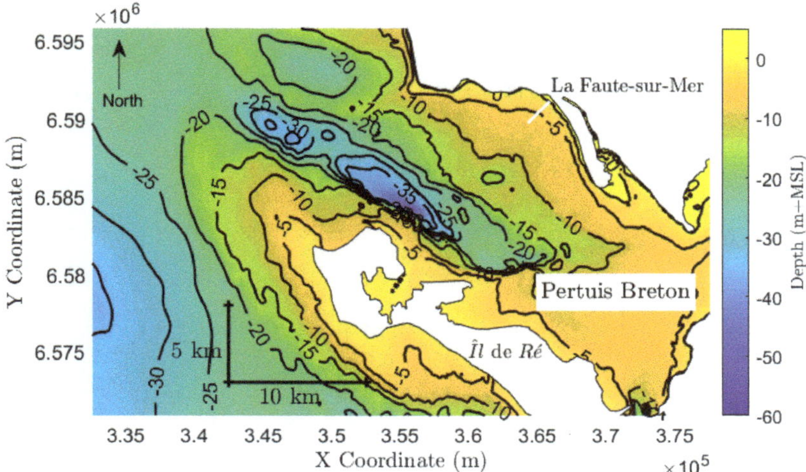

Figure 1. Bathymetric plot of the study area. Data sources: Service Hydrographique et Océanographique de la Marine (outer part of the estuary); Littoral Environnement et Sociétés (LIENSs) single beam sounder (shoreface); and Light Detection and Ranging (LIDAR originating from the Litto3D project) (supratidal zone). The white line shows the transect of instrument locations.

The area experiences a semi-diurnal tide, which ranges from less than 2 m (neap) to over 6 m (spring). The wave climate is generally quite energetic at the entrance of the embayment; with significant wave heights (H_{m0}) temporarily exceeding 8 m during annual winter storms [24]. However, this wave energy rapidly decreases inside the estuary due to refraction, diffraction and dissipation by depth-limited breaking and bottom friction. In 2010, a unique combination of storm surge and high spring tides resulted in severe flooding along the French Atlantic Coast. This extreme event resulted in loss of life and significant damage to property in la Faute-sur-Mer, with reported water levels in the village as high as 3 m [25]. More recently, the winter of 2013–2014 produced an unusually high-energy wave climate in the Bay of Biscay, consisting of six successive extreme events with offshore H_{m0} in the range of 5–12 m and peak period (T_p) exceeding 15 s. These extreme conditions resulted in the development of several washover deposits along the barrier bounding La Faute-sur-Mer to the West.

2. Methods

2.1. Field Experiment

The field campaign, conducted from 30 January to 11 March 2014, captured several of the before-mentioned events. In the present study we focus on an event which occurred on 2 February 2014 resulting in washover development. This was captured by means of an Acoustic Waves and Currents (AWAC) sensor on the shoreface, an Acoustic Doppler Current Profiler (ADCP) positioned 69 m seaward of the dune crest and a pressure transducer (PT) on the dune crest (Figure 2). It should be noted that the AWAC was displaced some 50 m—due to the severity of the weather conditions—and approached a steel mooring chain during the storm, which biased the compass. Thus, only 1D spectral analyses were performed on the measured data (see Section 2.3 for details on the data processing). In addition to the hydrodynamic data collected, a differential global positioning survey was carried out on the beach following the overwash event to capture the extent of the washover deposit.

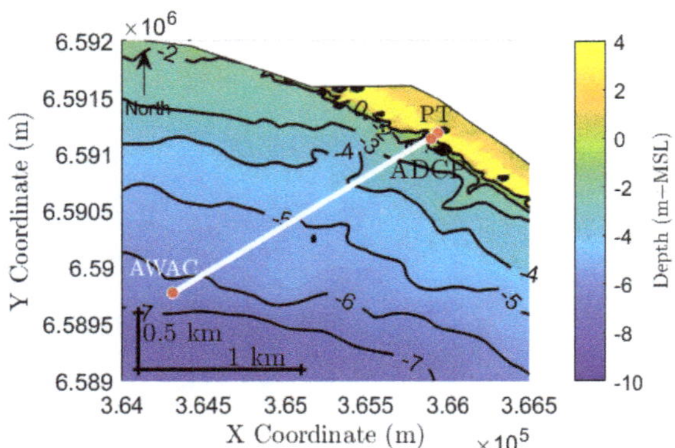

Figure 2. Instrument locations (red dots) in the nearshore. Solid white line corresponds to that in Figure 1, for reference.

2.2. Numerical Modeling

XBeach is an open-source, two-dimensional numerical model in the horizontal plane (2DH) which solves equations for wave propagation, mean flow including long waves, sediment transport and morphological changes. Here, we apply the Surfbeat mode (XBeachX release) which solves short-wave (gravity) motions using a reduced wave-action equation where the mean wave directions ($\overline{\theta}$) are first calculated and then the short-wave energy is propagated along these directions [26]:

$$\frac{\partial A}{\partial t} + \frac{\partial c_g cos\overline{\theta} A}{\partial x} + \frac{\partial c_g sin\overline{\theta} A}{\partial y} = -\frac{D_w + D_f}{\sigma}, \tag{1}$$

$$A(x, y, t, \theta_i) = \frac{S_w(x, y, t, \theta_i)}{\sigma(x, y, t)}, \tag{2}$$

$$\sigma = \sqrt{gk\ tanh\ kh}, \tag{3}$$

where the wave action, A is given by Equation (2), S_w is the wave energy density in each directional bin, σ is the intrinsic wave frequency given by Equation (3), h is the local water depth, k is the wave number, θ_i is the angle of incidence with respect to the x-axis; and D_w and D_f are dissipation terms which take into account wave breaking and bottom friction, respectively; and c_g is the group velocity. D_w is computed using the Daly et al. [27] parameterization, while D_f is determined by a specified short-wave friction coefficient, f_w [28].

On the other hand, flow motions at the scale of the wave group (infragravity motions) are solved directly using the nonlinear shallow water equations:

$$\frac{\partial \eta}{\partial t} + \frac{\partial h u^L}{\partial x} + \frac{\partial h v^L}{\partial y} = 0, \tag{4}$$

$$\frac{\partial u^L}{\partial t} + u^L \frac{\partial u^L}{\partial x} + v^L \frac{\partial u^L}{\partial y} - f v^L - v_h \left(\frac{\partial^2 u^L}{\partial x^2} + \frac{\partial^2 u^L}{\partial y^2} \right) = \frac{\tau_{sx}}{\rho h} - \frac{\tau^E_{bx}}{\rho h} - g \frac{\partial \eta}{\partial x} + \frac{F_x}{\rho h}, \tag{5}$$

$$\frac{\partial v^L}{\partial t} + u^L \frac{\partial v^L}{\partial x} + v^L \frac{\partial v^L}{\partial y} - f u^L - v_h \left(\frac{\partial^2 v^L}{\partial x^2} + \frac{\partial^2 v^L}{\partial y^2} \right) = \frac{\tau_{sy}}{\rho h} - \frac{\tau^E_{by}}{\rho h} - g \frac{\partial \eta}{\partial y} + \frac{F_y}{\rho h}, \tag{6}$$

where u^L and v^L are the Lagrangian velocities, F_x and F_y are wave-induced stresses (derived from the wave action model), v_h is the horizontal viscosity, f is the Coriolis coefficient, τ_{sx} and τ_{sy} are wind shear stresses; and τ_{bx}^{E} and τ_{by}^{E} are bed shear stresses (based on Eulerian velocities) determined here by a specified Chézy coefficient, C.

The model approach adopted in this study followed a somewhat uncommon nesting procedure within XBeach itself. It comprised a large-domain (45 × 25 km) 2DH hydrodynamic model of the embayment (the entire area shown in Figure 1) to simulate the transformation of the waves from deep water (seaward of the mouth of the embayment) to the shoreface (AWAC location) and a nested 1D morphodynamic model to simulate wave run-up and dune overwash at the coast of La Faute-sur-Mer. The objectives of the 2DH model were to determine the direction of wave attack in the shoreface—as the movement of the AWAC during the storm period disallowed the collection of directional data—and to generate surface elevation (zs) (including IG motions) and short-wave energy (E) time series to be used as boundary conditions for the nested, higher-resolution 1D model.

2.2.1. Hydrodynamic 2DH Model Setup

For the 2DH model, a constant grid spacing in the cross-shore direction, dx = 12.5 m and in the longshore direction, dy = 25 m were used; resulting in a 3600 × 1000 node grid (45 × 25 km domain). At the open boundary (seaward of Pertuis Breton in 30 m water depth), XBeach was forced with time-varying directional wave spectra and mean water levels obtained from the storm surge modeling system described by Bertin et al. [24] (Figure 3)—forced by fields of sea-level pressure and wind speed originating from the Climate Forecast System Reanalysis (CFSR) [29]. Consistent with typical winter storms [30,31], these deep-water wave conditions showed an energetic event with $H_{m0} > 5$ m and $T_p > 17$ s (Figure 3b,c) under spring-tide conditions. The mean wave direction (θ) varied between 260° and 270° with an average of 268° (Figure 3c). The 2DH model was then calibrated by optimizing the friction parameters; a combination of f_w = 0.06 and C = 75 $m^{1/2}s^{-1}$ provided the best agreement between the observed and modeled data. As the main purpose of the 2DH model was to investigate wave transformation across the embayment, morphology and sediment transport formulations were switched off.

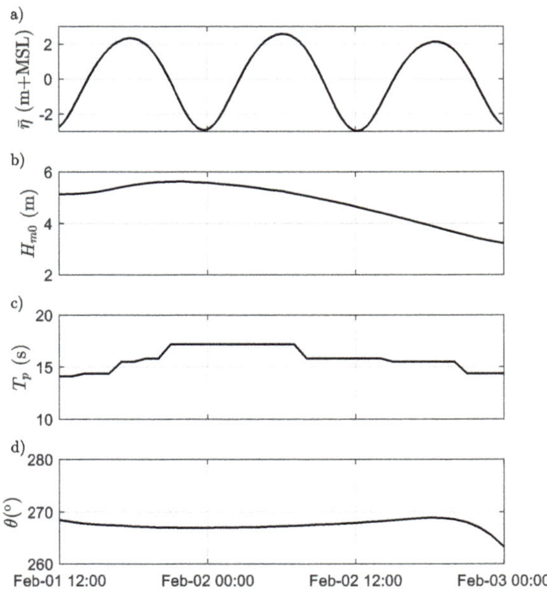

Figure 3. Offshore (**a**) mean water level; (**b**) significant wave height; (**c**) peak period; and (**d**) mean wave direction as modeled by WWIII (boundary conditions for the 2DH XBeach model) for February 1st to 3rd 2014.

2.2.2. Morphodynamic 1D Model Setup

The nested higher-resolution model was set up as a transect across the dune with its offshore boundary at the AWAC location (2 km domain shown by the white lines in both Figures 1 and 2). The cross-shore grid spacing (*dx*) was varied from 5 m offshore to 2 m at the coast to ensure that the steeply-sloping dune was correctly captured. The model was forced with long-wave surface elevation (*zs*) and short-wave energy, (*E*) time series obtained from the 2DH hydrodynamic model to simulate wave transformation to the nearshore (ADCP location) and dune (PT location). In order to model the eroding dune and washover development, sediment transport and morphology formulations were switched on. With respect to physical parameters, a median grain size (*D50*) = 0.0035 m was specified; the ratio of breaking waves to local water depth was set to 0.4; and the critical bed slope for wet areas was set to 1.1. Additionally, a transport term is added in XBeach which is proportional to the difference between the actual slope and a prescribed value (hereinafter referred to as "bermslope") to better represent conditions where the dune is known to maintain a relatively steep profile. The coefficient of this term is set to 10 times the usual bed-slope term, which produces a strong local onshore transport when the actual slope is less than the prescribed value. This term is only applied in a narrow region where the wave height to water depth ratio is > 1 [32]. A *bermslope* value of 0.12 provided the best agreement between the modeled and observed dune profiles.

2.3. Data Analysis

2.3.1. Separation of Gravity and Infragravity Waves

A spectral analysis was performed in order to identify the relative importance of infragravity waves with respect to gravity waves. All field pressure measurements were first corrected for the atmospheric pressure variations. Pressure energy density spectra were computed by applying the Welch's average periodogram with five Hamming-windowed segments (50% maximum overlap). These pressure spectra were then converted into surface elevation spectra, $C_{\eta\eta}$ using linear wave theory. The significant wave heights in both the infragravity ($H_{m0,IG}$) and gravity ($H_{m0,G}$) frequency bands were then determined as follows:

$$H_{m0,IG} = 4\sqrt{\int_{0.005}^{f_{split}} C_{\eta\eta}\, df}, \tag{7}$$

$$H_{m0,G} = 4\sqrt{\int_{f_{split}}^{0.5} C_{\eta\eta}\, df}, \tag{8}$$

where a split frequency (f_{split}) of 0.03 Hz is used to separate the infragravity and gravity bands. This choice of cut-off frequency—which is approximately equal to half the average peak frequency ($f_p = 1/T_p$) of the offshore waves (Figure 3c)—is based on the tendency that, in deep water, the majority of gravity-wave energy is found at frequencies > $f_p/2$, while the majority of IG-wave energy lies at frequencies < $f_p/2$ [33].

XBeach uses a representative single frequency for the gravity-band wave energy and therefore does not produce the gravity-band spectra. Modeled $H_{m0,G}$ was obtained directly from computed short-wave energy; while $H_{m0,IG}$ was obtained from the simulated long-wave surface elevation time series using Equation (7). In addition to significant wave heights, the modeled mean water levels ($\overline{\eta}$), maximum run-up (R_{max}) and post storm bed level (*zb*) were also assessed and compared to observations.

2.3.2. Separation of Incoming and Outgoing Signals

To assess the reflection of IG waves at the dune, the modeled and observed total long-wave surface elevation signals (zs) were separated into incoming (zs_{in}) and outgoing (zs_{out}) components as follows [34]:

$$zs_{in} = \frac{zs + \sqrt{\frac{h}{g}}u}{2}, \tag{9}$$

$$zs_{out} = \frac{zs - \sqrt{\frac{h}{g}}u}{2}. \tag{10}$$

This method, developed for normally-incident shallow-water waves, requires co-located measurements of surface elevation (zs) and currents (u)—by means of an ADCP, for example—in order to decompose the total signal into seaward (outgoing) and shoreward (incoming) propagating components [34]. These signals were then used to calculate the incoming and outgoing IG significant wave heights ($H_{m0,IG,in}$ and $H_{m0,IG,out}$) using Equation (7).

2.4. Error Metrics

The model-data comparisons were carried out by applying the Root-mean-square Error ($RMSE$), Relative Bias ($Rel.\ bias$) and Brier-Skill Score (BSS) error metrics; where Ψ is used as a stand-in for the parameter under consideration in a sample size N:

$$RMSE = \sqrt{\frac{1}{n}\sum_{i=1}^{N}\left(\Psi^{i}_{XBeach} - \Psi^{i}_{observed}\right)^2}, \tag{11}$$

$$Rel.\ bias = \frac{\sum_{i=1}^{N}\left(\Psi^{i}_{XBeach} - \Psi^{i}_{observed}\right)}{\sum_{i=1}^{N}\Psi^{i}_{observed}}, \tag{12}$$

$$BSS = 1 - \frac{\frac{1}{n}\sum_{i=1}^{N}\left(\Psi^{i}_{XBeach} - \Psi^{i}_{post\ storm}\right)^2}{\frac{1}{n}\sum_{i=1}^{N}\left(\Psi^{i}_{initial} - \Psi^{i}_{post\ storm}\right)^2}. \tag{13}$$

3. Results and Discussion

3.1. Wave Propagation Across the Embayment

In order to assess the accuracy of the 2DH model, the predicted $\overline{\eta}$, $H_{m0,G}$ and $H_{m0,IG}$ were compared to the AWAC observations for the February 2nd event (Figure 4). The $RMSE$ values for these parameters were 0.2 m, 0.17 m and 0.07 m, respectively. The underestimation of $H_{m0,G}$ at low tide is possibly due to the influence of local wind, which were not included in the hydrodynamic model. The three metrics showed $Rel.\ bias$ values of: −0.03, 0.02 and 0.03 for $\overline{\eta}$, $H_{m0,G}$ and $H_{m0,IG}$, respectively; suggesting relatively unbiased predictions of each parameter. To supplement the missing directional data in the nearshore, the model was used to compute the mean wave direction (θ). An average θ of 238° was found in the shoreface (Figure 4d); confirming that waves approached the coast of La Faute-sur-Mer with near-perpendicular incidence.

In general, the 2DH hydrodynamic model showed reasonable agreement with the observations at the AWAC location; indicating that it was a suitable tool to investigate the hydrodynamic processes at the scale of the Pertuis Breton Embayment. A further analysis of the model results shows energetic offshore gravity waves which decrease in magnitude rapidly at the mouth of the embayment (Figures 5 and 6). While entering the embayment, these waves refract towards the coastlines of La Faute-sur-Mer and Île de Ré as the water depth became increasingly shallower. While the gravity waves significantly decrease in amplitude from offshore to the AWAC location—by depth-induced breaking and bottom friction (Figure 5a)—the IG waves appear to propagate across the embayment relatively unchanged (Figure 5b). The result is an increase in the relative contribution of the IG waves ($H_{rms,IG}/H_{rms,G}$) in the nearshore (Figure 5c).

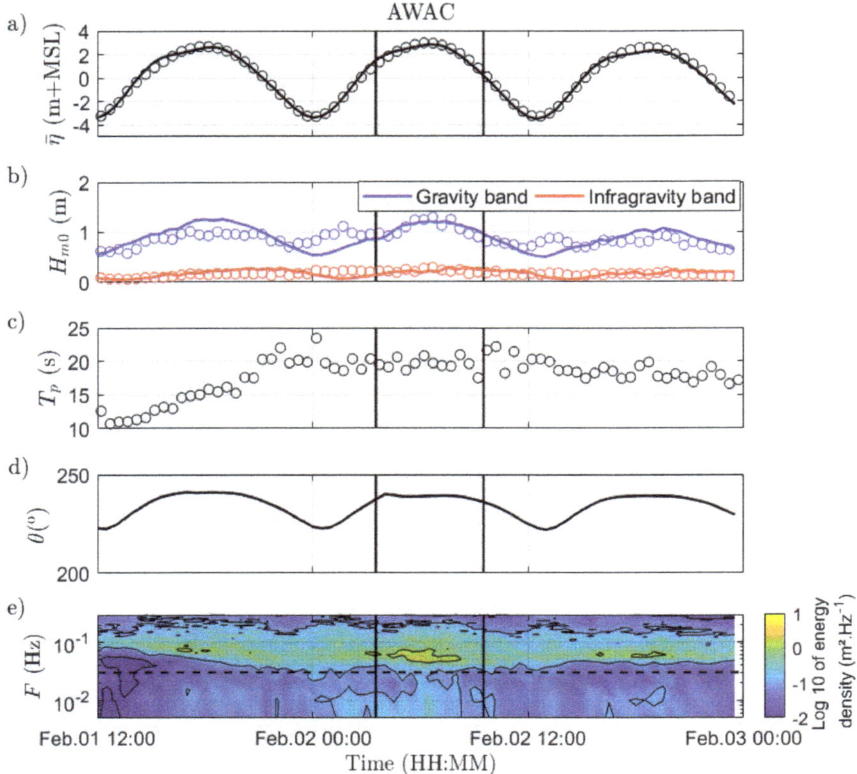

Figure 4. Modeled (solid) and observed (circles): (**a**) mean water levels; (**b**) significant wave heights; (**c**) peak periods; (**d**) mean wave direction; and (**e**) observed energy density, for the 2 February 2014 event (bounded by vertical lines) at the AWAC location. Contours in panel 'e' indicate areas of equal energy density and the dashed line separates gravity and IG frequencies.

A closer look at the transect in Figure 6 (offshore to the shoreface) confirms that $H_{m0,G}$ is significantly reduced (80%) compared to a 44% reduction in $H_{m0,IG}$, on average. These IG waves decayed at the entrance of the embayment (X ~ 340 to 350 km) concurrently with the higher frequency waves that force them; however, as the water depth became shallower, they experienced amplitude growth through the continuous transfer of energy from the gravity waves as they shoal over the mildly-sloping bathymetry (Figure 6b). On the other hand, the gravity waves continued to dissipate in shallow water (Figure 6a).

Ultimately, the results of the 2DH model were used as input for the nested 1D morphological model. This input comprised time series of: mean water levels ($\bar{\eta}$); short-wave energy (E), as a measure of the gravity waves; the long-wave surface elevation (zs), as measure of the infragravity component; and the mean wave direction (θ). The results of which are presented and discussed in the following section.

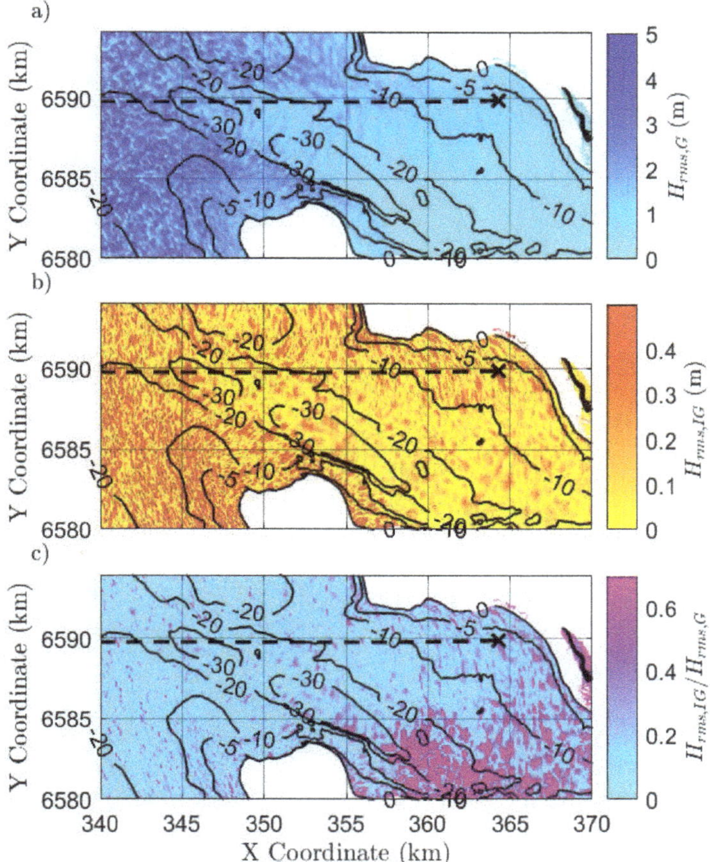

Figure 5. Snapshots of modeled instantaneous root-mean-square wave heights (H_{rms}) in both the (**a**) gravity and (**b**) IG bands; with (**c**) relative contribution of the IG waves ($H_{rms,IG}/H_{rms,G}$) across the embayment at the peak of the February 2nd event (at time = 06:00). Dashed lines correspond to transect in Figure 6 and '**X**' represents the AWAC location. Depth contours are provided for reference.

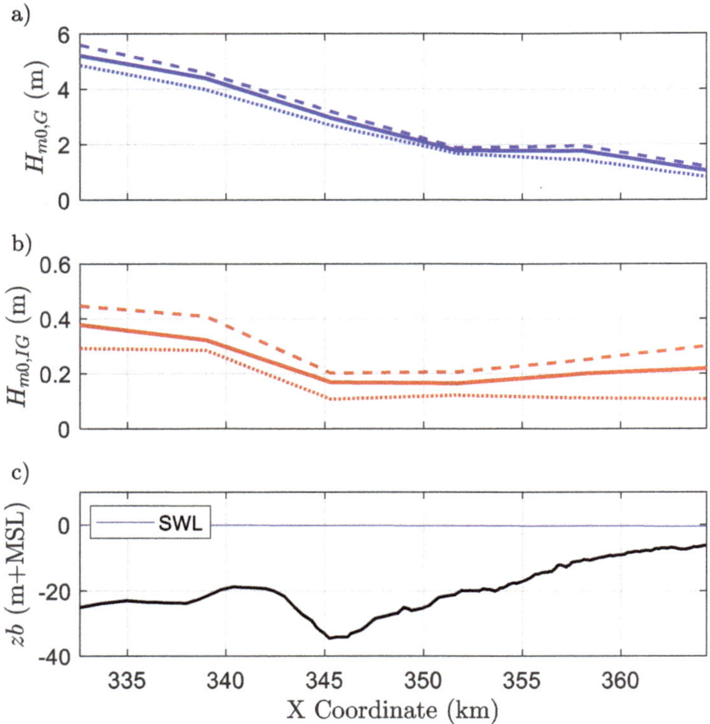

Figure 6. Transect across embayment showing the transformation of modeled maximum (dashed), mean (solid) and minimum (dotted): (**a**) gravity waves; and (**b**) infragravity waves, with (**c**) bed level for reference, for the February 2nd event (from time = 03:30 to 09:30).

3.2. Wave Run-up and Overwash of the Dune at La Faute-sur-Mer

In order to assess the performance of the 1D model, the predictions of $\overline{\eta}$ (Figure 7a), $H_{m0,G}$ and $H_{m0,IG}$ (Figure 7b) were compared to observations. *RMSE* values were 0.26 m, 0.23 m and 0.11 m for the three metrics, respectively with corresponding *Rel. bias* values of 0.09, 0.11 and −0.04, respectively. Overall, the model showed reasonable agreement with the observations of the ADCP which—together—indicate that IG waves were indeed significant, with $H_{m0,IG}$ exceeding 0.5 m during the storm (observed maximum of 0.64 m).

Though 1D, the model allowed for refraction (according to Snell's law) and the generation of longshore currents. Furthermore, as the area is quite uniform in the alongshore direction (Figure 2) the 1D model can be seen as representative of actual conditions; however, with the exception that it excludes high-frequency swash at the shoreline. XB-SB is also able to simulate the reflection of the IG waves at the dune (Figure 8), with the incoming component ($H_{m0,IG,in}$) and outgoing component ($H_{m0,IG,out}$) each estimated with reasonable accuracy: *RMSE* values of 0.06 m and 0.08 m, respectively; and *Rel. bias* values of ~0 and −0.09, respectively. The underprediction of $H_{m0,IG,out}$—though minor—does, however, result in the understimation of the bulk squared infragravity reflection coefficient ($R^2 = H^2_{m0,IG,out}/H^2_{m0,IG,in}$) with $RMSE = 0.35$ and $Rel.\ bias = -0.18$. This understimation of R^2 indicates that the model overestimated the nearshore IG wave dissipation [35].

The nested model was also run with morphology and sediment transport formulations disabled to assess the modeled run-up and overtopping of the dune without any erosion. The simulated run-up time series shows that at the peak of the storm (between 6:30 and 7:30), the run-up reaches and overtops the dune crest (elevation 3.92 m+MSL) (Figure 9). The maximum modeled run-up (R_{max}) of 4.09 m corresponded well with that observed by the PT (4.08 m) on the dune. As XBeach excludes high-frequency swash at the shoreline, these findings highlight the significance of low-frequency motions in dune overwash at La Faute-sur-Mer and similar coastlines.

With morphology and sediment transport enabled, the model accurately simulated the lowering and development of a washover deposit behind the dune (Figure 10) with a *BSS* of 0.78, which is considered good [36]. It is also worth noting that the model—without the *bermslope* option—completely overestimates the volume of erosion and is unable to reproduce the steep post-storm dune profile (Figure 10). The sensitivity of the model to the *bermslope* parameter value is also shown in Figure 10.

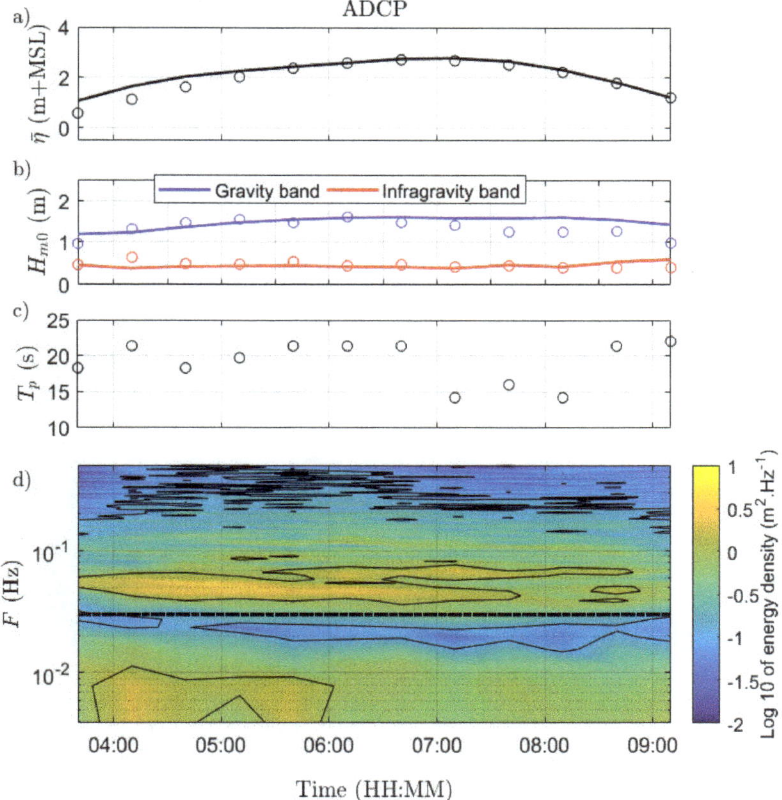

Figure 7. Modeled (solid lines) and observed (circles): (**a**) mean water level; (**b**) significant wave heights in the gravity and infragravity bands; (**c**) peak wave period; and (**d**) observed energy density, for the 2 February 2014 event taken at the ADCP location. Contours in panel 'd' indicate areas of equal energy density and the dashed line separates gravity and IG frequencies.

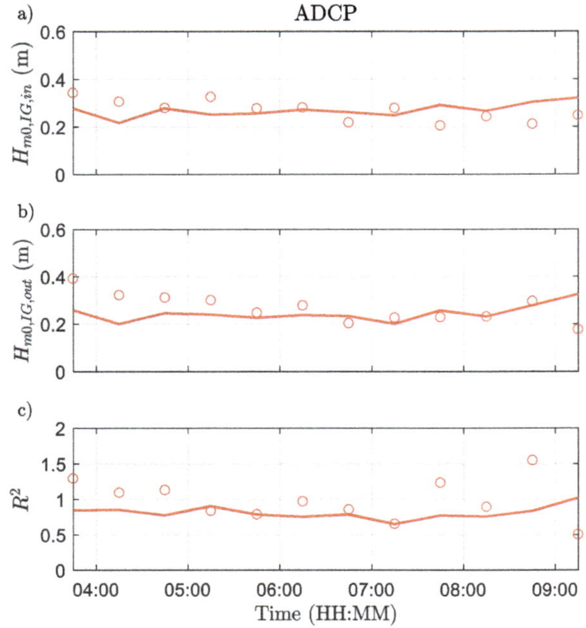

Figure 8. Modeled (solid lines) and observed (circles): (**a**) incoming and (**b**) outgoing IG significant wave heights; and (**c**) bulk squared IG reflection coefficient, for the 2 February 2014 event taken at the ADCP location.

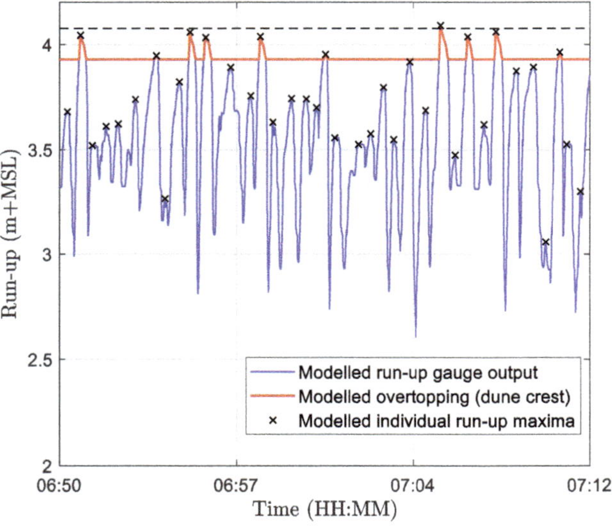

Figure 9. Modeled surface elevation at the shoreline (blue), run-up maxima (circles) for the 2 February 2014 event. Red line highlights the overtopping events. Dashed black line corresponds to the observed R_{max} The elevation of the dune crest is 3.92 m+MSL, for reference.

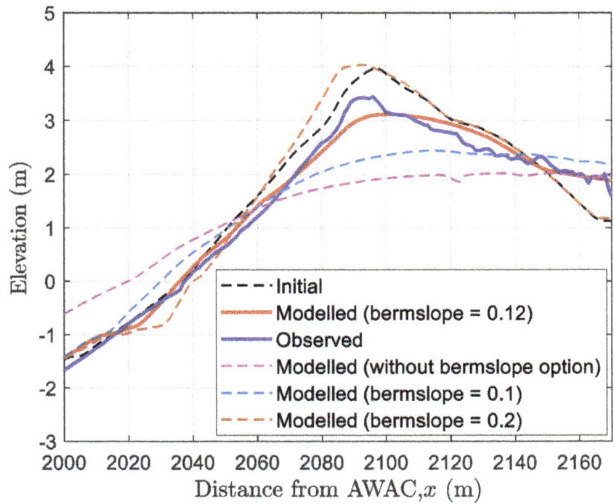

Figure 10. Modeled and observed dune erosion for the 2 February 2014 event.

A look at the transformation of H_{m0}, along the transect to the dune, shows that both gravity and IG-band waves are able to propagate over the mildy-sloping foreshore without further significant change (Figure 11). However, as the water depth becomes shallower at the dune toe (x = 2060 m), waves in both frequency bands experience amplitude growth, resulting in a 11% increase in $H_{m0,G}$ and a 109% increase in $H_{m0,IG}$—on average—compared to conditions at the AWAC location (x = 0 m); with the average ratio of IG- to gravity-band wave energy ($H_{m0,IG}/H_{m0,G}$) equalled to 0.42. While the amplitude growth of the gravity waves is due to shoaling in shallow water, the apparent doubling in amplitude of the IG wave at the dune toe is the combined result of shoaling and—to a greater extend—the superpositioning of the incoming and reflected IG waves. These findings further speak to the importance of accurately representing the infragravity wave component in shallow coastal environments.

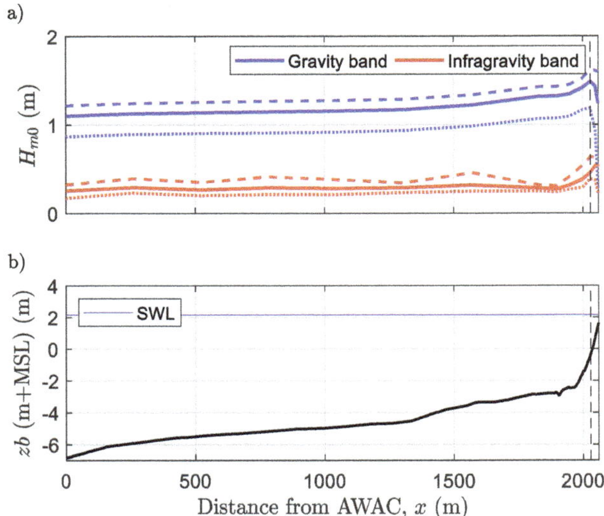

Figure 11. Transect across foreshore corresponding to Figure 2 showing the transformation of modeled maximum (dashed), mean (solid) and minimum (dotted): (**a**) gravity and infragravity waves, with (**b**) bed level for reference, for the February 2nd event (from time = 03:30 to 09:30). Dashed vertical lines indicate ADCP location.

4. Conclusions

In the present paper, we demonstrated that the XB-SB numerical model may be confidently applied—on a large scale (45 × 25 km domain)—to simulate the hydro-morphodynamcs of a complex embayment, under extreme conditions. Despite its limitations as a short-wave averaged but wave-group resolving model, the unconvential 2DH/1D approach was able to accurately reproduce both the large-scale hydrodynamics—wave heights and mean water levels across the embayment; and the local morphodynamics—steep post-storm dune profile and washover deposit.

The combined field data and numerical model results showed that offshore energetic gravity waves are significantly reduced by refraction, bottom friction and breaking over the shallow bathymetry; while infragravity waves are able to propagate into shallow water without being extensively changed. Thus, the relative contribution of the IG waves increased from less than 0.2 offshore to over 0.7 along the coast of La Faute-sur-Mer. Findings also showed that the noteworthy amount of IG-wave energy found at the dune toe—with the observed $H_{m0,IG}$ as high as 0.64 m—was the result of the superpositoning of the incoming and reflected IG waves. Given that the approach adopted here neglects the influence of gravity-band swash at the shoreline, our findings emphasize the significance of IG waves in washover development along the embayment; as XB-SB was able to accurately reproduce both the observed maximum water level at the dune crest and the extent of the resulting washover deposit.

Our work highlights the need for any method or model to account for IG-wave dynamics when applied to similar environnments. As many of the widely-used large domain phase-averaged models (e.g., SWAN/Delft3D) tend to exclude the often-dominant contribution of IG waves [37,38], the XB-SB coupled 2DH/1D approach described here may prove to be a suitable alternative.

Author Contributions: Conceptualization, C.L. and D.R.; methodology, C.L and X.B.; software, D.R.; validation, C.L.; formal analysis, C.L. and X.B.; investigation, X.B. and G.A.; resources, X.B. and D.R.; data curation, X.B.; writing—original draft preparation, C.L.; writing—review and editing, X.B. and D.R.; visualization, X.B. and C.L.; supervision, D.R. and X.B.; project administration, C.L.; funding acquisition, D.R.

Funding: This work is part of the *Perspectief* research programme *All-Risk* with project number B2 which is (partly) financed by NWO Domain Applied and Engineering Sciences, in collaboration with the following private and public partners: the Dutch Ministry of Infrastructure and Water Management (RWS); Deltares; STOWA; the regional water authority, Noorderzijlvest; the regional water authority, Vechtstromen; It Fryske Gea; HKV consultants; Natuurmonumenten; and waterboard HHNK.

Acknowledgments: Preliminary results of the research described here was first presented and published [39] at the Coastlab 2018 conference (May 22–26, Santander, Spain); the conference is acknowledged for providing the platform to present the described research.

Conflicts of Interest: The authors declare no conflict of interest.

References

1. Bertin, X.; Bakker, A.D.; Dongeren, A.V.; Coco, G.; Andre, G.; Ardhuin, F.; Bonneton, P.; Bouchette, F.; Castelle, B.; Craw, W.C. Infragravity Waves: From Driving Mechanisms to Impacts. *Earth-Sci. Rev.* **2018**, *177*, 774–799. [CrossRef]
2. Sallenger, A.H. Storm Impact Scale for Barrier Islands. *J. Coast. Res.* **2000**, *16*, 890–895.
3. Hunt, I.A. Design of Seawalls and Breakwaters. *J. Waterw. Harb. Div.* **1959**, *85*, 123–152.
4. Holman, R.A. Extreme Value Statistics for Wave Run-up on a Natural Beach. *Coast. Eng.* **1986**, *9*, 527–544. [CrossRef]
5. Hedges, T.S.; Mase, H. Modified Hunt's Equation Incorporating Wave Setup. *J. Waterw. Port Coast. Ocean Eng.* **2004**, *130*, 109–113. [CrossRef]
6. Stockdon, H.F.; Holman, R.A.; Howd, P.A.; Sallenger, A.H. Empirical Parameterization of Setup, Swash, and Runup. *Coast. Eng.* **2006**, *53*, 573–588. [CrossRef]
7. Mather, A.; Stretch, D.; Garland, G. Predicting Extreme Wave Run-up on Natural Beaches for Coastal Planning and Management. *Coast. Eng. J.* **2011**, *53*, 87–109. [CrossRef]
8. Soldini, L.; Antuono, M.; Brocchini, M. Numerical Modeling of the Influence of the Beach Profile on Wave Run-up. *J. Waterw. Port Coast. Ocean Eng.* **2013**, *139*, 61–71. [CrossRef]

9. Park, H.; Cox, D.T. Empirical Wave Run-up Formula for Wave, Storm Surge and Berm Width. *Coast. Eng.* **2016**, *115*, 67–78. [CrossRef]
10. Mayer, R.; Kriebel, D. Wave Run-up on Composite-slope and Concave Beaches. *Coast. Eng.* **1994**, *1*, 2325–2339.
11. Shand, R.D.; Shand, T.D.; McComb, P.J.; Johnson, D.L. Evaluation of Empirical Predictors of Extreme Run-up Using Field Data. In *Coasts and Ports 2011: Diverse and Developing: Proceedings of the 20th Australasian Coastal and Ocean Engineering Conference and the 13th Australasian Port and Harbour Conference*; Engineers Australia: Barton, Australia, 2011; p. 669.
12. Power, H.E.; Atkinson, A.L.; Hammond, T.; Baldock, T.E. Accuracy of Wave Run-up Formula on Contrasting Southeast Australian Beaches. In *Coasts and Ports 2013: 21st Australasian Coastal and Ocean Engineering Conference and the 14th Australasian Port and Harbour Conference*; Engineers Australia: Barton, Australia, 2013; pp. 1–6.
13. Stockdon, H.F.; Thompson, D.M.; Plant, N.G.; Long, J.W. Evaluation of Wave Run-up Predictions from Numerical and Parametric Models. *Coast. Eng.* **2014**, *92*, 1–11. [CrossRef]
14. Roelvink, D.; Reniers, A.; Dongeren, A.V.; De Vries, J.V.T.; Mccall, R.; Lescinski, J. Modeling Storm Impacts on Beaches, Dunes and Barrier Islands. *Coast. Eng.* **2009**, *56*, 1133–1152. [CrossRef]
15. Bellotti, G.; Archetti, R.; Brocchini, M. Experimental Validation and Characterization of Mean Swash Zone Boundary Conditions. *J. Geophys. Res. Oceans.* **2003**, *108*, 3250. [CrossRef]
16. Van Santen, R.; Steetzel, H.; De Vries, J.V.T.; Van Dongeren, A. Modeling Storm Impact on Complex Coastlines. Westkapelle, The Netherlands. *Coast. Eng. Proc.* **2012**, *1*, 52. [CrossRef]
17. Grzegorzewski, A.S.; Johnson, B.D.; Wamsley, T.V.; Rosati, J.D. Sediment Transport and Morphology Modeling of Ship Island, Mississippi, USA, During Storm Events. *Coast. Dyn.* **2013**, 1505–1516.
18. Van Dongeren, A.; Lowe, R.; Trang, D.M.; Roelvink, D.; Symonds, G.; Ranasinghe, R. Numerical Modeling of Low-frequency Wave Dynamics over a Fringing Coral Reef. *Coast. Eng.* **2013**, *73*, 178–190. [CrossRef]
19. McCall, R.T.; Masselink, G.; Poate, T.G.; Roelvink, J.A.; Almeida, L.P. Modeling the Morphodynamics of Gravel Beaches During Storms with Xbeach-g. *Coast. Eng.* **2015**, *103*, 52–66. [CrossRef]
20. Bertin, X.; Olabarrieta, M. Relevance of Infragravity Waves in a Wave-dominated Inlet. *J. Geophys. Res. Oceans* **2016**, *121*, 5418–5435. [CrossRef]
21. Lashley, C.H.; Roelvink, D.; Van Dongeren, A.; Buckley, M.L.; Lowe, R.J. Nonhydrostatic and Surfbeat Model Predictions of Extreme Wave Run-up in Fringing Reef Environments. *Coast. Eng.* **2018**, *137*, 11–27. [CrossRef]
22. Breilh, J.-F.; Bertin, X.; Chaumillon, E.; Giloy, N.; Sauzeau, T. How Frequent is Storm-induced Flooding in the Central Part of the Bay of Biscay? *Glob. Planet. Chang.* **2014**, *122*, 161–175. [CrossRef]
23. Nicolle, A.; Karpytchev, M. Evidence for Spatially Variable Friction from Tidal Amplification and Asymmetry in the Pertuis Breton (France). *Cont. Shelf Res.* **2007**, *27*, 2346–2356. [CrossRef]
24. Bertin, X.; Li, K.; Roland, A.; Bidlot, J.-R. The Contribution of Short-waves in Storm Surges: Two Case Studies in the Bay of Biscay. *Cont. Shelf Res.* **2015**, *96*, 1–15. [CrossRef]
25. Lumbroso, D.M.; Vinet, F. A Comparison of the Causes, Effects and Aftermaths of the Coastal Flooding of England in 1953 and France in 2010. *Nat. Hazards Earth Syst. Sci.* **2011**, *11*, 2321–2333. [CrossRef]
26. Roelvink, D.; McCall, R.; Mehvar, S.; Nederhoff, K.; Dastgheib, A. Improving Predictions of Swash Dynamics in Xbeach: The Role of Groupiness and Incident-band Runup. *Coast. Eng.* **2018**, *134*, 103–123. [CrossRef]
27. Daly, C.; Roelvink, D.; Van Dongeren, A.; De Vries, J.V.T.; McCall, R. Short Wave Breaking Effects on Low Frequency Waves. *Coast. Eng. Proc.* **2011**, *1*, 20. [CrossRef]
28. Ruessink, B.G.; Miles, J.R.; Feddersen, F.; Guza, R.T.; Elgar, S. Modeling the Alongshore Current on Barred Beaches. *J. Geophys. Res. Oceans.* **2001**, *106*, 22451–22463. [CrossRef]
29. Saha, S.; Moorthi, S.; Pan, H.L.; Wu, X.R.; Wang, J.D.; Nadiga, S.; Tripp, P.; Kistler, R.; Woollen, J.; Behringer, D.; et al. The NCEP Climate Forecast System Reanalysis. *Bull. Am. Meteorol. Soc.* **2010**, *91*, 1015–1058. [CrossRef]
30. Castelle, B.; Marieu, V.; Bujan, S.; Splinter, K.D.; Robinet, A.; Senechal, N.; Ferreira, S. Impact of the Winter 2013–2014 Series of Severe Western Europe Storms on a Double-barred Sandy Coast: Beach and Dune Erosion and Megacusp Embayments. *Geomorphology* **2015**, *238*, 135–148. [CrossRef]
31. Masselink, G.; Scott, T.; Russell, P.; Davidson, M.; Conley, D. The Extreme 2013/2014 Winter Storms: Hydrodynamic Forcing and Coastal Response Along the Southwest Coast of England. *Earth Surf. Process. Landf.* **2016**, *41*, 378–391. [CrossRef]

32. Roelvink, D.; Costas, S. Beach Berms as an Essential Link Between Subaqueous and Subaerial Beach/Dune Profiles. *IX J. Geomorfol. Litoral* **2017**, *17*, 79–82.
33. Roelvink, J.A.; Stive, M.J.F. Bar-generating Cross-shore Flow Mechanisms on a Beach. *J. Geophys. Res. Oceans (1978–2012)* **1989**, *94*, 4785–4800. [CrossRef]
34. Guza, R.; Thornton, E.; Holman, R. Swash on Steep and Shallow Beaches. In Proceedings of the 9th International Conference on Coastal Engineering, Houston, TX, USA, 3–7 September 1984.
35. Rijnsdorp, D.P.; Ruessink, G.; Zijlema, M. Infragravity-wave Dynamics in a Barred Coastal Region, a Numerical Study. *J. Geophys. Res. Oceans* **2015**, *120*, 4068–4089. [CrossRef]
36. Sutherland, J.; Peet, A.H.; Soulsby, R.L. Evaluating the Performance of Morphological Models. *Coast. Eng.* **2004**, *51*, 917–939. [CrossRef]
37. Buckley, M.; Lowe, R.; Hansen, J. Evaluation of Nearshore Wave Models in Steep Reef Environments. *Ocean Dyn.* **2014**, *64*, 847–862. [CrossRef]
38. Roeber, V.; Bricker, J.D. Destructive Tsunami-like Wave Generated by Surf Beat Over a Coral Reef During Typhoon Haiyan. *Nat. Commun.* **2015**, *6*, 7854. [CrossRef] [PubMed]
39. Lashley, C.H.; Bertin, X.; Roelvink, D. Field Measurements and Numerical Modeling of Wave Run-up and Overwash in the Pertuis Breton Embayment, France. In Proceedings of the 7th International Conference on the Application of Physical Modelling in Coastal and Port Engineering and Science (Coastlab18), Santander, Spain, 22–26 May 2018.

Journal of
Marine Science and Engineering

Article

Interaction of Swell and Sea Waves with Partially Reflective Structures for Possible Engineering Applications

Andrea Lira-Loarca [1,*], Asunción Baquerizo [1] and Sandro Longo [2]

1 Andalusian Institute for Earth System Research, University of Granada, Avenida del Mediterráneo s/n, 18006 Granada, Spain; abaqueri@ugr.es
2 Department of Engineering and Architecture, University of Parma, Parco Area delle Scienze, 181/A, 43124 Parma, Italy; sandro.longo@unipr.it
* Correspondence: aliraloarca@ugr.es; Tel.: +34-958-249-738

Received: 22 January 2019; Accepted: 30 January 2019; Published: 2 February 2019

Abstract: In this work, we investigate the interaction between the combination of wind-driven and regular waves and a chamber defined by a rigid wall and a thin vertical semi-submerged barrier. A series of laboratory experiments were performed with different values of incident wave height, wave period, and wind speed. The analysis focuses on the effect of the geometry of the system characterized in terms of its relative submergence d/h and relative width B/L. Results show that for the case of $d/h = 0.58$ a resonant effect takes place inside the chamber regardless of the wind speed. Wind-driven waves have a higher influence on the variation of the wave period of the waves seaward and leeward of the plate, as well as on the phase lag. Results show that the amplification or reduction of the wave energy inside the chamber is closely related to the wave period as compared to the 1st order natural period of the chamber.

Keywords: vertical barrier; semi-submerged; wind waves; experiments; laboratory

1. Introduction

The modeling of the interactions between forcing agents, e.g., waves and wind, and maritime structures is a challenging problem with important applications in coastal engineering. In the last decades, environmentally friendly coastal structures, such as partially submerged barriers, have become of great interest due to their frequent use in environmental protection, recreation, or wave energy extraction facilities. This type of barrier can reduce the wave energy inside a harbor or a marina while allowing sediment and water exchanges [1,2]. At the same time, these structures can be designed towards wave energy extraction when considering an oscillating water column (OWC) wave energy converter (WEC). Many experimental and theoretical studies have been performed to evaluate their efficiency and hydrodynamic behavior under regular and irregular waves.

Dean, (2008), Ursell and Dean, (1947) [3,4] first studied the wave reflection of incident progressive waves on a fully and partially submerged vertical barrier. Losada et al., (1992) [5] studied the linear theory for periodic waves impinging obliquely on a vertical thin barrier, using the eigenfunction expansion method. This study was then extended to analyze oblique modulated waves [6]. Linear theory was also applied to analyze the scattering of irregular waves impinging on fixed vertical thin barriers, with a good agreement between the analytical model and experimental data by other authors [7]. Koutandos et al., (2010) [1] presented an experimental study of waves acting on a partially submerged breakwater with four different configurations, including a single fixed barrier under regular and irregular waves in shallow and intermediate water. The results showed the effect of the various configurations on transmission, reflection, and energy dissipation, highlighting that the main governing design parameter could change depending on forcing conditions (for short waves the main parameter was the submergence of the barrier, for long waves it was the width of the chamber). Jalon et al., (2018) [8] presented an analytical model to optimize the design configuration of a vertical thin barrier concerning different criteria, such as harbor tranquility or wave energy extraction. Other types of structures, e.g., slotted-, porous-, and Jarlan-type barriers, have also been widely studied [9,10].

Regarding the experimental studies on vertical semi-submerged barriers, Kriebel et al., (1999) [11] studied their efficiency under regular and irregular waves in terms of the transmission and reflection coefficients, while Liu and Al-Banaa, (2004) [2] carried out experiments under solitary waves focusing on the wave forces acting on the barrier.

Nonetheless, in nature we find swell waves coexisting with wind-driven waves. Swell waves are described as long-period waves that have been traveling for long distances and in absence of wind can be locally analyzed by means of standard wave spectra. Wind waves are actively growing due to forcing action from the local wind, and are characterized as non-regular waves consisting of a spectrum in continuous evolution. However, most of the experimental studies on wave interaction with maritime structures such as the ones mentioned, are limited to regular and/or irregular waves generated by a paddle, without considering (1) the effect of wind forcing on both the incident and reflected swell wave trains, (2) the interaction of the driven sea waves with the structure and (3) the non-linear interaction between the different wave field components. Therefore, there is still need for data on the wave-structure interaction under wind-driven waves superimposed on swell, taking into account their intrinsic irregular and random complex nature.

The present study is based on a series of laboratory experiments of wave-structure interaction with paddle-generated regular waves in combination with wind-driven waves, due to wind blowing in the direction swell propagation. The structure consists of a thin vertical semi-submerged barrier delimiting a chamber along with an impermeable back wall. This type of structures can be used in maritime engineering applications such as harbor tranquility and energy extraction [7,8,12].

The manuscript is structured as follows. Section 2 describes the adopted methodology, the experimental facility and measurements. Section 3 contains the critical analysis of the data and the discussion. Conclusions are given in Section 4.

2. Methodology

2.1. Experimental Setup and Data Analysis

Experiments were conducted in the Atmosphere-Ocean Interaction Flume (CIAO) of the Andalusian Institute for Earth System Research, Granada, Spain. It is 16 m long, 1 m wide and is designed for an optimal water depth of 70 cm (Figure 1). This facility is dedicated to the study of the coupled processes between the ocean and the atmosphere. It is equipped with (i) two opposite piston-type wavemakers with active absorption system that allow the generation of regular and irregular waves with periods between 1–5 s, and therefore in the range of shallow to intermediate conditions, and heights up to 25 cm; (ii) a closed-circuit wind generation system (wind tunnel) with mean wind speed at start of the test section—representative of U_{10} conditions in prototype—up to 12 m/s that generate waves with an effective fetch length up to 12 m, and (iii) a current-generation system for currents of up to 75 cm/s. The system allows the generation of waves and currents following or opposing the wind direction, with all the possible combinations. The system allows for the active absorption of incoming waves by the secondary paddle and the compensation of re-reflection by the generating paddle. More information on this system and its performance can be found on Lykke Andersen et al., (2016) [13]. For a detailed description of the CIAO facility the reader is referred to Clavero et al., (2013), Nieto et al., (2015) [14,15].

A rigid thin barrier, with a thickness of 1.9 cm and a submergence d was installed vertically (see Figure 2). This barrier was located at 11 m from the wavemaker and spaced a distance B (width of the chamber) from a vertical rigid back wall. The x- and z-axes refer to the stream-wise and vertical direction, respectively, with the origin at the mean water level in the section where the thin barrier is located.

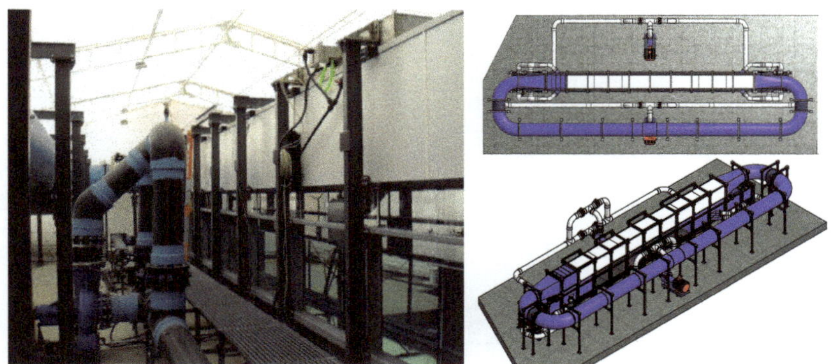

Figure 1. Atmosphere-Ocean Interaction Flume (CIAO). Schematics provided by VTI.

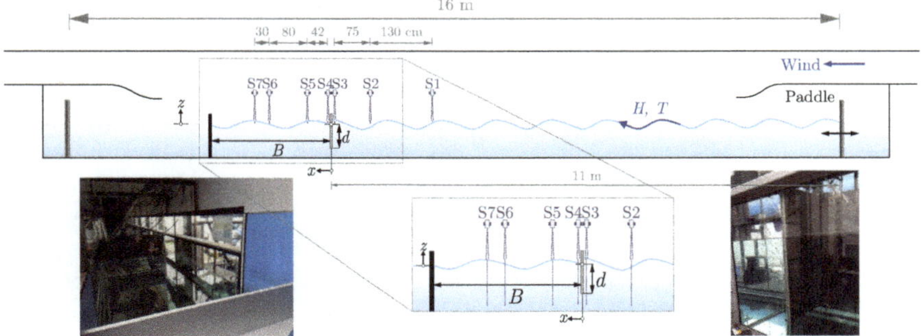

Figure 2. The experimental setup of the flume. The system consists of a semi-submerged rigid thin barrier separated a distance B (chamber width) from a vertical rigid back wall. S1–S7 are ultrasound distance gauges for free surface elevation measurements. Tests were done with a combination of regular waves (paddle-generated) and wind-driven waves traveling in the same direction.

Water level measurements were obtained using seven UltraLab ULS 80D acoustic wave gauges installed at different sections along the flume, with a maximum repetition rate equal to 75 Hz, a declared space resolution of 0.18 mm and a reproducibility of $\pm 0.15\%$. The overall accuracy is 0.5 mm. Free surface measurements were taken for 185 s.

Measurements of wind speed were collected before the installation of the system (vertical plate or back wall) using a Pitot tube at $x = 0$ m (plate section) and at different heights above the water level. The signal was acquired for ≈ 60 s with a data rate of 1 kHz. Figure 3 shows the wind speed profiles at different rotation rates of the fan used in the tests, with U_{ref} representing the average speed in the vertical profile [16].

A combination of paddle (with H^{input}, T^{input}, representing the r.m.s. wave height and period given as inputs to wave generation software), and wind waves (U_{ref}, representing the reference wind speed) were generated for a chamber with a constant width $B = 2.5$ m and different plate submergences d. Each regular wave experiment was performed (i) in the absence of wind, and (ii) in combination with wind waves using different U_{ref}. Regular waves were generated using only one of the wavemakers with the active absorption on.

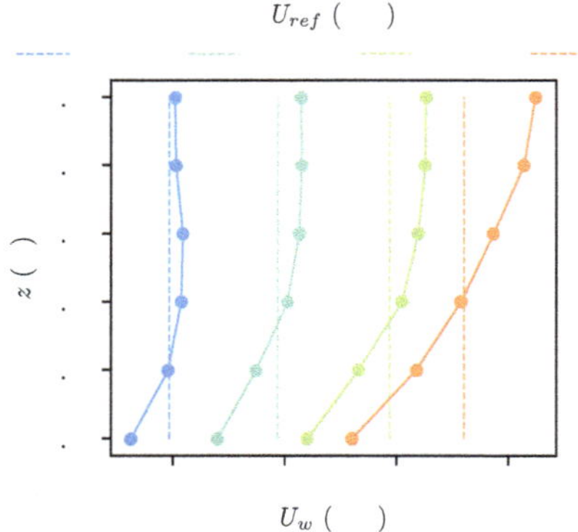

Figure 3. Wind speed profiles (symbols) and reference speed U_{ref} (dashed vertical lines) for the different rotation rates of the fan.

Table 1 lists the values of the parameters of the experiments. H_{rms0} and T_{z0} represent the r.m.s. wave height and the mean wave period obtained by the statistical analysis of the surface elevation data in the section of the wave gauge S1 ($x = -2.1$ m) for the experiment without wind ($U_{ref} = 0$ m/s). The statistical analysis consists on applying the zero-upcrossing technique to the surface elevation time series obtaining the individual wave heights and periods of the signal. The mean wave period T_{z0} is always coincident with the input period T^{input}, whereas H_{rms0} is not coincident with the imposed value H^{input}. L_0 represents the wavelength (experiment without wind) obtained by means of the linear dispersion equation $\omega_0^2 = gk_0 \tanh(k_0 h)$ where $\omega_0 = \frac{2\pi}{T_{z0}}$ is the angular frequency, $k_0 = \frac{2\pi}{L_0}$ is the wavenumber and $h = 0.7$ m is the water depth and T_{z0} is the mean wave period obtained from the experiment without wind. The two first natural oscillation periods of the system were approximated using Merian's formula $T_n = \frac{2B}{n\sqrt{gh}}$ with $n = 1, 2, ...,$ and are equal to $T_1 = 1.9$ s and $T_2 = 0.95$ s, respectively. Figure 4 presents the signal of the wave gauge S1 for the different experiments and different wind conditions.

Table 1. Parameters of the experiments. H^{input} and T^{input} are the input values given to the wave generation software. H_{rms0}, T_{z0} and L_0 represent the r.m.s. wave height, mean wave period and wavelength, respectively, obtained by the statistical analysis of the surface elevation data of S1 for the experiment without wind. The colum H_{rms0} depicts the mean value $\pm$ the error —when applicable— between two repetitions of the same case. d/h is the relative submergence, B is the chamber width and U_{ref} represents the reference wind speed.

Test	H^{input} (cm)	H_{rms0} (cm)	$T^{input} = T_{z0}$ (s)	d/h	B/L_0	T_{z0}/T_1	U_{ref} (m/s)
R1a	7	6.4 ± 0.5	1.85	0.33	0.6	0.97	0, 3.95, 5.88, 7.88, 9.21
R1b	7	5	1.85	0.58	0.6	0.97	0, 3.95, 5.88, 7.88, 9.21
R1c	7	8.9 ± 0.15	1.85	0.71	0.6	0.97	0, 3.95, 5.88, 7.88, 9.21
R2a	6	6.3	1.65	0.58	0.7	0.87	0, 3.95, 5.88, 7.88, 9.21
R3a	7	2.7	2.78	0.58	0.4	1.46	0, 3.95, 5.88, 7.88, 9.21
R4a	7	6.7	4.76	0.58	0.2	2.5	0, 3.95, 5.88, 7.88, 9.21
R5a	6	2.8 ± 0.25	3	0.58	0.3	1.58	0, 3.95, 5.88, 7.88, 9.21
R6a	6	5.8 ± 1.4	4.7	0.58	0.2	2.47	0, 3.95, 5.88, 7.88, 9.21

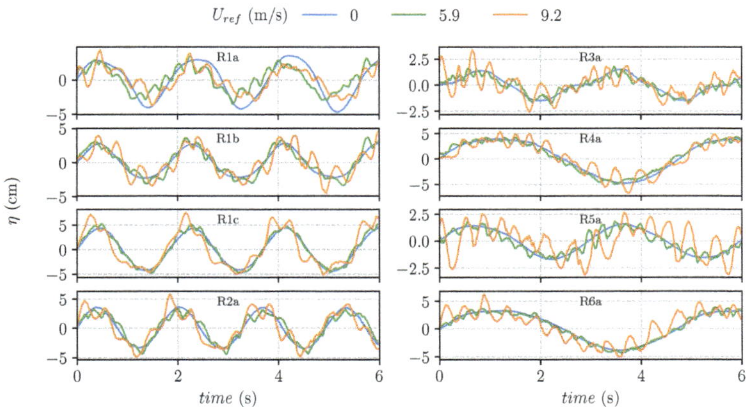

Figure 4. Free surface elevation of the wave gauge S1 for the experiments presented on Table 1. In each panel, different colors of the lines refer to different wind speeds.

In addition to the statistical analysis by means of the zero-upcrossing technique, data were processed by spectral analysis using the python Scipy Signal processing functions. The spectra are obtained with the Welch method, using a Hanning window with 50% overlap resulting in 17 degrees of freedom and a spectral resolution of 0.039 Hz.

The model investigated in this work can be replicated in real-world applications guaranteeing Froude similarity, commonly adopted for the design of maritime structures. We notice that using a Froude-based similarity and working with the same fluid viscosity for both model and prototype, inevitable violates the Reynolds number scaling. In this case, a partial dynamic similarity is achieved using Froude scaling, with length and pressure scale of λ and a time scale of $\sqrt{\lambda}$. Assuming a geometrical scale of $1/14$, experiment R1a, e.g., scales to an incident wave height of $H^{input} = 1$ m, $T^{input} = 7$ s, $L = 59$ m in water depths of $h = 10$ m.

2.2. Expected Wave Characteristics

Jalon et al., (2018) [8] presented an analytical model to optimize the design configuration of a vertical thin barrier concerning different criteria, such as harbor tranquility or wave energy extraction using a prototype setup as the once presented in this work. Their results showed the presence of nodal and antinodal frequencies of the spectra on both the leeward and seaward side depending on the geometrical configuration and the incident wave characteristics focusing on the relative submergence (d/h) and relative width (B/L) parameters. For values of $d/h > 0.25$, maximum values of the reflection coefficient where found for very small values of B/L and slightly larger than 0.5 for which also maximum capture coefficients were found.

For B/L close to 0.6, the reflection coefficient increases with d/h (experiments R1a-R1c—Table 1) up to values of $d/h \approx 0.7$ when it starts decreasing again. In these cases, the highest capture coefficient is obtained for $d/h \approx 0.6$. At $x = 0$, their results showed the presence of quasi-nodes (quasi-antinodes) in the seaward (leeward) sides for the configurations resembling experiments R1a and R2a and the quasi-antinodes on both sides experiment R1c.

For d/h close to 0.6, and varying B/L (experiments R1b, R3a, R4a) the reflection coefficient increases with B/L while the transmission coefficient presents similar values for B/L of 0.2 (R4a) and 0.4 (R3a) and larger values for $B/L = 0.6$ (R1b). At $x = 0$, their results showed the presence of quasi-antinodes (quasi-nodes) in the seaward (leeward) sides for the configurations resembling experiments R3a and R4a and the opposite for experiment R1b.

3. Results and Discussion

The key element of the analysis is the influence of the geometry of the system and of the wind-generated waves on the measured variables. Therefore, we analyzed the results with respect to the relative submergence d/h and to the different U_{ref}. During the experiments, the relative submergence d/h was modified by varying the submergence of the plate d whereas the water depth h was kept constant. To test different values of the relative width B/L, the chamber width B remained fixed and different configurations of wave periods with and without the combination of wind were tested, changing the wavelength of the incident wave L_0.

All the results presented henceforth are dimensionless with respect to the corresponding values of the measurements in absence of wind and at Section S1.

Figures 5–8 show the results for the different R1-experiments, corresponding to the same initial theoretical values $H^{input} = 7$ cm, $T^{input} = 1.85$ s and wind speeds, but with a different relative submergence d/h (see Table 1). The mean wave period of the regular waves ($T_{z0} = 1.85$ s) is very close to the 1st order natural period of the chamber, with $T_{z0}/T_1 = 0.97$.

Figure 5 depicts, in the first and second row, the dimensionless r.m.s. wave height and wavelength for the gauges S2 (seaward region) and S5, S6 and S7 (leeward region) for each experiment. In the leeward region, the ratio H_{rms}/H_{rms0} is related to the transmission coefficient. The third row shows the mean phase lag defined as $\Phi = \omega t_{max}$, with ω being the angular frequency. t_{max} represents the time difference of the maximum surface elevation position between the signal of each sensor and the signal of sensor S1 within a wave period. For its calculation, the signal of each sensor is divided into waves of period T^{input} and the position of the maximum surface elevation is found.

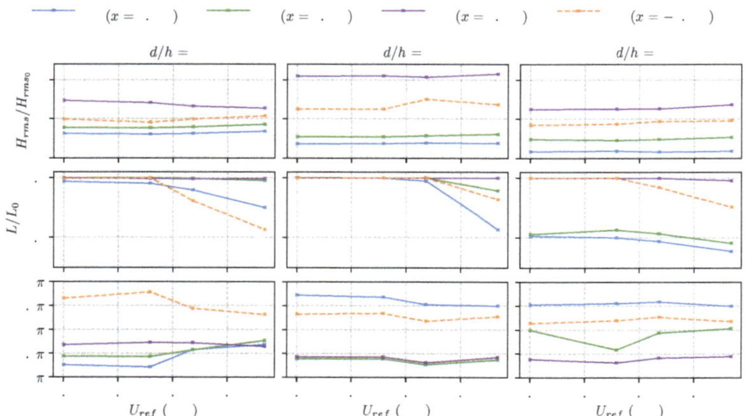

Figure 5. R1a-b-c-experiments with constant input values $H^{input} = 7$ cm, $T^{input} = 1.85$ s and different relative submergences d/h. The first row shows the r.m.s. wave height H_{rms}, the second row shows the wavelength L, the third row shows the phase lag Φ. Each panel shows the results for S5-S6-S7 sensors (violet, green, and blue lines, respectively) in the leeward region, and for sensor S2 (orange dashed line) in the seaward region. Values are dimensionless with respect to H_{rms0}, L_0 measured in Section S1 without wind.

It can be observed that for each individual sensor and experiment the wave height is constant regardless of the wind speed, showing that wind-driven waves have almost no influence in the amplification or reduction of the periodic wave inside the chamber, when $T_{z0}/T_1 \approx 1$. In the case of the highest plate submergence, $d/h = 0.71$ (R1c), the minimum value of the wave height is observed. This result was expected, given the high reflection of the incident wave and the low energy transmitted inside the chamber. The highest values are obtained for the experiment R1b with $d/h = 0.58$. In this case,

S5 gives a value of $H_{rms}/H_{rms0} \approx 2$, showing a resonant behavior inside the chamber. However, a high variation between the different sensors inside the chamber is observed in this test, with values in the range 0.5–2 between the sensors S7 and S5. It can be attributed to the partial standing pattern inside the chamber that exhibits a spatial variation of the total wave amplitude with values depending on the relative distance of each gauge to quasi-nodes and quasi-antinodes. Also, the phase lag shows a similar variability, with $0.2\pi < \Phi < 0.5\pi$ for sensor S5 and $1.5\pi < \Phi < 1.7\pi$ for sensor S7. We notice the presence of quasi-antinodes and surface elevations in phase opposition ($\Phi \approx \pi$) which is critical for the study of the loads acting on the plate [8]. These results are in agreement with the results in Jalon et al., (2019) [8], where the same structural configuration is studied analytically. The maximum values of the capture coefficient (a proxy for the transmission coefficient), with the presence of quasi-antinodes, are obtained for $B/L \approx 0.6$ (R1a-b-c-experiments).

The main influence of wind-driven waves is observed in the estimations of L/L_0. As observed, L/L_0 decreases for increasing wind speeds and is more evident for Section S2, where the wind is expected to have a higher influence. A change of the phase lag can be observed for varying U_{ref} for the tests R1a and R1b, whereas for tests R1c, with $d/h = 0.71$, the phase lag is almost constant; however, as already highlighted, it varies considerably for different sections. This can be attributed to the fact that higher reflections are likely to be associated with higher submergences showing partial standing oscillations. Therefore, a reduction of the non-linear interaction between wind waves and the periodic component is expected.

Figure 6 depicts the dimensionless surface elevation η/A_{rms0}, where A_{rms0} is the r.m.s. wave amplitude obtained from S1 in the absence of wind. It can be observed that the highest relative amplitude inside the chamber is given for tests with $d/h = 0.33 - 0.58$, with almost double of the seaward wave amplitude at the Section S5 (which is very close to a quasi-antinode), for all wind speeds. The deformation of the regular wave induced by wind action in the seaward region is transmitted inside the chamber, where an asymmetry between crests and troughs is observed. This could be due to the amplification modes of the system and to the re-reflection, both controlled by geometry. In passing, we notice that free surface oscillations are strongly affected by the forcing term and by the damping, the latter mainly due to dissipation [17]. For limited dissipation, a blow up of the oscillations is expected for multiple harmonic in the forcing term. In this respect, the free surface oscillations in the chamber can vary significantly if energy is extracted.

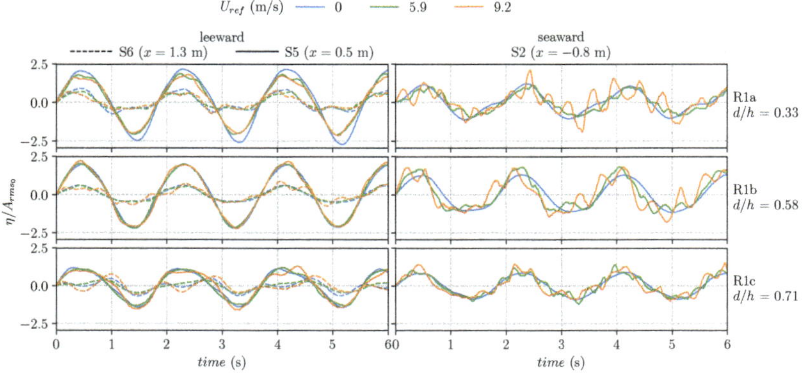

Figure 6. R1a-b-c-experiments with constant nominal values $H^{input} = 7$ cm, $T^{input} = 1.85$ s and different relative submergences d/h. Free surface elevation is dimensionless with respect to the amplitude measured at Section S1 in the absence of wind. The left column shows the results for the two sensors S5–S6 in the leeward region (continuous and dashed lines, respectively), the right column shows the results for sensor S2 in the seaward region. In each panel, different colors of the lines refer to different wind speeds.

A similar behavior can be observed for all three d/h values tests in Section S2 in the seaward region, where the wind generates short waves traveling alongside the long regular wave. This effect is less noticeable in the test with $d/h = 0.71$ (R1c), where a higher reflection of the periodic wave components is expected.

These results are mirrored in the power spectrum $S' = S(f)f_p/m_0$ shown in Figure 7, dimensionless with respect to the peak frequency f_p and to the zeroth-order moment m_0 obtained from data at Section S1 in the absence of wind. At high frequencies, the scenario is controlled by wind-driven waves, energy increases, and the resonant frequencies become less noticeable. Inside the chamber the peak frequency in all sections almost corresponds to the frequency of the regular waves, which is very close to the natural period of the chamber, $T_{z0}/T_1 = 0.97$. The energy is higher for sensor S5 closest to the plate and the energy in the chamber increases with respect to the seaward region, which is associated with the peak frequency of regular waves and to the first resonant period, more so for the case with the highest relative submergence $d/h = 0.71$; there is an energy decrease at higher frequencies with respect to the seaward side. This is attributed to non-linear interaction between wind-driven waves, periodic waves and resonant waves controlled by the geometry of the system and to the filtering effect of the screen, more efficient for high frequency components. For clear cut results, it will be necessary to further explore the wave profile and its variation with different geometrical configurations and with different forcing conditions.

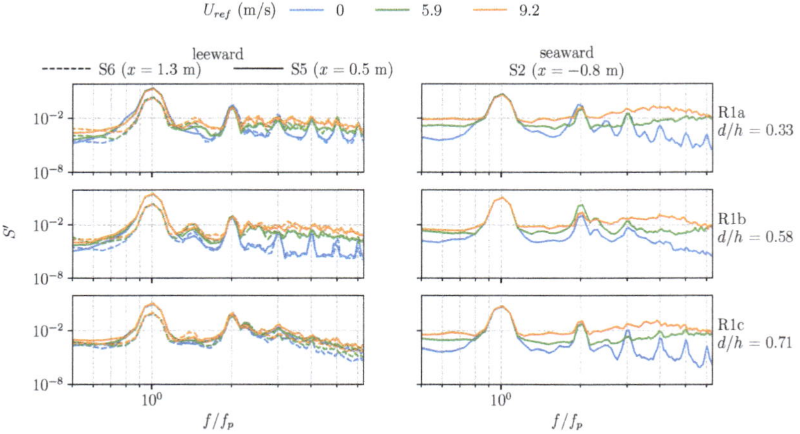

Figure 7. Dimensionless power spectrum S'. For caption see Figure 6.

Figure 8 shows the dimensionless phase-averaged surface level for the R1a-b-c-experiments, with and without wind-driven waves and for different sensors. It can be observed that both the predominant regular wave phase, and the wave profile, change from the seaward to the leeward region. In the case of the highest relative submergence $d/h = 0.71$ (R1c), the transmission of the longer periodic wave is expected to be lower, and the changes could be due to the wind acting on the free surface. Sensor 5, at section $|x|/L = 0.12$, shows a wave profile similar to the incident wave, with an increment in wave height due to its position in a quasi-antinode. Sensor 6 shows a reduction of the wave height and some changes in wave profile. As expected, due to the shading effect of the plate, the influence of wind-driven waves is limited to the seaward region, with a more evident variation of the wave profile for the case of $d/h = 0.58$ (R1b) and $U_{ref} = 5.9$ m/s.

To analyze the effect of the dimensionless parameter B/L in the resonant behavior of the chamber for different wind conditions, different regular wave (paddle-generated) and wind-wave configurations were tested for a fixed relative submergence $d/h = 0.58$. This analysis is especially relevant for the design of OWC devices that have performance directly related to the oscillation of the water column.

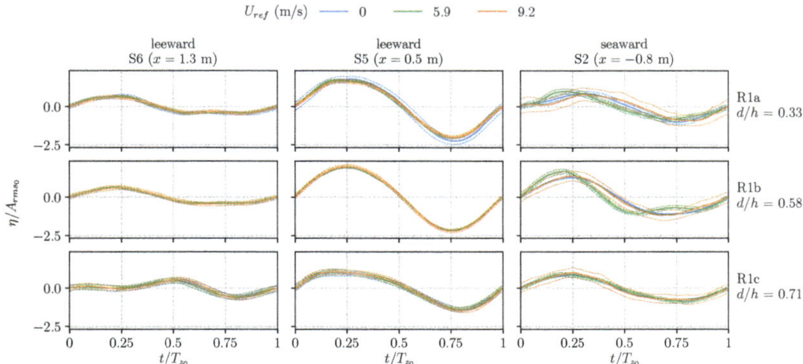

Figure 8. Phase-averaged dimensionless free surface elevations. For caption see Figure 6. The dashed lines correspond to the phased-averaged value ± one standard deviation.

Figure 9 shows the same variables shown in Figure 5, but the comparison is made between different regular wave experiments with the same relative submergence $d/h = 0.58$. Each column represents two sets of experiments with initial theoretical values of H^{input} = 6–7 cm and similar values of T_{z0}/T_1.

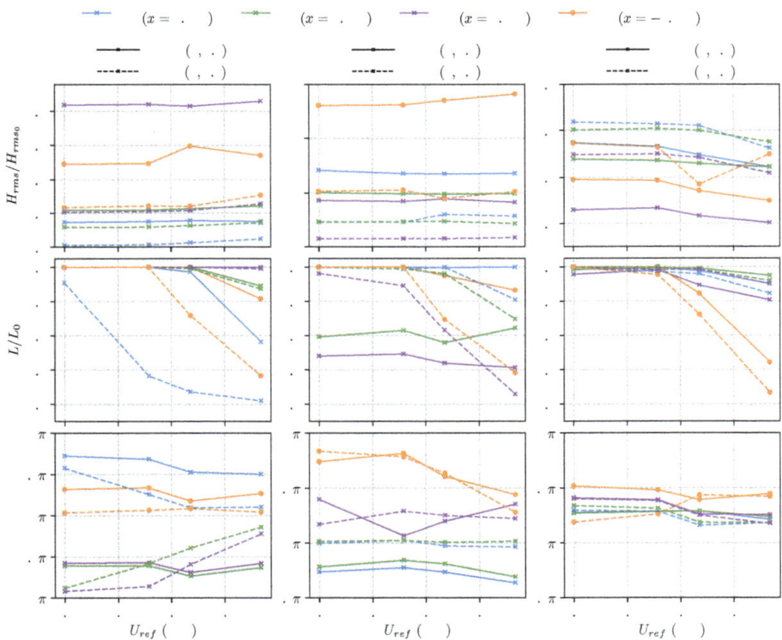

Figure 9. Experiments with the same relative submergence $d/h = 0.58$. For caption see Figure 5.

Values of $H_{rms}/H_{rms0} \approx 3$ can be observed in the leeward region for tests with $T_{z0}/T_1 \approx 1.5$, while at Section S2 in the seaward region, results $H_{rms}/H_{rms0} \approx 5-6$. We notice that the scaling value of wave height H_{rms0} is significantly reduced with respect to the nominal value (Table 1). For example, R3a was carried out using a nominal wave height of $H^{input} = 7$ cm while the statistical analysis of S1 for this experiment without wind gave a value of $H_{rms0} = 2.7$ cm. In this case, at Section S2, the closest to the plate in the seaward region, very high values of r.m.s. wave height were obtained (higher reflection), with consequent very high values of H_{rms}/H_{rms0}. It can also be observed that for experiments R4a and R6a, corresponding to the experiments with higher wave periods of the regular wave, there is a noticeable stronger influence of wind-driven waves than for experiments with lower wave periods, characterized by a constant value of H_{rms}/H_{rms0} for the different wind speeds. A reduction of the wavelength L/L_0 can be observed for increasing wind speed for almost all tests. In the seaward region, there is an almost five-fold reduction with the highest wind. In the leeward region, this influence changes depending on T_{z0}/T_1, e.g., for $T_{z0}/T_1 \approx 2.5$ there is a negligible change of the wavelength with wind speed.

Another relevant indicator of the system behavior is the phase lag. For tests with $T_{z0}/T_1 \leq 1.5$ the phase lag varies greatly from one sensor to the other for the same experiment, whereas for tests with $T_{z0}/T_1 \approx 2.5$ results $\pi \leq \Phi \leq 1.5\pi$ regardless of wind speed. Hence, in some tests, the relation between the period of the incident wave height and the resonant wave periods of the system plays a major role than the wind speed.

Figure 10 shows H_{max} and H_{rms} of sensors S2, S5, and S6. Each point represents a different configuration of regular wave (paddle-generated) and wind speed and they were ordered according to the relation T_z/T_1 (x-axis) where T_z is the mean wave period of each sensor and case and T_1 the 1st order natural period of the chamber. The first row presents the results for S2 (seaward region) where the highest values of H/H_{rms0} are achieved. This is due to the fact that as previously explained, in this position the regular periodic wave component is predominant over the shorter wind waves, therefore the highest values of wave height are obtained. For the sensors S5-S6 in the leeward region, an amplification of wave energy is observed for values T_z/T_1 close to 1 for all wind speeds. For most of the experiments in the absence of wind, the same values of H_{max} and H_{rms} are obtained while in tests with wind, H_{max}/H_{rms0} presents higher values than H_{rms}/H_{rms0}. This is expected and due to the fact that the r.m.s. is a statistic that takes into account all the wave heights in the signal, therefore, taking into account the shorter wind-driven waves, its value is expected to be lower than H_{max}.

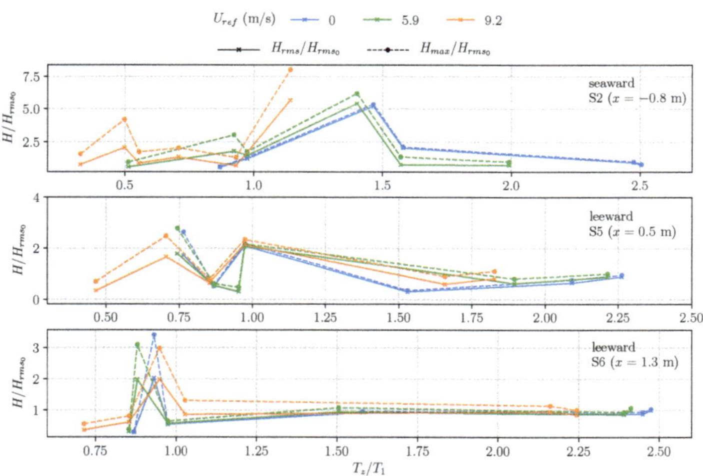

Figure 10. Dimensionless maximum and r.m.s. wave heights for different sensors (row-wise). In each panel, different points correspond to the experiments with the same relative submergence d/h and different configurations of regular waves (paddle-generated) with and without the combination of wind waves. In each panel, the different colored lines represent the results for experiments with and without wind.

4. Conclusions

In the present study, the hydrodynamic interaction of regular waves, in combination with short wind-driven waves, with a vertical fixed semi-submerged barrier was experimentally investigated. The study was focused on the influence of the seaward wave characteristics and certain geometric characteristics of the structure, with special attention on the role of wind waves providing new insight of the influence of wind-driven waves on the incident wave characteristics and in the wave field inside the chamber as it still remains a topic to be further investigated in maritime and coastal engineering. A series of conclusions can be derived.

When analyzing a regular wave configuration with a wave period similar to the 1st order natural period of the chamber, an amplification of wave energy in the seaward region is measured for a relative submergence of $d/h = 0.58$, in agreement with the results obtained by Jalon et al., (2019) [8]. In the case of higher submergences, the fixed plate operates in a highly reflective manner and a reduction of the wave energy inside the chamber can be observed. Therefore, the submergence of the plate is a key parameter for the design of these types of structures, either for harbor protection (energy reduction), or WEC devices (energy amplification).

Wind-driven waves have a high influence on the wave symmetry and phase lag between the seaward and leeward regions. This is an important factor to take into account since it could lead to strong forces on the barrier when the phase lag equals π.

The influence of wind waves is more noticeable for the experiments where $T_{z0}/T_1 \geq 1.5$, in which the short-crested waves proved to have a higher influence on the resulting wave periods and heights in the seaward region. Therefore, the influence of wind waves on the behavior of the structure depends greatly on the period of the predominant periodic component (swell).

The overall behavior of the system depends on the relative values of periodic waves, wind waves, and resonant periods of the chamber. The analysis has been conducted without extracting energy from the system, since no OWC device was inserted in the experimental layout. In real conditions, with an OWC device active, we expect some minor modifications, with a general reduction of the wave height in both leeward and seaward regions. In this respect, the present analysis can be confidently applied for verifying the structures in the most critical conditions.

As a future extension, this research could involve (i) the analysis of irregular waves, where is expected that wind waves effects depend on the frequency range; (ii) the analysis of grouping effects, where a sequence of grouped waves can force the system with presently unnoticed effects.

Author Contributions: All authors conceptualized the study. A.L.-L. did the experimental work and analysis of data with the guidance of A.B and S.L. All authors contributed to the preparation of the manuscript.

Acknowledgments: A.L.L. is supported by the research group TEP-209 (Junta de Andalucía) and project AQUACLEW. Project AQUACLEW is part of ERA4CS, an ERA-NET initiated by JPI Climate, and funded by FORMAS (SE), DLR (DE), BMWFW (AT), IFD (DK), MINECO (ES), ANR (FR) with co-funding by the European Union (Grant 690462). During the preparation of the manuscript, A.L.L. was doing a research stay at the University of Parma (PhD cotutelle agreement) funded by the Campus of International Excellence of the Sea (CEIMAR) and the University of Granada.

Conflicts of Interest: The authors declare no conflict of interest. The funders had no role in the design of the study; in the collection, analyses, or interpretation of data; in the writing of the manuscript, or in the decision to publish the results.

References

1. Koutandos, E.; Prinos, P.; Gironella, X. Floating breakwaters under regular and irregular wave forcing: reflection and transmission characteristics. *J. Hydraul. Res.* **2005**, *43*, 174–188. [CrossRef]
2. Liu, P.L.F.; Al-Banaa, K. Solitary wave runup and force on a vertical barrier. *J. Fluid Mech.* **2004**, *505*, 225–233. [CrossRef]
3. Dean, W.R. On the reflexion of surface waves by a submerged plane barrier. *Math. Proc. Camb. Philos. Soc.* **2008**, *41*, 231–238. [CrossRef]

4. Ursell, F.; Dean, W.R. The effect of a fixed vertical barrier on surface waves in deep water. *Math. Proc. Camb. Philos. Soc.* **1947**, *43*, 374–382. [CrossRef]
5. Losada, I.J.; Losada, M.A.; Roldán, A.J. Propagation of oblique incident waves past rigid vertical thin barriers. *Appl. Ocean Res.* **1992**, *14*, 191–199. [CrossRef]
6. Losada, M.A.; Losada, I.J.; Roldán, A.J. Propagation of oblique incident modulated waves past rigid, vertical thin barriers. *Appl. Ocean Res.* **1993**, *15*, 305–310. [CrossRef]
7. Losada, I.J.; Losada, M.A.; Losada, R. Wave spectrum scattering by vertical thin barriers. *Appl. Ocean Res.* **1994**, *16*, 123–128. [CrossRef]
8. Jalón, M.; Lira-Loarca, A.; Baquerizo, A.; Losada, M. An analytical model for oblique wave interaction with a partially reflective harbor structure. *Coast. Eng.* **2019**, *143*, 38–49, doi:10.1016/j.coastaleng.2018.10.015. [CrossRef]
9. Bennett, G.S.; McIver, P.; Smallman, J.V. A mathematical model of a slotted wavescreen breakwater. *Coast. Eng.* **1992**, *18*, 231–249. [CrossRef]
10. Isaacson, M.; Premasiri, S.; Yang, G. Wave interactions with vertical slotted barrier. *J. Waterw. Port Coast. Ocean Eng.* **1998**, *124*, 118–126. [CrossRef]
11. Kriebel, D.; Sollitt, C.; Gerken, W. Wave forces on a vertical wave barrier. In Proceedings of the 26th International Conference on Coastal Engineering, Copenhagen, Denmark, 22–26 June 1998; pp. 2069–2081.
12. Karimirad, M. *Offshore Energy Structures*; Springer International Publishing: Cham, Switzerland, 2014; p. 298.
13. Lykke Andersen, T.; Clavero, M.; Frigaard, P.; Losada, M.; Puyol, J. A new active absorption system and its performance to linear and non-linear waves. *Coast. Eng.* **2016**, *114*, 47–60. doi:10.1016/j.coastaleng.2016.04.010. [CrossRef]
14. Clavero, M.; Aguilera, L.; Nieto, S.; Longo, S.; Losada, M.A. Experimental study of the interaction atmosphere-ocean. In Proceedings of the XII Coast and Ports Spanish Conference, Cartagena, Spain, 7–8 May 2013.
15. Nieto, S.; Lira-Loarca, A.; Clavero, M.; Losada, M.A. Canal de Interacción Atmósfera-Océano (CIAO). In Proceedings of the XIII Coast and Ports Spanish Conference, Avilés (Asturias), Spain, 24–25 June 2015.
16. Addona, F. Swell and Wind-Waves Interaction under Partial Reflection Conditions. Ph.D. Thesis, University of Parma, Parma, Italy and University of Granada, Granada, Spain, 2018.
17. Longo, S.; Chiapponi, L.; Liang, D. Analytical study of the water surface fluctuations induced by grid-stirred turbulence. *Appl. Math. Model.* **2013**, *37*, 7206–7222. [CrossRef]

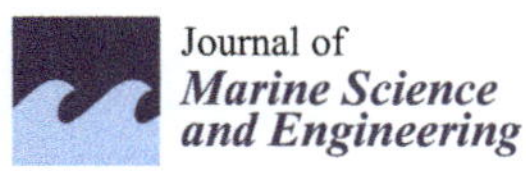
Journal of
Marine Science and Engineering

Article

Development of an Outdoor Wave Basin to Conduct Long-Term Model Tests with Real Vegetation for Green Coastal Infrastructures

Jochen Michalzik *, Sven Liebisch and Torsten Schlurmann

Leibniz Universität Hannover, Ludwig-Franzius-Institute for Hydraulic, Estuarine and Coastal Engineering, 30167 Hannover, Germany; liebisch@lufi.uni-hannover.de (S.L.); schlurmann@lufi.uni-hannover.de (T.S.)

* Correspondence: michalzik@lufi.uni-hannover.de; Tel.: +49-511-762-14625

Received: 12 November 2018; Accepted: 10 January 2019; Published: 18 January 2019

Abstract: The demand for physical model tests with real vegetation is increasing due to the current trend to elucidate the performance and durability of green coastal infrastructures to ensure and promote ecosystem services. To address this demand, a new outdoor wave basin (OWB) was built in August 2017 at the Ludwig-Franzius-Institute in Hannover, Germany. This paper reviews the general characteristics and the ongoing development of the new OWB. First insights into the long-term development of the ecosystem services of different grass revetments are discussed in terms of their ecological value and safety standards of sea dikes. Focus is placed on the resistance and ecological value of different grass mixtures that are typically applied on sea dikes situated along the North Sea. Further research concepts are briefly described to highlight how experiments in the new OWB may contribute to the current understanding and design recommendations of green coastal infrastructures. The operation of the OWB enables the performance of long-term experiments over seasonal growth stages of coastal vegetation using either fresh or seawater with wave load stresses and varying sea water levels. The first conducted experiments with different grass revetment combinations mimic typical storm surge conditions with a constant wave load (with a duration of up to 10 hours every second week) on a natural dike.

Keywords: outdoor wave basin; long-term development; vegetation development; ecosystem services; nature-based

---

1. Introduction

Traditionally coastal protection measures have been designed and optimised in terms of safety levels and costs, with minor consideration for the environment. Practical guidelines for coastal structures (e.g., the Shore Protection Manual [1], the Coastal Engineering Manual [2] and the Overtopping Manual [3]) are widely applied and conventionally rely on so-called hard solutions (e.g., dikes, seawalls or breakwaters). As the importance of sustainable development is increasingly recognised, greater emphasis is placed on designing environmentally friendly coastal structures. As such, soft solutions (e.g., dunes, seagrass meadows, saltmarshes and mangrove forests) are increasingly considered. The combination of coastal habitat (e.g., salt marshes) with hard solutions—known as hybrid infrastructure—is another way to implement soft solutions [4–7]. The potentials and performances of soft solutions are strikingly outlined [8] but still ineffectively addressed in terms of reliable and practical design guidelines [4,9] that consider ecological and engineering aspects in a mutual approach.

To incorporate coastal safety and other ecosystem services of green infrastructure in practical design guidelines, a deeper understanding of the interaction between soft solutions and hydraulic boundary conditions is essential.

In the past, investigations were conducted to estimate the dissipation of hydrodynamic energy by vegetation for various conditions and functions [10]. For example, the investigation of the failure of grass cover layers at seaward and shoreward dike slopes by Piontkowitz [11] or through the Sheldebak test in 1994 [12]. However, most of the studies used either artificial vegetation or focused on certain life stages of the particular vegetation by conducting only short-term physical model tests [13–16]. As pointed out by the ecosystem engineering approach to the management (EAM) of infrastructure by Hastings et al. [17] and the Building with Nature (BwN) approach by De Vriend [18] and proven from field and laboratory studies by Narayan et al. [8], one of the major knowledge gaps is the long-term development of nature-based sustainable solutions addressing performance and design of green coastal infrastructures. Physical model tests under controlled boundary conditions are required to study the ability of vegetation to adapt when exposed to hydraulic stresses. Recently, Silinski et al. [19] showed that the wave dissipation due to specific vegetation properties is affected by wave exposure. In this case, the ecosystem services of tidal marsh vegetation vary significantly between locations with different hydraulic boundary conditions. These findings should be considered in the design of ecosystem-based coastal protection measures with real vegetation. Thus, increasing demand for the highly interdisciplinary physical model tests with real vegetation and guidelines for the conduction of these experiments arises. Lara et al. [20] reviewed these issues and pointed out the complexity by defining a guideline for experiments with real vegetation in laboratories. Concerning the performance of long-term physical model tests, it is of utmost importance to mimic the natural conditions over seasonal cycles and vegetation growth phases. As such field studies are considered the most suitable option to deliver the most reliable results. However, results of field studies are also subjected to uncertainties due to transient boundary conditions, lack of control and limitations of monitoring techniques [21–23]. Controlled boundary conditions are essential to develop profound knowledge with a detailed understanding of the interactions between vegetation and biotic and abiotic factors. Thus, an outdoor wave basin is the most consistent and cost effective method for the development of green coastal infrastructures with ecosystem services of regional vegetation [20].

1.1. Impact of Waves on Coastal Ecosystems

Numerous field studies show that the development of coastal ecosystem and its services are affected by wave load [24–29]. Firstly, the rough environmental conditions in the wave impact zone result in a decreasing ecological niche leading to a limitation of the species richness in the coastal ecosystems [24]. Secondly, wave loads influence the coastal ecosystem characteristics like shoot density, expansion and survival rates of the shoots [27]. Bos et al. [27] showed a decrease in the survival rate of eelgrass as exerted stresses, due to wave loading, increased. Additionally, the wave loads change the zonation of vegetation within the coastal ecosystem [28,30]. The vertical zonation can occur during different life stages of the vegetation and result in a species-specific occupation of different water-depth zones inside the coastal ecosystem. Finally, the direct and indirect effects of wave load influence the individual mechanical characteristics of the vegetation [26]. Coops and Van der Velde [22] investigated the effect of wave load on the stem height of two seagrass species. Investigating four different helophytes, the direct and indirect effects of wave load were identified by Coops et al. [31]. The study provided evidence for the relation between increased biomass productions as wave exposure decreases. The findings of Blanchette [32] are in agreement with the findings of Coops et al. [31] showing the same relation for intertidal plants and their adaption to wave load. A decrease in stem height of a salt marsh leads to reduced wave attenuation [16] and a decrease of the effective ecosystem services of the coastal ecosystem [19,32]. Taking into account that wave loads exceed the boundaries of the coastal ecosystem during storm surges the review of the aquatic and littoral vegetation has to be extended for terrestrial vegetation. Therefore, the direct effects of wave loads are interpreted as mainly mechanical induced stress due to the frequent drag forces in combination with the excessive bending of the stems during wave run-up. The indirect effects of wave loads are interpreted as a loss of available plant nutrients in the topsoil. The findings of 80 studies are reviewed by Biddington [33], confirming the expected analogies between aquatic, littoral and terrestrial vegetation in their capability for adapting to mechanical stress. The review by Biddington [33] also highlighted various adaptations of plant growth, development and physiology by showing a wide range of mechanical induced stress with numerous adaptation strategies of the vegetation. Thus, the impact of wave loads on the coastal ecosystem can be characterized as dynamic and complex interactions affecting the ecosystem services of the environment.

1.2. Methodological Approach

Construction and design processes of coastal infrastructure along the European coastlines adhere reliably protect and safeguard the coastal hinterland from wave attack and storm surges. As highlighted by Schoonees et al. [4], traditional coastal structure provide limited ecosystem services in any proper design or maintenance approach. According to the ecosystem engineering approach, a high potential for increasing the ecosystem services of coastal infrastructures while preserving or possibly even enhancing the existing safety standards can be found in coastal infrastructure with vegetation [17]. To improve the design and maintenance of those compound coastal infrastructures, a profound understanding of the complex long-term performance and durability of vegetation development under altered metocean and hydrodynamic stresses, for example, waves, storm surges or sea level changes, are inevitable [34,35]. Thus, a new outdoor wave basin (OWB) has been developed and installed in the Ludwig-Franzius-Institute, Leibniz University Hannover, Germany in order to formulate practical design guidelines, considering ecological and engineering aspects in a mutual approach. As a pilot research study in the new OWB a typical sea dike in prototype scale has been erected and tested under realistic, long-term wave loading for innovative monitoring approaches within the Ecodike project (see Section 3).

The aim of the pilot project is to enhance the ecosystem services of dikes and revetments while preserving or possibly enhancing the existing safety standards. A controlled investigation of the effect of wave load on the vegetation development and resistance for sea dikes is required in order to reliably determine wave run-up reduction and mitigation of overtopping as grass revetment. Consequently, a concrete basin was built in August 2017 enabling multiple and faster model constructions due to the use of heavy machines inside the basin.

2. Outdoor Wave Basin

2.1. Dimensions and Construction

The OWB has been constructed adjacent to the large wave flume (GWK) of the Coastal Research Centre (Forschungszentrum Küste, FZK) and the 3D-wave-current-basin (WSBM) in the Northern city limit of Hannover, Germany. Due to a direct connection to the neighboured shipping channel an additional natural water reservoir is available besides the groundwater connection via a drain sump. The entrance of the OWB is designed for easy access by large trucks and to enable direct unloading of bulk material inside the basin. Furthermore, it is possible to use wheel loaders inside the OWB with up to 40 t for a fast model construction. The OWB is 24.10 m long and 14.2 m wide, containing a 12.55 m long and 14.2 wide deep section (see blue rectangle in Figure 1B). The maximum water levels are 0.8 m at the wave maker and 1 m in the deep section. The surrounding walls have a height of 1 m at the wave maker position and 2 m at the deep section which enables water levels up to 2 m for certain model setups. The OWB is also designed for tests with seawater. Those experiments would be conducted using a flexible PVC water tank as a closed system with water treatment next to the OWB (see Section 2.4). For test campaigns which require remote sensing, a 10 m high observation tower is available next to the OWB (see Figure 1A). The empty basin is shown together with the general wave maker position marked with a green rectangle in Figure 1B.

Figure 1. New outdoor wave basin during construction with observation tower (**A**) and empty basin after finishing the concrete works (**B**) in August 2017.

The deep section on the right-hand side in Figure 2 is 0.2 m lower than the marked wave maker position. This allows space for toe protection for model dikes (see Section 3) or the root development for model setups with imbedded vegetation.

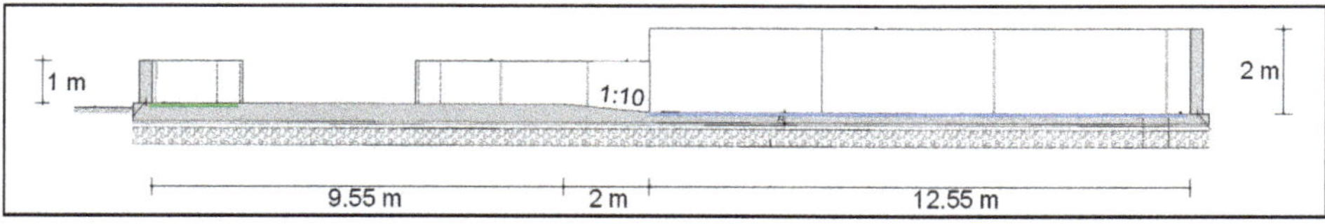

Figure 2. Profile of the outdoor wave basin with wave maker position (green) and deep section (blue).

2.2. Wave Maker

The DHI piston type wave maker system consists of three independent wave maker elements with an adjustable wave paddle width of 3 to 5 m. Thus, waves can either be generated over the total width or if necessary only in a selected section of the basin. The piston type wave maker elements are each driven by a hydraulic cylinder and are capable to generate a maximum stroke of 0.60 m. The wave maker motion and acceleration are controlled with a PC with Ethernet connection to the wave maker. The signals for the generation of irregular waves are computed from empirical energy density spectra (e.g., JONSWAP). The maximum water level at the wave maker (0.6 m) allows the generation of wave heights up to H_s = 0.25 m with periods of up to T_p = 3 s (see Figure 3A). An active wave absorption technique up to second order is about to be integrated in order to enable enduring modelling environments, thereby avoiding abnormal sea state conditions in the basin. Figure 3B shows the wave maker elements with a width of 7 m, which are fixed by their own weight and additional concrete blocks.

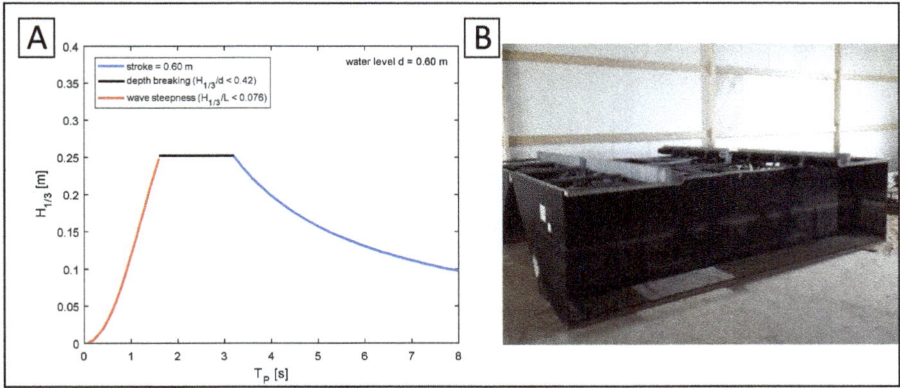

Figure 3. (**A**) Wave parameters for OWB (theoretical values, 90% practical reachable) and (**B**) Two DHI piston type wave paddle elements with a maximum width of 7 m and height of 1 m.

2.3. Water Transfer System

Filling the basin to the maximum water level of 0.6 m (i.e., 0.8 m in the deep section) takes between 2- and 4 hours, depending on the type of model constructed inside the basin. To obtain the maximum water level, 240 m^3 of water are required. The OWB is filled either through freshwater from the neighbouring inland waterway (MLK—Mittellandkanal) or by groundwater from an adjacent drain sump. To avoid disturbance due to critical biological and chemical composition or water temperatures, a multiparameter water quality probe is deployed for monitoring during the experiments. The OWB is designed to also accommodate experiments with seawater. Therefore, a highly durable flexible PVC water tank will be made available in the coming months to store seawater next to the OWB. Experiments with seawater have to be conducted in a closed system to prevent corrosion of the general water transfer system at the existing infrastructure. Since the capacity of the flexible PVC water tank is limited, a water treatment can be installed to ensure the water quality during but also between the experiments.

2.4. Measuring Equipment and Data Acquisition

A multitude of measuring systems for the investigation of hydraulic and coastal engineering parameter is available in the OWB. Since the outdoor facility operates under real climate conditions, robust measuring devices and sensor equipment are installed permanently. In addition, a meteorological station, pressure sensors, a multiparameter water quality probe and velocimeters are deployed on a permanent basis. Additionally, 3D laser scans, ultrasonic sensors or hyperspectral measurements are conducted in episodic campaigns during experiments. A control room for data acquisition and wave generation are available in front of the OWB.

3. Pilot Project—Ecodike

The main objective of the pilot research project, named *Ecodike*, is to investigate how the ecological value of dikes and revetments as coastal protection structures can be enhanced. In collaboration with A. Graunke and N. Wrage-Mönnig (Grassland and Fodder Sciences, Faculty of Agriculture, University of Rostock) a test vegetation was developed for this project. An inventory of existing and new data of the vegetation types and their traits such as root depth, growth height or trampling resistance was carried out for the main natural ecosystems on the German coast. Suitable dike vegetation was identified on plant traits and properties that enhance erosion resistance. Multivariate statistics were used to classify and rank the main abiotic conditions for certain species or traits. In conclusion, species that are potential candidates for covering revetments and dikes were identified based on to their natural environment and were defined as different test vegetation.

In total six different test vegetation were designed, ranked by increasing ecological values (e.g., species richness and biodiversity) from test vegetation 1 to 6. Test vegetation 1 is a traditional grass mixture for dikes which contain only four grass species [36]. These number of grass species are substituted and increased systematically: test vegetation 2 contain 4 different species, while test vegetation 6 consist of 18 different herb and grass species. The ecosystem services offered by vegetation 6 are not limited to coastal protection, but also offer an increased quality of coastal habitats which promote a balanced ecosystem.

In this pilot project four of these six test vegetations were selected based on their suitability as grass cover for sea dikes. Physical model tests are conducted to evaluate the practical safety standards of these test vegetations and its application as revetments for sea dikes. An innovative monitoring approach, using full-waveform laser scans and hyperspectral measurements, is used to monitor the long-term vegetation development following seasonal variations in the growth phases under wave load. To identify the effects of wave load on the long-term vegetation development of the test vegetation, the seaside slope of a typical sea dike, containing a sand core and a 0.3 m clay layer, was constructed in the OWB. The model dike has a height of 1.5 m, a width of 14.2 m and a slope of 1:6 (see Figure 4A). According to the guidelines for experiments with real vegetation defined by Lara et al. 2016, original clay material for dike construction was transferred from the coast of the North Sea to Hannover to create natural conditions for the test vegetation. The model dike is divided into two main sections in order to investigate the vegetation development with and without stresses induced by wave loads. The two main sections are further divided into four subsections, each containing one of the four selected test vegetations. Within the last vegetation period in the overall dry year of 2018, grass revetments have been episodically exposed to wave load with wave heights ranging between 0.10 and 0.25 m and wave periods of 1 to 3 s for up to 7 to 10 h every second week. The tested grass mixtures consist of different types of unique grasses and herbs; and as such resemble a rather flexible, elastic vegetation coverage with only little structural stiffness on a 1:6 slope. Based on visual observations and preliminary measurements, no distinct differences in wave run-up and reflection between the four vegetation segments could be identified.

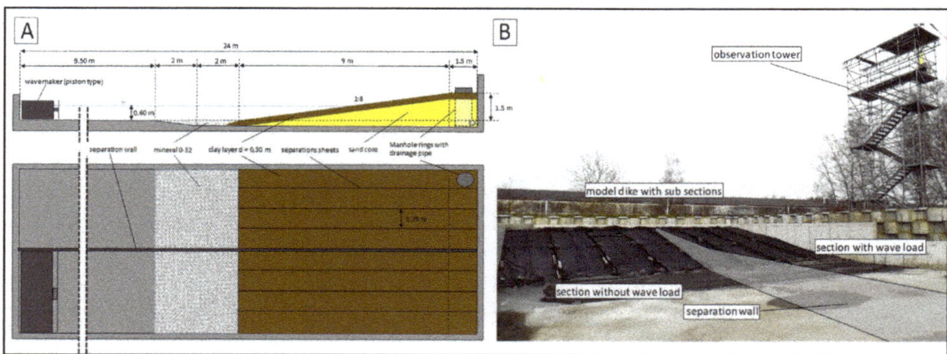

Figure 4. Model setup for the Ecodike project (**A**) and constructed model dike with two sections containing four subsections covered by protection-sheet during the winter months in February 2018 (**B**).

Duration of tests are divided into 10 min segments with 5 min breaks to avoid the development of any lateral patterns inside the basin. Additionally, the test vegetation is exposed to salt stress by artificially watering the dike with seawater to mimic environmental conditions on the North Sea. Furthermore, short test sets with regular waves are conducted to estimate wave run-up heights and maximum flow velocities on the dike for different stages in the anticipated lifetime of the target vegetation. After the vegetation period, the vegetation development (e.g., root depth and density, coverage index, species composition) and the resistance of the dike is quantified and analysed with shearing stress and pull-out tests. The combination of full-waveform laser scans and hyperspectral measurements, as well as the dike resistance parameter based on the measured vegetation development with and without wave load, lead to new insights in the long-term vegetation development. The specific adaptation processes of the vegetation to wave load will be identified to predict the associated long-term dike resistance. For example, an adaptation by increasing the subsurface biomass can lead to a decrease in the aerial biomass and positively influence their mechanical characteristics, thus possibly resulting in an increased dike resistance [19]. Thus, an innovative monitoring approach is developed based on the long-term vegetation development on the model dike. Figure 4B shows the prepared model dike covered by a sheet to protect the slope against erosion and settlement of unwanted vegetation during the winter months.

3.1. Preliminary Results—Erosion Resistance after One Vegetation Period

Preliminary insights during the initial phase after one vegetation period (i.e., March to September) indicate distinct differences between the test vegetations performance on the sea dike depending on the specific type of grass revetment of the test vegetation. Since the long-term test programme is not completed yet, only test vegetation 1 and 6 (i.e., with the lowest and the highest ecological value, respectively) are compared. Erosion only occurs at the dike sections of test vegetation 6 with high ecological values. Erosion depths have been measured at the wave impact zone about 0.2 times H_s below the still water level using a surface profilometer. The accumulated mean erosion depth after four wave load scenarios with a total duration of 20 hours and significant wave heights H_s = 0.25 m, are hardly significant for test vegetation 1. In contrast, test vegetation 6 experienced erosion of 3 cm on average. It should be noted that the test vegetation 6 has lower coverage indices due to a significantly lower seed per area ratio of 1 g/m^2 during the initial sowing process. Therefore, this test vegetation is more likely to show erosion compared to test vegetation 1 which has low species richness but significantly higher seed per area ratio of 18 g/m^2. A green value method was developed to differentiate between clay covered and uncovered by vegetation using RGB-images. Coverage indices were subsequently analysed at the wave impact zone. The mean coverage indices of the test vegetation 1 are 95% and for test vegetation 6 55%. Yet, it is expected that the coverage indices of test vegetation 6 are likely to increase in the upcoming seasonal vegetation growth phase with significant less erosion potential for any of the upcoming 20 h of wave load scenarios.

3.2. Outlook

After the test programme, the development of test vegetation with and without wave load is compared to better understand the limiting effects of wave load on the vegetation development and its ecosystem services (e.g., achievable coverage indices). Further tests of the different aspects of the achieved grass revetment resistance with and without wave load (e.g., shearing strength, pull-out tests and root depth and density) will help to understand the long-term development of the ecosystem service of the grass revetment. These measurements are conducted after completing the test programme to avoid disturbing the vegetation development. The initial damages created by conducting the grass revetment resistance measurements are likely to initiate erosion development [37,38].

4. Conclusions

The construction and the ongoing development of the new OWB enables new and innovative ways to investigate green coastal infrastructures with ecosystem services. The long-term experiments are extending the current knowledge on the long-term development of the ecosystem services while leading towards more ecologically friendly and sustainable coastal protection measures. The pilot research project provides first insights into the long-term development of the ecosystem services of different grass revetments (see Section 3). By focusing on the resistance and the ecological value of these grass mixtures, the compatibility of a high ecological value and safety standard of sea dikes are investigated. Future research considering the complex interactions between ecological and engineering aspects of the coastal infrastructure will help to develop reliable and practical design guidelines. The performance and design of green coastal infrastructures will be addressed in future projects by enhancing the state of the art for green coastal infrastructures to improve the understanding of the long-term development of nature-based, sustainable solutions. The upcoming projects can be used to enhance the knowledge of a multitude of issues resulting from the ecosystem-based coastal solutions. Long-term physical model tests under controlled boundary conditions will be used to identify differences in adaptation of vegetations to hydraulic stresses.

Author Contributions: For this research article the specifying of the individual contributions of the authors is as follows: conceptualization, J.M., S.L. and T.S.; methodology, J.M.; investigation, J.M.; resources, T.S.; writing—original draft preparation, J.M. and S.L.; writing—review and editing, J.M., S.L. and T.S.; visualization, J.M.; supervision, S.L. and T.S.; project administration, S.L. and T.S.; funding acquisition, T.S.

Funding: The EcoDike-project is funded by the German Federal Ministry of Education and Research (BMBF) (Fkz. 03F0757F). The support is highly appreciated.

Conflicts of Interest: The authors declare no conflict of interest. The funders had no role in the design of the study; in the collection, analyses or interpretation of data; in the writing of the manuscript or in the decision to publish the results.

References

1. CERC. *Shore Protection Manual (SPM)*, 4th ed.; 2. U. S. Army Engineer Waterways Experiment Station, Ed.; U. S. Government Printing Office: Washington, DC, USA, 1984.
2. USACE. *Coastal Engineering Manual*, 1110th–2nd ed.; U.S. Army Corps of Engineers, Ed.; USACE: Washington, DC, USA, 2002.
3. EurOtop. *Manual on Wave Overtopping of Sea Defences and Related Structures. An Overtopping Manual Largely Based on European Research, but for Worldwide Application*, 2nd ed. 2016. Available online: https://www.researchgate.net/publication/313501579_EurOtop_Manual_on_wave_overtopping_of_sea_defences_and_related_sturctures__An_overtopping_manual_largely_based_on_European_research_but_for_worlwide_application_2nd_edition (accessed on 18 January 2019).
4. Schoonees, T.; Gijón Mancheno, A.; Scheres, B.; Bouma, T.J.; Silva, R.; Schlurmann, T.; Schüttrumpf, H. *Hard Structures for Coastal Protection, towards Greener Designs*; Springer Nature Switzerland AG: Basel, Switzerland, 2019; under review.
5. David, G.; Schulz, N.; Schlurmann, T. Assessing the Application Potential of Selected Ecosystem-Based, Low-Regret Coastal Protection Measures. In *Ecosystem-Based Disaster Risk Reduction and Adaptation in Practice*; Springer International: Berlin/Heidelberg, Germany, 2016; pp. 457–482.
6. Van der Nat, A.; Vellinga, P.; Leemans, R.; van Slobbe, E. Ranking coastal flood protection designs from engineered to nature-based. *Ecol. Eng.* **2016**, *87*, 80–90. [CrossRef]
7. Silva, R.; Lithgow, D.; Esteves, L.S.; Martínez, M.L.; Moreno-Casasola, P.; Martell, R.; Pereira, P.; Mendoza, E.; Campos-Cascaredo, A.; Winckler Grez, P.; et al. Coastal risk mitigation by green infrastructure in Latin America. *Proc. Inst. Civ. Eng. Marit. Eng.* **2017**, *170*, 39–54. [CrossRef]
8. Narayan, S.; Beck, M.W.; Reguero, B.G.; Losada, I.J.; Van Wesenbeeck, B.; Pontee, N.; Sanchirico, J.N.; Ingram, J.C.; Lange, G.M.; Burks-Copes, K.A. The effectiveness, costs and coastal protection benefits of natural and nature-based defences. *PLoS ONE* **2016**, *11*, 1–17. [CrossRef] [PubMed]
9. Capobianco, M.; Stive, M.J.F. Soft Intervention Technology as a Tool for Integrated Coastal Zone Management Soft intervention technology as a tool for integrated zone management. *J. Coast. Conserv.* **2000**, *6*, 33–40. [CrossRef]
10. Järvelä, J. Flow resistance of flexible and stiff vegetation: A flume study with natural plants. *J. Hydrol.* **2002**, *269*, 44–54. [CrossRef]
11. Piontkowitz, T. Failure of Grass Cover Layers at Seaward and Shoreward Dike Slopes. 2009. Available online: http://resolver.tudelft.nl/uuid:7c5a02bb-1358-4448-ac71-41bb5e06f4af (accessed on 25 September 2018).

12. TAW. Technical Report Erosion Resistance of Grassland as Dike Covering Technical Report. 1997, pp. 1–49. Available online: http://resolver.tudelft.nl/uuid:446b0289-dad6-4c87-b8cd-bb81128a5770 (accessed on 10 September 2018).
13. Silinski, A.; Heuner, M.; Schoelynck, J.; Puijalon, S.; Schröder, U.; Fuchs, E.; Troch, P.; Bouma, T.J.; Meire, P.; Temmerman, S. Effects of wind waves versus ship waves on tidal marsh plants: A flume study on different life stages of scirpus maritimus. *PLoS ONE* **2015**, *10*, e0118687. [CrossRef] [PubMed]
14. Silinski, A.; Heuner, M.; Troch, P.; Puijalon, S.; Bouma, T.J.; Schoelynck, J.; Schröder, U.; Fuchs, E.; Meire, P.; Temmerman, S. Effects of contrasting wave conditions on scour and drag on pioneer tidal marsh plants. *Geomorphology* **2016**, *255*, 49–62. [CrossRef]
15. Strusínska-Correia, A.; Husrin, S.; Oumeraci, H. Tsunami damping by mangrove forest: A laboratory study using parameterized trees. *Nat. Hazards Earth Syst. Sci.* **2013**, *13*, 483–503. [CrossRef]
16. Möller, I.; Kudella, M.; Rupprecht, F.; Spencer, T.; Paul, M.; van Wesenbeeck, B.K.; Wolters, G.; Jensen, K.; Bouma, T.J.; Miranda-Lange, M.; et al. Wave attenuation over coastal salt marshes under storm surge conditions. *Nat. Geosci.* **2014**, *7*, 727–731. [CrossRef]
17. Hastings, A.; Byers, J.E.; Crooks, J.A.; Cuddington, K.; Jones, C.G.; Lambrinos, J.G.; Talley, T.S.; Wilson, W.G. Ecosystem engineering in space and time. *Ecol. Lett.* **2007**, *10*, 153–164. [CrossRef]
18. de Vriend, H.; van Koningsveld, M.; Aarninkhof, S. 'Building with nature': The new Dutch approach to coastal and river works. *Proc. Inst. Civ. Eng. Civ. Eng.* **2014**, *167*, 18–24. [CrossRef]
19. Silinski, A.; Schoutens, K.; Puijalon, S.; Schoelynck, J.; Luyckx, D.; Troch, P.; Meire, P.; Temmerman, S. Coping with waves: Plasticity in tidal marsh plants as self-adapting coastal ecosystem engineers. *Limnol. Oceanogr.* **2018**, *63*, 799–815. [CrossRef]
20. Lara, J.L.; Maza, M.; Ondiviela, B.; Trinogga, J.; Losada, I.J.; Bouma, T.J.; Gordejuela, N. Large-scale 3-D experiments of wave and current interaction with real vegetation. Part 1: Guidelines for physical modeling. *Coast. Eng.* **2016**, *107*, 70–83. [CrossRef]
21. Yang, S.L.; Li, H.; Ysebaert, T.; Bouma, T.J.; Zhang, W.X.; Wang, Y.Y.; Li, P.; Li, M.; Ding, P.X. Spatial and temporal variations in sediment grain size in tidal wetlands, Yangtze Delta: On the role of physical and biotic controls. *Estuar. Coast. Shelf Sci.* **2008**, *77*, 657–671. [CrossRef]
22. Coops, H.; Van der Velde, G. Effects of waves on helophyte stands: Mechanical characteristics of stems of Phragmites australis and Scirpus lacustris. *Aquat. Bot.* **1996**, *53*, 175–185. [CrossRef]
23. Paul, M.; Amos, C.L. Spatial and seasonal variation in wave attenuation over Zostera noltii. *J. Geophys. Res. Ocean.* **2011**, *116*, 1–16. [CrossRef]
24. Nishihara, G.N.; Terada, R. Species richness of marine macrophytes is correlated to a wave exposure gradient. *Phycol. Res.* **2010**, *58*, 280–292. [CrossRef]
25. Puijalon, S.; Léna, J.-P.; Rivière, N.; Champagne, J.-Y.; Rostan, J.-C.; Bornette, G. Phenotypic plasticity in response to mechanical stress: Hydrodynamic performance and fitness of four aquatic plant species. *New Phytol.* **2008**, *177*, 907–917. [CrossRef]
26. Coops, H.; Van Den Brink, F.W.B.; Van Der Velde, G. Growth and morphological responses of four helophyte species in an experimental water-depth gradient. *Aquat. Bot.* **1996**, *54*, 11–24. [CrossRef]
27. Bos, A.R.; Van Katwijk, M.M. Planting density, hydrodynamic exposure and mussel beds affect survival of transplanted intertidal eelgrass. *Mar. Ecol. Prog. Ser.* **2007**, *336*, 121–129. [CrossRef]
28. Coops, H.; Van der Velde, G. Impact of hydrodynamic changes on the zonation of helophytes. *The Netherlands J. Aquat. Ecol.* **1996**, *30*, 165–173. [CrossRef]
29. Heuner, M.; Silinski, A.; Schoelynck, J.; Bouma, T.J.; Puijalon, S.; Troch, P.; Fuchs, E.; Schröder, B.; Schröder, U.; Meire, P.; et al. Ecosystem engineering by plants on wave-exposed intertidal flats is governed by relationships between effect and response traits. *PLoS ONE* **2015**, *10*, e0138086. [CrossRef] [PubMed]
30. Sundermeier, A.; Schröder, U.; Wolters, B. Zum Einfluss des Wellenschlags auf Röhricht an der Unteren Havel-Wasserstraße. In Proceedings of the Bundesanstalt für Gewässerkunde, Veranstaltungen 2/2007; Bundesanstalt für Gewässerkunde: Koblenz, Germany, 2007; pp. 65–70. Available online: https://www.bafg.de/DE/08_Ref/U3/07_Monit_beweis/02_Roehrichtentw/R%C3%B6hrichtetw.pdf?__blob=publicationFile (accessed on 18 January 2019).
31. Coops, H.; Boeters, R.; Smit, H. Direct and indirect effects of wave attack on helophytes. *Aquat. Bot.* **1991**, *41*, 333–352. [CrossRef]

32. Blanchette, C.A. Size and Survival of Intertidal Plants in Response to Wave Action. *Ecology* **1997**, *78*, 1563–1578. [CrossRef]
33. Biddington, N.L. The effects of mechanically-induced stress in plants—A review. *Plant Growth Regul.* **1986**, *4*, 103–123. [CrossRef]
34. Szmeja, J.; Galka, A. Phenotypic responses to water flow and wave exposure in aquatic plants. *Acta Soc. Bot. Pol.* **2008**, *77*, 59–65.
35. Eisenmann, J. *Weidenspreitlagen an Binnenwasserstrassen—Untersuchungen zur Geotechnischen Standsicherheit*; Universität für Bodenkultur Wien: Wien, Austria, 2015.
36. EAK. *Empfehlungen für die Ausführung von Küstenschutzwerken*; Ausschuss für Küstenschutzwerke, Ed.; Korrigiert; Kuratorium für Küstenforschung im Küsteningenieurwesen (KFKI): Hamburg, Germany, 2002; Volume 65, ISBN 9783804210561.
37. Klein Breteler, M.; Bottema, M.; Mourik, G.C.; Capel, A. Resilience of dikes after initial damage by wave attack. *Coast. Eng.* **2012**, *2012*, 1–14. [CrossRef]
38. Van Steeg, P.; Labrujere, A.; Mom, R. Transition structures in grass covered slopes of primary flood defences tested with the wave impact generator. In Proceedings of the 36th IAHR World Congress 2015, Hague, The Netherlands, 28 June–3 July 2015; pp. 1–12.

Journal of
Marine Science and Engineering

MDPI

Communication

Non-Hydrostatic Modeling of Waves Generated by Landslides with Different Mobility

Ryan P. Mulligan *, W. Andy Take and Gemma K. Bullard

Department of Civil Engineering, Queen's University, Kingston, ON K7L 3N6, Canada
* Correspondence: ryan.mulligan@queensu.ca; Tel.: +1-613-533-6503

Received: 18 June 2019; Accepted: 8 August 2019; Published: 10 August 2019

Abstract: Tsunamis are generated when landslides transfer momentum to water, and these waves are major hazards in the mountainous coastal areas of lakes, reservoir, and fjords. In this study, the influence of slide mobility on wave generation is investigated using new: (i) experimental observations; (ii) theoretical relationships; and (iii) non-hydrostatic numerical predictions of the water surface and flow velocity evolution. This is accomplished by comparing landslides with low and high mobility and computing the momentum flux from landslides to water based on data collected in laboratory experiments. These slides have different materials, different impact velocities, different submarine runout distances, and generate very different waves. The waves evolve differently along the length of the waves' flume, and the experimental results are in close agreement with high-resolution phase-resolving simulations. In this short communication, we describe new research on landslide generated waves conducted at Queen's University, Canada, and presented at Coastlab18 in Santander, Spain.

Keywords: landslide waves; tsunamis; laboratory experiments; momentum balance; numerical wave modeling

---

1. Introduction

Landslide mobility (LM) describes the velocity and distal reach of the debris from a slide. This is a complex phenomenon of interacting processes that depends on many landslide parameters including the density, porosity, internal shear strength, basal friction, particle collisions, pore pressures, and possible fluidization. Despite this complexity, a slide with a higher LM would have a higher velocity and a longer runout distance at the base of a slope compared with a slide with the same volume but a lower LM. These highly mobile slides that travel long distances are, therefore, more hazardous (e.g., [1]). When a landslide impacts a body of water, momentum is transferred from the sliding mass to the water, generating a wave or series of waves [2]. Unlike the case of tsunamis generation by earthquakes where vertical and horizontal motions of the seafloor cause water displacement [3,4], subaerial landslides generate waves primarily from the streamwise momentum flux or the along-channel component of mass flow into water [5,6]. Other experimental studies (e.g., [7,8]) have found relationships between the dependence of the wave energy and maximum wave amplitude on the slide impact velocity and the hill slope angle.

Highly mobile landslides have the potential to be a more severe hazard due to the high speed and longer runout distances that can be affected [9]. The analytical work of [2,10] attempt to capture the proportion of the landslide contributing to the wave using the time and length scales of impact. These scale parameters describe how quickly the submerged mass decelerates after impacting the reservoir, which is a measure of landslide mobility. The influence of changing landslide mobility was indirectly investigated by [11] using positively buoyant particles and a range in landslide porosity. The study found that increasing the landslide porosity results in a decrease in the maximum near-field amplitude due to a larger deceleration upon impact. This potentially lends support to the hypothesis that increasing landslide mobility will result in larger wave amplitudes, however, the influence of landslide mobility has yet to be quantified directly.

In the present study, we investigate the influence of landslide mobility on wave generation by postulating that a highly mobile slide carries higher momentum when it is transferred to water and therefore, generates a larger wave. This is accomplished by comparing the waves generated by cases for very different slide materials that include granular material (low LM) and water (high LM). These are selected to be representative of a dry soil or rock landslide (using the granular particles) a mudflow or saturated debris flow (using water). We perform laboratory experiments (Section 2), calculate the theoretical momentum balance (Section 3) and conduct non-hydrostatic numerical simulations of surface elevation and flow velocity evolution (Section 4) to investigate the differences between the landslides with different mobility and the waves they generate. The goal of this paper is to highlight the importance of landslide mobility on the near field wave properties and wave evolution into the far field by a discussion of recent advances in understanding landslide-generated tsunamis from large scale experiments, consideration of the momentum transfer, and a numerical phase-resolving wave propagation model.

2. Laboratory Experiments

Experiments were performed in a landslide flume consisting of an 8.23 m long slope inclined at 30° to the horizontal to gravitationally accelerate landslides into a 33.0 m long and 2.1 m wide horizontal wave flume described in detail by [12], shown in Figure 1. Tests with different mobility were selected as end-member examples in the present study. However, a considerable number of experiments have been conducted by [13], and some of these tests are described by [14]. Material is released from a source volume box at the top of the slope, accelerates down the landslide slope, impacts the water with thickness s and velocity v_s where it generates waves, propagates along the flume and runs up the slope at the end of the flume. In the present study, a triangular source volume V of 3 mm diameter ceramic beads was used as the more collisional landslide material with lower mobility, and a similar volume of water was used as the highly mobile material. The mean water depth h was very similar for both tests, differing by 0.01 m. The landslide and near field wave properties were observed using a system of high-speed cameras, and these slides generate very different waves, as illustrated by the images of the near-field waves in Figure 2. The observational methods and accuracy of data obtained in the experiments are described in detail by [13,14]. The water surface was also measured using nine capacitive probes (P1–P9) along the flume that sample at 100 Hz, and the maximum wave amplitudes were 0.09 m and 0.29 m at P1 for the low LM and high LM tests, respectively. The fluid velocity was measured 0.012 m above the flume bottom using an acoustic velocimeter (Nortek Vectrino Profiler) operating at 100 Hz at station P2.

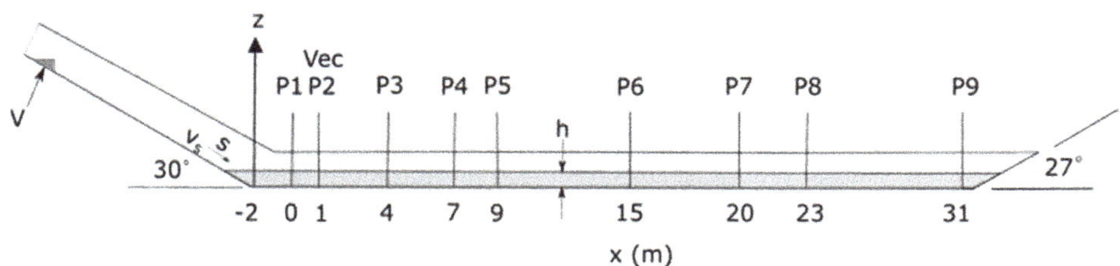

Figure 1. Landslide flume experimental apparatus, indicating the locations of wave probes (P1–P9) and acoustic velocity sensor (Vec). The numerical modeling domain begins at $x = 0$, with boundary conditions from observations at P1.

Figure 2. Images of the waves generated by landslides with different mobility: (**a**) low mobility; (**b**) high mobility.

3. Theoretical Momentum Balance

The generation and propagation of tsunamis are highly complex. However, by considering the momentum balance under idealized conditions, theoretical relationships for the maximum wave amplitude in the near-field zone can be derived [2]. Following this approach, the landslide momentum flux ($\boldsymbol{M_s}$) can be estimated as

$$M_s = \frac{\rho_s s v_{s\cos\alpha}}{\Delta t_e} \tag{1}$$

where the sliding mass with bulk density ρ_s impacts the water on slope α at velocity v_s parallel to the slope. It has thickness s on impact with the water, and the resulting wave is generated over the effective time Δt_e that momentum is imparted from the slide to the water [12]. The fluid momentum flux ($\boldsymbol{M_f}$) is given by:

$$M_f = \frac{\rho g\left(h a_m + \frac{1}{2}{a_m}^2\right)}{L} \tag{2}$$

This depends on the maximum wave amplitude a_m above the still water level, the initially still water depth h, the fluid density ρ, gravitational acceleration g, and submarine runout length L over which the wave is generated. The idealized momentum flux between the slide (M_s) and the fluid (M_f) is expressed per unit length of the slide and per unit width of the flume in units of kg/(ms^2), and this is equivalent to force per unit area (N/m^2). In this approach, idealized theoretical equations are developed for landslides with a significant horizontal velocity component on moderate slopes, and are not valid for vertical rockfalls. The vertical fall velocity of a landslide is also important, and this is accounted for in the development of a limiting relationship based on the fluid continuity equation [2].

In the present study, we examined the balance of momentum between the landslide and the fluid at the wave probe closest to the impact region (P1) for the low LM and high LM experiments. The terms in Equations (1) and (2) were evaluated using the bulk landslide properties in Table 1 that were measured using cameras and wave probes. This table provides the experimental conditions and indicates the differences in forcing for the end-member mobility cases, and each experiment results in very different wave characteristics. The time series of water surface elevations at P1 are shown in Figure 3, with the maximum wave amplitude of 0.29 m for the high LM experiment and a significantly smaller value of 0.09 m for the low LM experiment. The runout distances were 3.0 and 1.6 m, and the effective times were 0.5 and 1.0 s for the high and low LM cases, respectively. The value of the momentum terms for these experiments and a larger dataset with varying slide volume and reservoir depth are shown in Figure 4. The momentum contained in both the landside and the fluid was greater for the high LM slides compared to the low LM landslides. Overall there was a high correlation for the high mobility slides (R = 0.89 for 12 tests) and for the low mobility slides (R = 0.92 for 4 tests), and these values are high considering that other momentum terms, such as dissipation are neglected.

Table 1. Experimental conditions for examples of landslides with different mobility.

Mobility	ρ_s (kg/m^3)	V (m^3)	v_s(m/s)	s (m)	h (m)	a_m (m)
low LM (granular)	1400	0.34	4.00	0.040	0.30	0.09
high LM (water)	1000	0.30	5.38	0.043	0.31	0.29

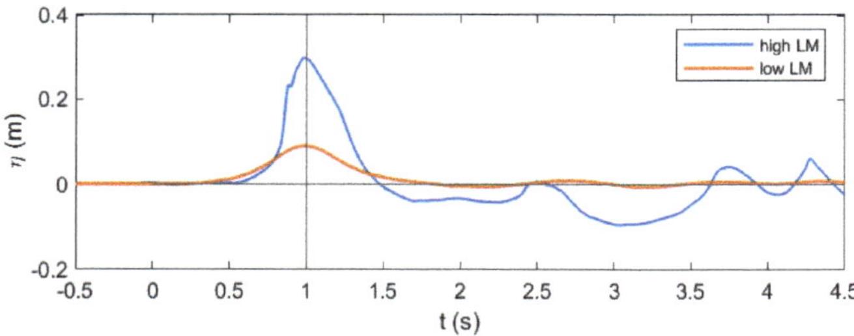

Figure 3. Example time series of water surface elevation near the impact site for landslides with low and high mobility, adjusted such that the maximum amplitude occurs at $t = 1$ s. These waves were observed at wave probe P1 and correspond to the experimental conditions listed in Table 1.

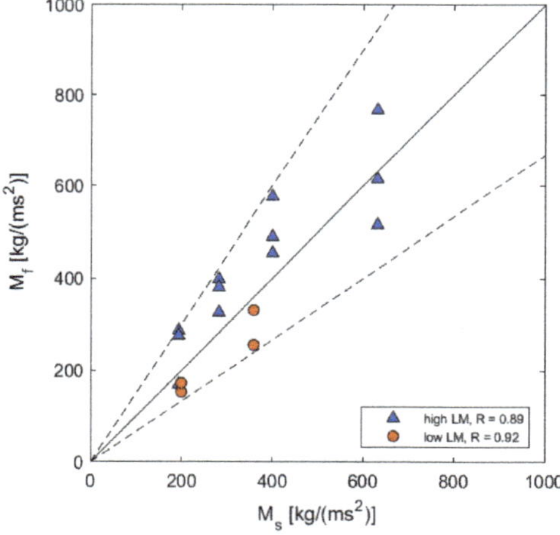

Figure 4. Balance of momentum terms for a set of experiments with waves generated by landslides with low and high mobility. The solid line represents perfect correlation, the dashed lines are ±15% deviation, and correlation coefficients for each type of slide are shown.

The results indicate that the significantly larger waves in the high LM cases were generated by higher slide impact velocities and the longer submarine runout distances. This greater length provides a larger near-field region over which the momentum is transferred to the water before the wave speed c exceeds the decelerating submarine part of the slide and the wave propagates along the flume.

4. Numerical Modelling

4.1. Model Set-up

The numerical model is SWASH (Simulating WAves till SHore), a non-hydrostatic wave-flow model that uses a finite-difference approach to solving the fluid momentum and continuity equations [15] and has been used to simulate surface waves observed in laboratory [16] and field [17] conditions. This phase-resolving wave model includes discretization of the oscillating water column into vertical layers and numerically solves for the vertical pressure gradient, important for simulating propagation and dispersive characteristics of the wave field [18,19]. Aligned with a previous study [20] for waves generated by granular landslides, the model was applied in the present study to landslide-generated waves, with new results corresponding to data obtained from new experiments. Here we applied boundary conditions developed from water surface elevation observations from the low LM (granular particles, representing a dry soil or rock landslide) and high LM (water, representing a mudflow or saturated debris flow) experiments. The measurements were made at the first wave probe near the source region, and the model was used to simulate propagation and transformation along the flume. The model grid is shown in Figure 1, with the boundary condition applied at x = 0 m. The boundary conditions were provided by the water surface elevation measurements at wave probe 1 (P1), and only η was as input, and not fluid velocity at or below the surface. Comparison between measurements and model results were made at the other wave probe locations (P1–P9). The horizontal grid spacing was 0.10 m, and the model used 20 layers in the vertical direction. Important processes in the model that control the wave transformation and evolution are horizontal mixing, vertical mixing, breaking, and bottom friction.

4.2. Model Results

The simulated water surface elevations are compared with laboratory wave probe observations in Figures 5 and 6 for the low LM and high LM cases, respectively. The results indicate that the different landslide mobility generated waves with very different initial shapes that also evolved differently as they propagate away from the landslide source region. The low LM case had a smaller initial wave (Figure 5a) that propagates away from the source with little change in amplitude or shape along the flume (Figure 5b–f). The model results indicate that for the low LM slide, only bottom friction influences the wave evolution, with equivalent results obtained when other processes were turned off (horizontal mixing, vertical mixing, breaking). In contrast, the much larger wave in the high LM case (Figure 6a) significantly changes in amplitude and shape along the flume (Figure 6b–f). Despite the general agreement, this case was more difficult to simulate, and the amplitude, timing, and shape were not in perfect agreement with the observations. Small differences in the observed and simulated water surface were evident, particularly in the high mobility case, likely due to the high degree of non-linearity and the complexity of wave breaking. In the high mobility case, more complex fluid processes were actively changing the wave (horizontal mixing, vertical mixing, wave breaking, bottom friction) and were needed to simulate the wave evolution.

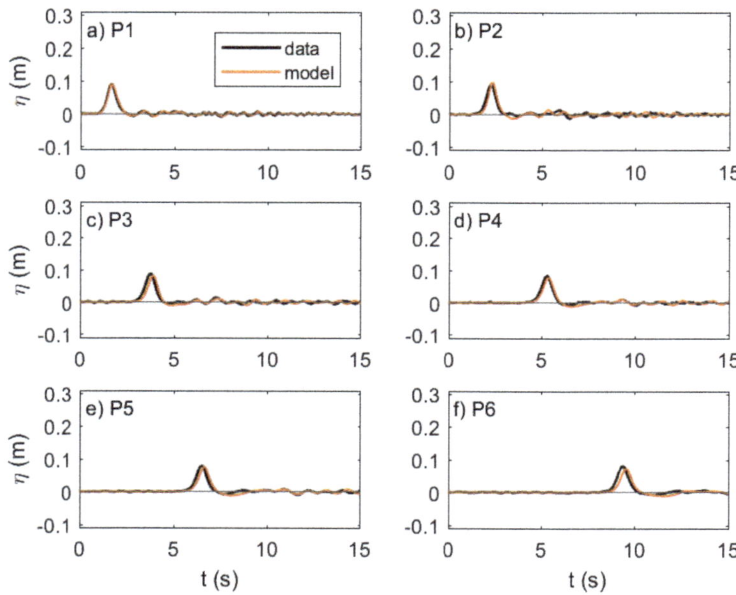

Figure 5. Water surface elevation time series for the low landside mobility case from experimental observations and numerical simulation. Time $t = 0$ s is referenced to the slide impact time. Wave probe sites shown in Figure 1 are: (**a**) P1; (**b**) P2; (**c**) P3; (**d**) P4; (**e**) P5; and (**f**) P6.

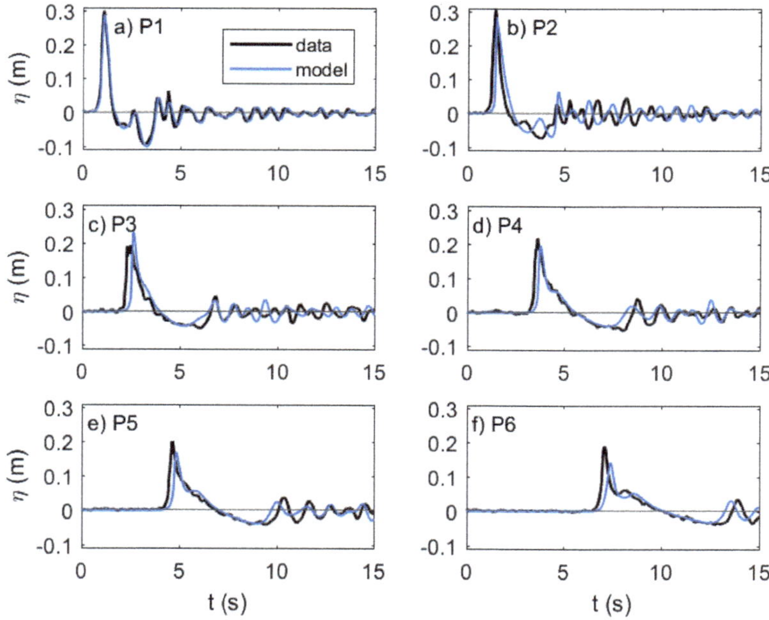

Figure 6. Water surface elevation time series for the high landside mobility case from experimental observations and numerical simulation. Time $t = 0$ s is referenced to the slide impact time. Wave probe sites shown in Figure 1 are: (**a**) P1; (**b**) P2; (**c**) P3; (**d**) P4; (**e**) P5; and (**f**) P6.

Near-bed velocity measurements were made at P2, and the raw data has been smoothed using a low-pass Butterworth filter. The smoothed observations are shown in Figure 7a with the model results for the lowest vertical layer. The data and model are in fairly close agreement, particularly with the timing of the fluid velocity for the first and largest wave in the train. The amplitude was over-predicted by the model by about 20%, but only for a brief time at the leading edge of the wave. A second peak in velocity at t = 2.8 s was due to the arrival of the submarine bubble plume flow at the sensor. Since only the water surface elevation was prescribed as the boundary condition, this was not simulated by the model. Further experiments to collect velocity measurements and continue numerical modeling is the subject of ongoing research. The results shown in Figure 7a have small oscillations following the higher velocity fluctuation of the leading wave. It is important to note that in time, small waves follow the leading wave and therefore, occur after it (e.g., t > 6 s). These waves, measured at site P2 near the landslide slope, were not influenced by reflection from the far end of the flume as it takes reflected waves 30–35 s to arrive back to this location. The simulated horizontal velocity field is shown in Figure 7b over the vertical profile, indicating the high positive along-channel flow at the wave crest followed by the longer duration and weaker current associated with the trough.

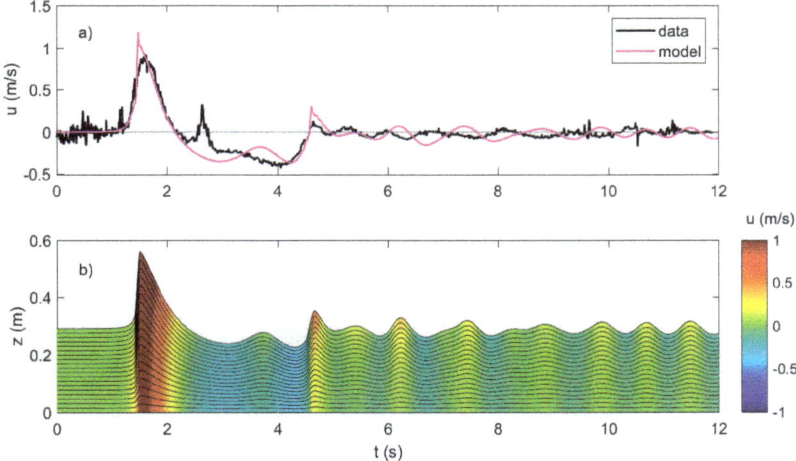

Figure 7. Along-flume fluid velocity time series at probe P2 during the wave passage for the high landside mobility case: (**a**) observed and predicted near-bottom velocity; (**b**) simulated water surface elevation and color contours of horizontal velocity, with black lines indicating the vertical grid.

5. Summary and Conclusions

This short communication provides new insight for predicting landslide waves by combining results from experimental observations, theoretical relationships, and non-hydrostatic numerical modeling. We investigated the role of landslide mobility, the combination of different materials and landslide behavior that controls the speed and impact properties, on the generation of waves by comparing two cases with different mobility. The larger wave was forced by the higher impact velocity and the longer submarine runout distance from the slide with high mobility, as greater momentum was transferred from the slide to the water over this larger length scale. The results indicated that the landslide with low mobility had a smaller initial wave that propagates away from the source with little change in amplitude or shape along the flume, and in contrast, the much larger wave generated by the high mobility landslide significantly evolved in amplitude and shape with distance from the region of impact. Overall, landslide mobility plays a major role in determining the size of the near-field wave and therefore, also influences its evolution with distance from the impact site due to breaking and fluid mixing. Using the combination of large-scale laboratory experiments and other numerical methods, the complex process of wave generation by landslides can be even more accurately quantified and better understood. Future research could focus on the near-field velocity with new experiments and numerical modeling, as this would lead to a more refined boundary condition for simulating landslide waves in coastal environments.

Author Contributions: Conceptualization, R.P.M., W.A.T., G.K.B.; methodology, R.P.M., W.A.T., G.K.B.; laboratory experiments, G.K.B.; numerical modelling, R.P.M.; writing, reviewing and editing, R.P.M., W.A.T., G.K.B.; funding, R.P.M., W.A.T.

Funding: This research was funded by the Natural Science and Engineering Research Council of Canada (NSERC) Discovery Grant programs of the first author [RGPIN/04043-2018] and the second author [RGPIN/04245-2015]. The design and construction of the large-scale landslide flume was funded by the Canada Foundation for Innovation.

Conflicts of Interest: The authors declare no conflicts of interest.

References

1. Legros, F. The mobility of long-runout landslides. *Eng. Geol.* **2002**, *63*, 301–331. [CrossRef]
2. Mulligan, R.P.; Take, W.A. On the transfer of momentum from a granular landslide to a water wave. *Coast. Eng.* **2017**, *125*, 16–22. [CrossRef]
3. Tanioka, Y.; Satake, K. Tsunami generation by horizontal displacement of ocean bottom. *Geophys. Res. Lett.* **1996**, *23*, 861–864. [CrossRef]
4. Fujiwara, T.; Kodaira, S.; No, T.; Kaiho, Y.; Takahashi, N.; Kaneda, Y. The 2011 Tohoku-Oki earthquake: Displacement reaching the trench axis. *Science* **2011**, *334*, 1240. [CrossRef] [PubMed]
5. Zweifel, A.; Hager, W.H.; Minor, H.-E. Plane impulse waves in reservoirs. *J. Waterw. Port Coast. Ocean Eng.* **2006**, *132*, 358–368. [CrossRef]
6. Heller, V.; Hager, W.H. Wave types of landslide generated impulse waves. *Ocean Eng.* **2011**, *38*, 630–640. [CrossRef]
7. Fritz, H.M.; Hager, W.H.; Minor, H.E. Near field characteristics of landslide generated impulse waves. *J. Waterw. Port Coast. Ocean Eng.* **2004**, *130*, 287–302. [CrossRef]
8. Ataie-Ashtiani, B.; Nik-Khah, A. Impulsive waves caused by subaerial landslides. *Environ. Fluid Mech.* **2008**, *8*, 263–280. [CrossRef]
9. Iverson, R.M.; George, D.L.; Allstadt, K.; Reid, M.E.; Collins, B.D.; Vallance, J.W.; Schilling, S.P.; Godt, J.W.; Cannon, C.M.; Magirl, C.S.; et al. Landslide mobility and hazards: Implications of the 2014 Oso disaster. *Earth Planet. Sci. Lett.* **2015**, *412*, 197–208. [CrossRef]
10. Zitti, G.; Ancey, C.; Postacchini, M.; Brocchini, M. Impulse waves generated by snow avalanches falling into lakes. In Proceedings of the 36th IAHR World Congress, The Hague, The Netherlands, 28 June–3 July 2015.
11. Lindstrøm, E.K. Waves generated by subaerial slides with various porosities. *Coast. Eng.* **2016**, *116*, 170–179. [CrossRef]
12. Miller, G.S.; Take, W.A.; Mulligan, R.P. Tsunamis generated by long and thin granular landslides in a large flume. *J. Geophys. Res. Oceans* **2017**, *122*, 653–668. [CrossRef]
13. Bullard, G.K. Wave Characteristics of Tsunamis Generated by Landslides of Varying Size and Mobility. Ph.D. Thesis, Queen's University, Kingston, ON, Canada, 2018; 145p.
14. Bullard, G.K.; Mulligan, R.P.; Take, W.A. An enhanced framework to quantify the shape of impulse waves using asymmetry. *J. Geophys. Res. Oceans* **2019**, 124. [CrossRef]
15. Zijlema, M.; Stelling, G.; Smit, P.B. SWASH: An operational public domain code for simulating wave fields and rapidly varied flows in coastal waters. *Coast. Eng.* **2011**, *58*, 992–1012. [CrossRef]
16. Rijnsdorp, D.P.; Smit, P.B.; Zijlema, M. Non-hydrostatic modelling of infragravity waves under laboratory conditions. *Coast. Eng.* **2014**, *85*, 30–42. [CrossRef]
17. Gomes, E.R.; Mulligan, R.P.; Brodie, K.L.; McNinch, J.E. Bathymetric control on the spatial distribution of wave breaking in the surf zone of a natural beach. *Coast. Eng.* **2016**, *116*, 180–194. [CrossRef]
18. Smit, P.; Zijlema, M.; Stelling, G. Depth-induced wave breaking in a non-hydrostatic, near-shore wave model. *Coast. Eng.* **2013**, *76*, 1–16. [CrossRef]

19. Smit, P.; Janssen, T.; Holthuijsen, L.; Smith, J. Non-hydrostatic modeling of surf zone wave dynamics. *Coast. Eng.* **2014**, *83*, 36–48. [CrossRef]
20. Mulligan, R.P.; Take, W.A.; Miller, G.S. Propagation and runup of tsunamis generated by gravitationally accelerated granular landslides. In Proceedings of the Coastlab16, Ottawa, ON, Canada, 10–13 May 2016.

Article

An Operational Wave System within the Monitoring Program of a Mediterranean Beach

Andrea Ruju *,†, Marinella Passarella, Daniele Trogu, Carla Buosi, Angelo Ibba and Sandro De Muro

Department of Chemical and Geological Sciences, University of Cagliari, Cittadella Universitaria, 09042 Monserrato (CA), Italy; marinella.passarella@unica.it (M.P.); daniele.trogu@outlook.com (D.T.); cbuosi@unica.it (C.B.); aibba@unica.it (A.I.); demuros@unica.it (S.D.M.)

* Correspondence: rujua@unica.it; Tel.: +39-070-6757778

† Current address: France Energies Marines, Bâtiment Cap Océan, Technopôle Brest Iroise, 525 Avenue Alexis de Rochon, 29280 Plouzané, France

Received: 20 December 2018; Accepted: 30 January 2019; Published: 2 February 2019

Abstract: This work assesses the performance of an operational wave system in the Mediterranean Sea by comparing computed data with measurements collected at different water depths. Nearshore data measurements were collected through a field experiment carried out at Poetto beach (Southern Sardinia, Italy) during spring 2017. In addition to coastal observations, we use intermediate and deep water wave data measured by two buoys: one situated North-West of Corsica and the other in the Gulf of Lion. The operational wave system runs once a day to predict the wave evolution up to five days in advance. We use a multi-grid approach in which a large grid extends over the entire Mediterranean basin and a fine grid covers the coastal seas surrounding the islands of Sardinia and Corsica. The comparison with measurements shows that the operational wave system is able to satisfactorily reproduce the wave evolution in deep and intermediate waters where the relative error of the significant wave height is 17%. The error exceeding 25% in coastal waters suggests that the use of a finer grid and the coupling with an atmospheric model able to catch local effects is advisable to accurately address nearshore wave processes driven by coastal wind forcing.

Keywords: operational system; wave forecast; wave modelling; Mediterranean Sea; monitoring program; beach management

---

1. Introduction

The growing importance of economic activities in coastal areas highlights the need for the development and implementation of monitoring programs as a support for management and coastal hazard assessment. In this context, remote and in situ observations allow the monitoring of deep water and coastal dynamics and provide crucial data to track their evolution [1]. However, measurements have two main downsides: first, they usually lack large spatial coverage since data are available at specific locations and in the second place they are unable to foresee evolutions beyond the time of the latest measurement. Operational wave systems based on forecast simulations are able to predict wave conditions a few days in advance and provide conditions over a wide geographical area, thus extending and complementing the information from measurements [2]. Nowadays, monitoring programs including the combination of measurements and simulations seem to represent the best choice in terms of data reliability, spatial/time coverage and costs [3]. For this reason they continue to receive an increasing attention from coastal scientists and managers, providing benefits for beach management, risk assessment and coastal planning [4–6].

In this work, we assess the performance of an operational wave system in the Mediterranean Sea by comparing computed wave data with three measured wave datasets, one of them obtained in the nearshore and the others in intermediate and deep waters. The nearshore data were collected in the Gulf of Cagliari, Southern Sardinia, where this study focuses. In recent decades, the coastal areas of the Gulf of Cagliari have been experiencing an intense urbanization with the development of the city of Cagliari and its hinterland (about 500,000 inhabitants). Residential development in addition to industrial activities have led to significant human pressure on the two beach systems, Giorgino and Poetto, included in the metropolitan area of Cagliari. A monitoring program, including periodic field surveys and video image acquisition, has been set up and implemented in the last five years at Poetto beach by the Coastal and Marine Geomorphology Group (CMGG) of the University of Cagliari [7].

As a specific task of the beach monitoring program, a field experiment was carried out in spring 2017 to collect nearshore wave and current data. We use these nearshore data, together with intermediate and deep water buoy data proceeding from the French agencies Cerema and Meteo France, to assess the reliability of an operational wave system running for the entire duration of the experiment. A spectral wave model coupled with an atmospheric model is used to simulate wave propagation over the Mediterranean Sea, with an increased spatial resolution in the coastal waters surrounding the islands of Sardinia and Corsica. This operational wave system, replicable in similar contexts, can serve as a key tool for decision makers and stakeholders in various coastal and maritime sectors. For example, the maritime transport needs accurate wave forecasts in order to secure operations at sea and prevent potential pollution. Moreover, a detailed wave information is valuable to predict vulnerability to storm impacts including flooding risk. For instance, widely-used formulations for maximum runup prediction on beaches rely upon incoming wave parameters measured or computed in coastal waters outside the surf zone [8,9]. Thus, wave operational systems appear to be of importance for port authorities, civil protection and coastal managers to anticipate and respond to extreme wave conditions that could interfere in port operations or damage beach infrastructures, potentially leading to safety risks. Another important application concerns scientific and economic aspects linked to the coastal and marine planning projects, operational assistance in coastal development, and monitoring of coastal processes such as beach erosion and sediment transport.

The ultimate goal of this work is to test the viability of incorporating an operational wave system in a beach monitoring program for a Mediterranean urban beach, including remote as well as in situ measurements. Besides, our work aims to extend the current knowledge of the capabilities of operational wave systems in forecasting wave conditions in the Mediterranean Sea. In this study, special attention is devoted to the identification of the model limitations in reproducing coastal wave conditions generated over limited Mediterranean fetches (on the order of hundreds of km) and under the local influence of varying sea breezes. Wave modelling under these environmental conditions has been recognized as particularly challenging by the literature [10,11], thus requiring further research.

The paper is structured as follows. Section 2 describes the monitoring program along with the operational wave system. Section 3 reports the system validations against measured data. Section 4 discusses the results and Section 5 draws some conclusions.

2. Study Area and Methods

2.1. Study Area

Giorgino and Poetto beaches are located in the innermost part of the Gulf of Cagliari, respectively, west and east of the metropolitan area of Cagliari. Recent economic development (with the coexistence of industrial, fishing, touristic and recreational activities) has increased the vulnerability of Giorgino and Poetto beaches to coastal risks and in particular to flooding and erosion. For this reason, these microtidal wave-dominated beaches have been extensively studied through the BEACH (Beach Environment, management And Coastal Hazard) [7,12] and NEPTUNE (Natural Erosion Prevision Through Use of Numerical Environment) [13] projects carried out since 2013 by the Coastal and Marine Geomorphology Group (CMGG) of the University of Cagliari. De Muro et al. [14] presented an integrated sea-land approach for mapping sedimentological processes of Poetto beach; whereas the geomorphological processes and anthropogenic modifications at Giorgino beach were studied by De Muro et al. [13].

Giorgino beach is an 11 km long embayment where the residential development has modified the coastal morphology and the fluvial system. The presence of industrial and port activities (such as dredging, dumping, movement of ships, anchorages, loading berths, fishing, discharge of untreated sewage) has been identified as the main cause of turbidity, pollution, toxicity and erosion [13]. Poetto beach is an urban sandy beach with a length of 8 km and a maximum width of about 100 m; it is the most populated beach in Sardinia (about 100,000 visitors per day in the summer [15]). See Figure 1 that displays a view of a moderate swell on Poetto beach from the video camera system installation. A nourishment project was carried out in 2002 on the western side of the beach area. The nourishment has significantly modified the textural, compositional and morphological features of the backshore, shoreline and shoreface [14]. Besides the mentioned beach nourishment, the intensification of anthropic pressure through an increasing number of residential users and tourists together with frequent beach cleaning operations (e.g., removal of seagrass berm) and other maintenance procedures (e.g., bulldozing) have a significant impact on the beach system affecting beach morphodynamic processes.

Figure 1. Snapshot of the Poetto beach captured by the video camera system during the moderate swell (significant wave height H_s = 1.27 m) occurred on 26 April 2017.

2.2. The Monitoring Program

The project NEPTUNE, led by the University of Cagliari, aimed at providing support for coastal management through the set-up of a monitoring program to track the main processes of the beach systems included in the metropolitan area of Cagliari. The measurements involved periodic topographic and bathymetric surveys, seabed quality assessment through bioindicators [16,17] and hydrodynamic observations. Due to the recognized importance of seagrasses in the mitigation of flood and erosion risks in microtidal Mediterranean beaches [18–20], special attention was devoted to the relationship between wave hydrodynamics and spatial distribution of the meadow of the endemic Mediterranean seagrass *Posidonia oceanica* [21,22]. Moreover, a video camera system was implemented to monitor coastal processes of Poetto beach including, for instance, beach morphodynamics and wave runup [23] and for the assessment of human impacts (especially beach cleaning and bulldozing). The video system application, operative nowadays, allows the observation of a broad range of nearshore hydrodynamic and morphological processes in support of coastal management and engineering [24,25]. Video monitoring techniques offer the chance to provide detailed coastal state information with high resolution in time and space in a relatively cheaper and safer way than deploying in situ instruments [24]. Through this system, it is also possible to provide direct support to coastal managers. In addition, video images can integrate a warning system for forecasting beach flooding, in order to prevent the damage of coastal infrastructures [26].

2.3. Field Wave Data

Field measurements were carried out over a 6-week period at Poetto beach, from 30 March to 16 May 2017. Presented within this work are data collected by an Acoustic Wave And Current (AWAC) profiler deployed in the nearshore at a water depth of 18 m. The deployment location was chosen by scuba divers in a sandy intramatte region between patches of *Posidonia oceanica* meadow. The AWAC profiler measured directional spectra and bulk wave parameters with a time interval of two hours with the purpose of synchronizing incoming wave data with video data from the video camera system located on land (see Figure 1). The spectral estimates were obtained by applying Fourier transforms of time series of 1024 s sampled at 2 Hz. Besides marine data, a meteorological station located on land in the proximity of the beach recorded wind speed and other atmospheric variables. Readers are referred to Passarella [23] to obtain more details about this experiment.

In addition to coastal observations, we present here wave data measured by two wave buoys: one located in intermediate depths and the other moored in deep waters. Since during the period of the experiment there was no availability of measurements from wave buoys located in the vicinity of Sardinian coasts, we used data from the French network. The buoy located in intermediate waters is La Revellata directional buoy (managed by Cerema) that is moored in 130 m depth in the sea North-West of the island of Corsica (42.57° N, 8.65° E). Since at this water depth the low-frequency band of the wave spectrum ($f < 0.8$ Hz) is affected by the bottom, we refer to this buoy as located in intermediate water depths. The deep water device is an omnidirectional wave buoy managed by Meteo France. It is located in the Gulf of Lion (42.06° N, 4.64° E) in approximately 2300 m depth. It collects bulk wave parameters (significant wave height and period) every hour. Moreover, the buoy is equipped with an anemometer that measures wind speed and wind direction at 3 m above the sea level. Figure 2 shows the geographic setting with the locations of the two wave buoys and the AWAC.

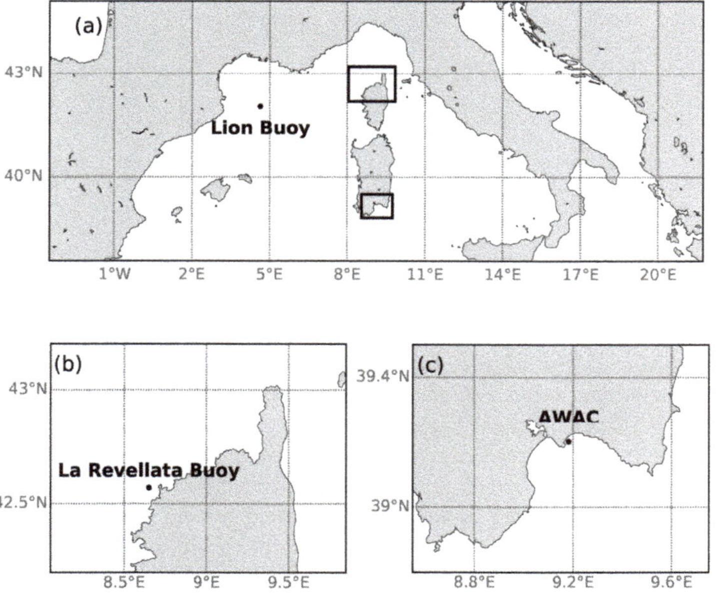

Figure 2. Geographic setting. (**a**) Western Mediterranean Sea; (**b**) Northern Corsica; (**c**) Southern Sardinia. The black dots indicate the location of the wave devices.

2.4. Operational Wave System

The operational system runs once a day to predict the wave evolution in the entire Mediterranean basin for 5 days ahead. This system includes the coupling of an atmospheric model with a wave propagation model.

The atmospheric evolution is predicted by the Global Forecast System (GFS) run by the National Center for Environmental Prediction (NCEP). The GFS is a numerical model that simulates the development of the atmosphere surrounding the entire terrestrial globe. NCEP runs the model every 6 h providing the full range of physical variables (such as temperature, pressure, wind speed, cloud cover, precipitation, etc.) that drive and characterize the atmospheric evolution up to 16 days in advance. The GFS model is set up with an horizontal spatial resolution of 0.25° (∼27 km at Mediterranean latitudes) whereas, in the vertical direction, the atmosphere is divided into 64 layers. The model output is setup with a time step of 3 h.

In the operational system, the wind field computed by the GFS model represents the forcing for the WaveWatch III (WWIII) wave model [27]. The WWIII model is run once a day and is set up to predict the wave evolution up to 5 days in advance with a time step of 3 h. WWIII is a depth-averaged and phase-averaged numerical model able to simulate the entire set of processes involved in the propagation of sea wave energy, from its generation in deep waters up to its dissipation in shallow waters. This makes WWIII computationally efficient and thus particularly suitable for wave simulations over large spatial domains such as those of ocean basins characterized by a wide range of water depths [28,29]. Moreover, the inclusion of shallow water physics packages in the recent versions of WWIII (such as version 4.18 used in this work) has extended the field of application of this model towards coastal waters. At the same time, it is worth mentioning the limit of WWIII in simulating the wave propagation pattern induced by the diffraction process around small islands [30] or artificial structures [31,32]. However, since diffraction is not expected to significantly affect wave propagation within the shoreface of Poetto beach, the use of WWIII in coastal waters appears justified in this study.

With the main purpose of simulating wave processes with different spatial scales, we adopt a multi-grid approach in which a large grid with a spatial resolution of 0.167° (∼18 km) extends over the entire Mediterranean basin, and a fine grid with a spatial resolution of 0.017° (∼1.8 km) covers the coastal seas surrounding the islands of Sardinia and Corsica. The fine grid extends over a region between the shoreline and 22 km offshore where boundary conditions are imposed accordingly to the wave field computed over the coarse grid. To avoid pronounced grid-boundary concavities located at the middle of large gulfs, the grid width extension is increased in these areas (see the panel (b) of Figure 3 that plots the grid configuration in the Gulf of Cagliari). Both coarse and fine grid simulations run over the bathymetry obtained using the ETOPO1 (Earth topography 1 arc minute) dataset [33]. The wave model is set up with 24 frequency bins and 24 directions. The frequency range covers from 0.04 to 0.74 Hz, whereas wave directions are 15° spaced. Nonlinear wave interactions in deep waters were computed with the Discrete Interaction Approximation method [34]. Triad wave interactions, accounting for nonlinear wave processes in intermediate and shallow waters, are also included. Table 1 lists the parameterizations used for the main source terms, see the WWIII manual [35] for an exhaustive explanation of the source terms.

Table 1. Source term treatment in WWIII.

	$S_{in} + S_{ds}$	S_{nl}	S_{bot}	S_{db}	S_{bs}	S_{tr}
Parameterizazion	ST4	NL1	BT4	DB1	BS1	TR1

During the period of the field experiment, wave model simulations started every morning once the first daily GFS simulation had been made available by NCEP. As initial condition, the model uses a restart file from the simulation of the previous day. The simulations are 117 h long from 00:00 of the first day until 21:00 of the fifth day. Under the current configuration, the multi-grid simulations run in parallel using 2 cores (2.6 GHz) with an average simulation rate close to 90 simulated hours per hour. The output parameters, requested at intervals of three hours, are provided across both the large scale and small scale grids and at the points corresponding to the locations of the Lion buoy, La Revellata buoy and the AWAC. We examine in this paper the following parameters: wind speed at 10 m above the mean sea level U_{10}, significant wave height H_s, peak wave period T_p, mean wave period T_m, mean wave direction θ_m, wave direction at spectral peak θ_p, wave directional spread σ and full 2D wave spectrum $E(f,\theta)$. The comparison between simulations and measurements is carried out by means of single point statistical indicators such as correlation coefficient R, bias B, normalized root mean square error $RMSE$, relative error $RelE$ and scatter index SI. They are defined as:

$$R = \frac{\sum_{i=1}^{N}(M_i - \overline{M})(O_i - \overline{O})}{\sqrt{\sum_{i=1}^{N}(M_i - \overline{M})^2 \sum_{i=1}^{N}(O_i - \overline{O})^2}}, \tag{1}$$

$$B = \frac{1}{N}\sum_{i=1}^{N}(M_i - O_i), \tag{2}$$

$$RelE = \frac{1}{N}\sum_{i=1}^{N}|\frac{M_i - O_i}{O_i}|, \tag{3}$$

$$RMSE = \sqrt{\frac{1}{N}\sum_{i=1}^{N}(M_i - O_i)^2}, \tag{4}$$

$$SI = \frac{RMSE}{\overline{O}}, \tag{5}$$

where N is the number of measurements, M is the model result, O is the observation and the overbar indicates the sample mean.

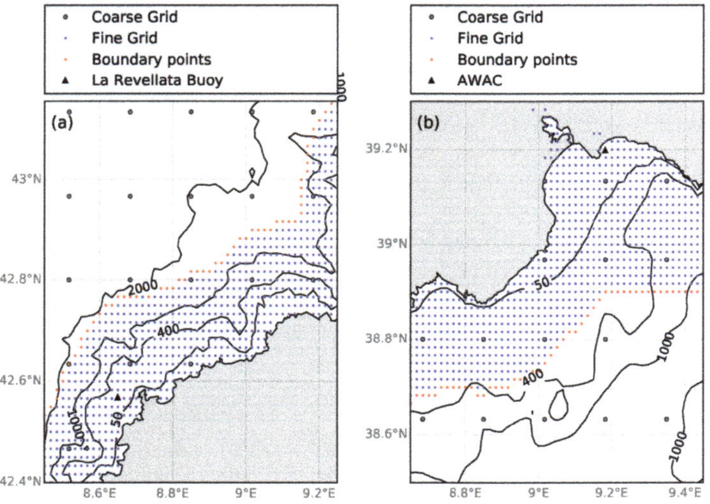

Figure 3. Details of computational grids. (**a**) Corsica; (**b**) Gulf of Cagliari. Thick lines are the coastline, thin lines represent the isobaths.

3. Results

Figures 4 and 5 show the comparison between measured and computed data of significant wave height H_s at the Lion buoy (deep water), La Revellata buoy (intermediate water) and AWAC (nearshore) at Poetto. Comparisons of wind data time series are displayed in Figure 6. Wind measurements are extrapolated to an elevation equal to that of the GFS output (10 m above the mean sea level) by assuming that a logarithmic velocity profile develops within the atmospheric boundary layer:

$$\frac{U_z}{U_{10}} = 1 + \frac{C^{0.5}}{0.4} ln(\frac{z}{10}), \tag{6}$$

$$C = 0.001(0.067U_{10} + 0.75), \tag{7}$$

where U_{10} is the wind speed at 10 m above the mean sea level, C is a friction coefficient obtained with the formulation proposed by Garratt [36]. The statistical indicators of parameters displayed in Figures 5 and 6 are reported in Table 2. The operational wave system is able to satisfactorily reproduce the evolution of the significant wave height in deep/intermediate waters where the relative error is 17% both at Lion and La Revellata buoys. In addition, the wind evolution is well addressed by the GFS with a relative error (15%) on the same order of magnitude as that of significant wave height at the Lion buoy. Positive bias values suggest that both the predicted wind speed and wave height are slightly overestimated in deep waters. On the other hand, small negative bias values are observed at La Revellata buoy. The correlation coefficients of both wind and significant wave height (0.61 and 0.84) at Poetto beach are lower than those obtained in deep waters (0.91 and 0.93). The relative error of significant wave height exceeding 25% is likely to be related to the inaccuracies in wind speed predictions (relative error of 34%). In contrast with deep waters, nearshore data at Poetto beach are characterized by a pronounced short-period variability with daily peaks in wind speed and significant wave height possibly due to sea breezes (see Section 4). Although the system reproduces the general wind pattern, some peaks in wind speed are underestimated; see for instance the days following 24 April in which the anemometer measured wind speeds on the order of 20 kts but the GFS predicts speeds lower than 15 kts. As a result, significant wave heights during the same week are underestimated (although with less discrepancies than wind) by the wave model.

Table 2. Results of statistical analysis of spectral bulk wave parameters and wind speed U_{10} at Lion buoy, La Revellata buoy (only wave data) and Poetto locations.

	$\overline{O}$	*R*	*B*	*RMSE*	*RelE*	*SI*
Lion H_s	1.34 m	0.93	0.16 m	0.35 m	0.17	0.21
Lion U_{10}	15.82 kts	0.91	0.40 kts	2.98 kts	0.15	0.16
La Revellata H_s	1.00 m	0.91	−0.12 m	0.29 m	0.17	0.23
La Revellata T_m	4.33 s	0.84	0.02 s	0.75 s	0.14	0.17
La Revellata T_p	5.87 s	0.73	−0.09 s	1.16 s	0.14	0.20
La Revellata σ	38.11°	0.24	−7.53°	15.29 °	0.30	0.40
Poetto H_s	0.35 m	0.85	0.00 m	0.14 m	0.28	0.34
Poetto T_m	2.97 s	0.77	0.35 s	0.90 s	0.22	0.30
Poetto T_p	5.16 s	0.50	−0.34 s	1.93 s	0.27	0.34
Poetto σ	56.59°	0.49	−24.98°	29.81°	0.46	0.53
Poetto U_{10}	9.28 kts	0.62	−1.77 kts	4.27 kts	0.34	0.40

Figures 7 and 8 plot the time series of measured and predicted bulk wave parameters at La Revellata buoy and in the nearshore at the AWAC location, respectively. The evolution of significant wave height, already presented in Figure 5, is kept here as a reference. La Revellata buoy (Figure 7) measured maximum values of the mean period between 6 and 7 s while the peak period reached 10 s under energetic conditions. The model is able to satisfactorily catch the time evolution of mean and peak period (*relE* of both mean and peak period is 0.14 with absolute *B* values lower than 0.1 s). The mean wave direction of propagation ranges from West-South-West to North and seems well reproduced by the model. The wave directional spread evolution is poorly caught by the model (*RelE* = 0.30) that, in addition, significantly underestimate its values ($B = -7.53°$).

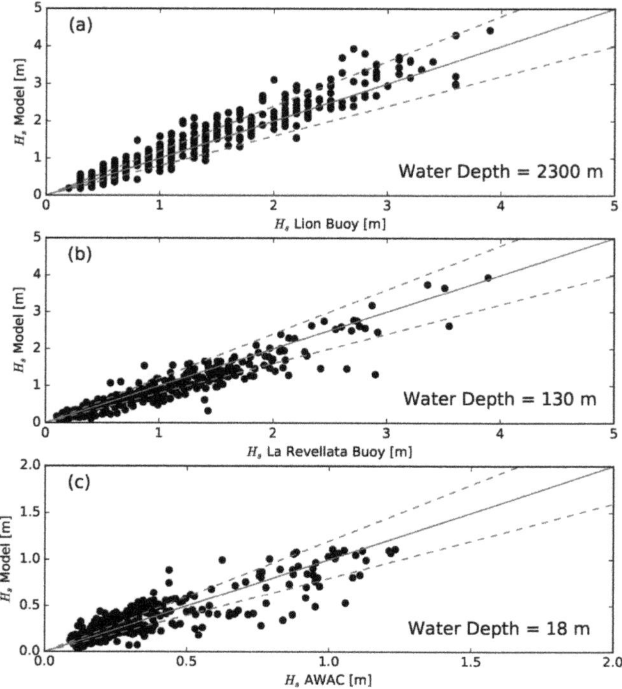

Figure 4. Scatter plots of computed versus observed H_S for the period of the experiment (from 31 March to 15 May 2017). (**a**) Lion buoy; (**b**) La Revellata; (**c**) AWAC at Poetto beach. Solid line, perfect agreement; dashed line, 20% error bound.

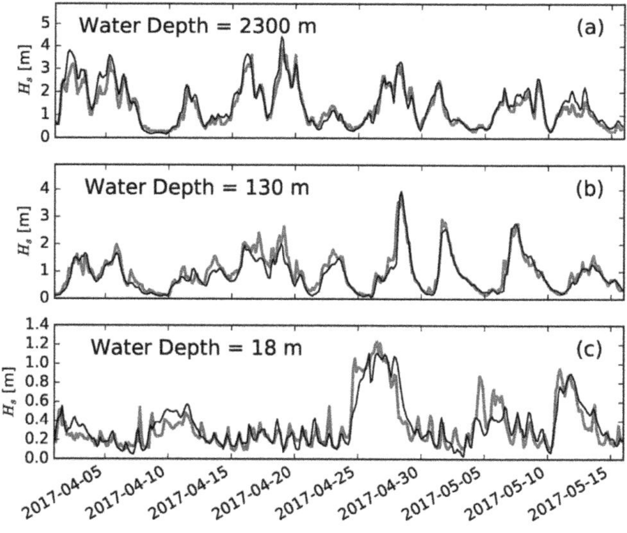

Figure 5. Comparisons between predictions and measurements of H_S time series at Lion buoy (**a**), La Revellata buoy (**b**) and AWAC at the Poetto beach (**c**). Thick gray line: observations; thin black line: predictions.

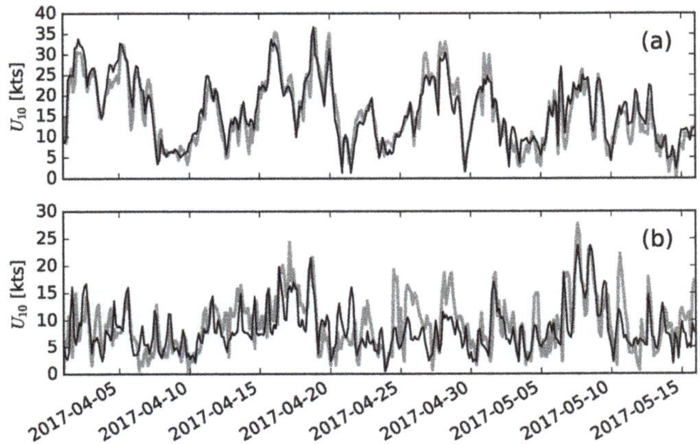

Figure 6. Comparisons between predictions and measurements of wind speed U_{10} time series at Lion buoy (**a**) and at the Poetto beach (**b**). Thick gray line: observations; thin black line: predictions.

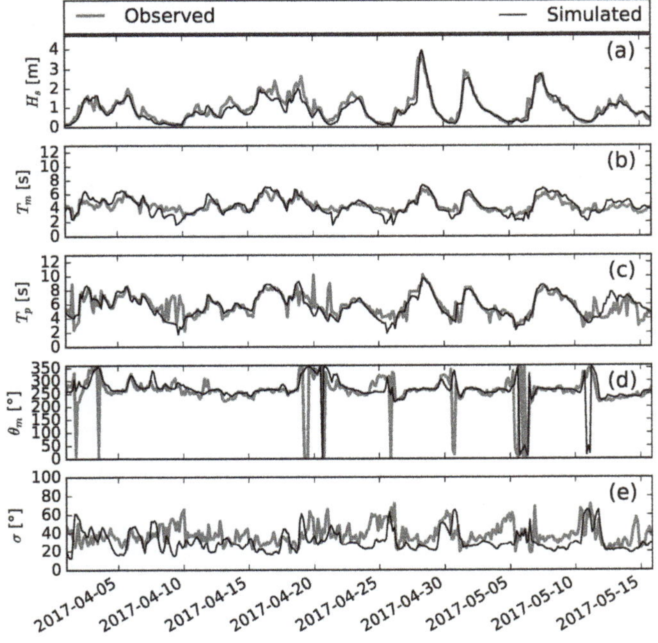

Figure 7. Comparisons between measurements and predictions of spectral bulk wave parameters at La Revellata buoy. H_s: significant wave height (**a**); T_m: mean period (**b**); T_p: peak period (**c**); θ_m: mean wave direction (**d**); σ: wave directional spread (**e**).

In coastal water at Poetto beach (Figure 8), mean periods (panel b) rise to values on the order of 5 and 6 s during relatively energetic conditions. At the same time, the peak period (panel c) reaches 7 to 8 s under energetic wave states with some abrupt variations due to shifts of the spectrum peak between sea and swell bands. While the wave model seems to correctly address the peak period maximum values ($relE$ = 0.27, bias B is -0.34 s), the mean period remains overestimated especially under energetic conditions ($relE$ = 0.22, B = 0.35 s). We will comment about this later in Section 4.

The AWAC recorded incident waves with a mean direction of propagation ranging almost exclusively between East and South-East. On the other hand, the mean wave direction predicted by the model tends to be more South and South-East with some notable inversions to offshore propagation. These inversions with waves coming from the North-West are noticeable also in the measured peak wave direction and are likely to be associated with offshore winds, weak incoming wave energies and small wave periods. Panel (f) of Figure 8 shows that the system significantly underestimates the values of the wave directional spread in nearshore waters (bias $B = -25°$). Large directional spread values on the order of 80° are recorded when peak wave direction turns to North-West (see panel e) implying the superposition of offshore and onshore propagating spectral components. In addition, the system predicts directional spread values in phase with offshore peak wave directions but with values hardly exceeding 60°. Besides the discrepancies in magnitude, the system is able to catch the main trend of directional spread variation with low values appearing when the mean wave direction is mainly South-East oriented (onshore).

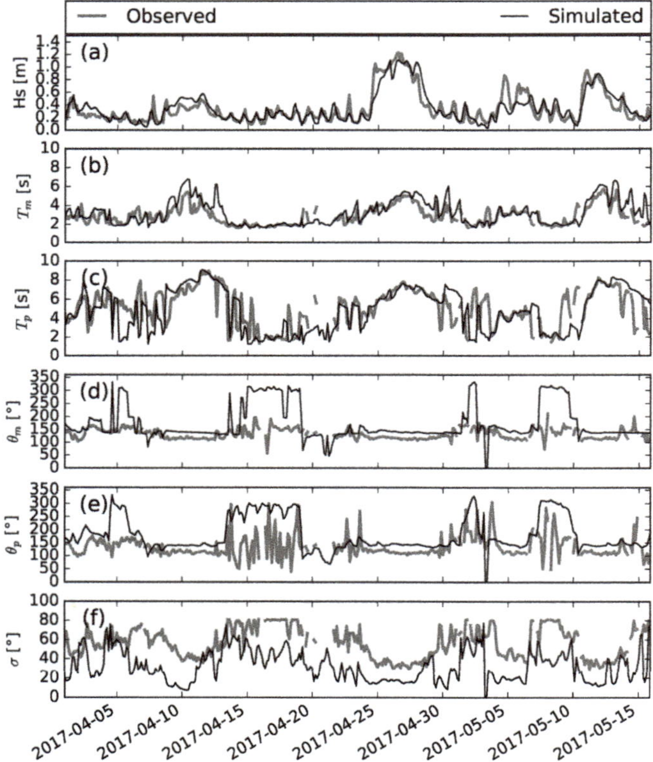

Figure 8. Comparisons between measurements and predictions of spectral bulk wave parameters at Poetto beach. H_s: significant wave height (**a**); T_m: mean period (**b**); T_p: peak period (**c**); θ_m: mean wave direction (**d**); θ_p: peak wave direction (**e**); σ: wave directional spread (**f**).

Although the operational wave system gives a prediction five days ahead, up to this point this work has made the assessment of the results for the first day of prediction only (with a temporal extension from 0 to 21 h from the beginning of the simulation). The quality of the prediction is expected to decrease as the time from the start of the simulation (lead-time) increases [6]. Figure 9 shows how the relative error at the Lion buoy rises from 17% of the first day to 27% of the fourth day of simulation (from 72 to 93 h from the start). Under the same lead-time range, the relative error passes from 17 to 28% at La Revellata buoy and from 28 to 35% at Poetto beach. This deterioration of the wave forecast accuracy with the lead-time is likely to be related to larger uncertainties in the wind forcing fields [10,37]. Despite the accuracy loss, the relative error remains under moderate values, confirming the reliability of the prediction for the time window covered by the simulation. The bias B values at the three locations keep the same sign (positive at Lion buoy and negative at la Revellata) along the days of the time window without showing a clear trend. The analysis of the prediction accuracy as a function of lead-time is not further pursued in this work.

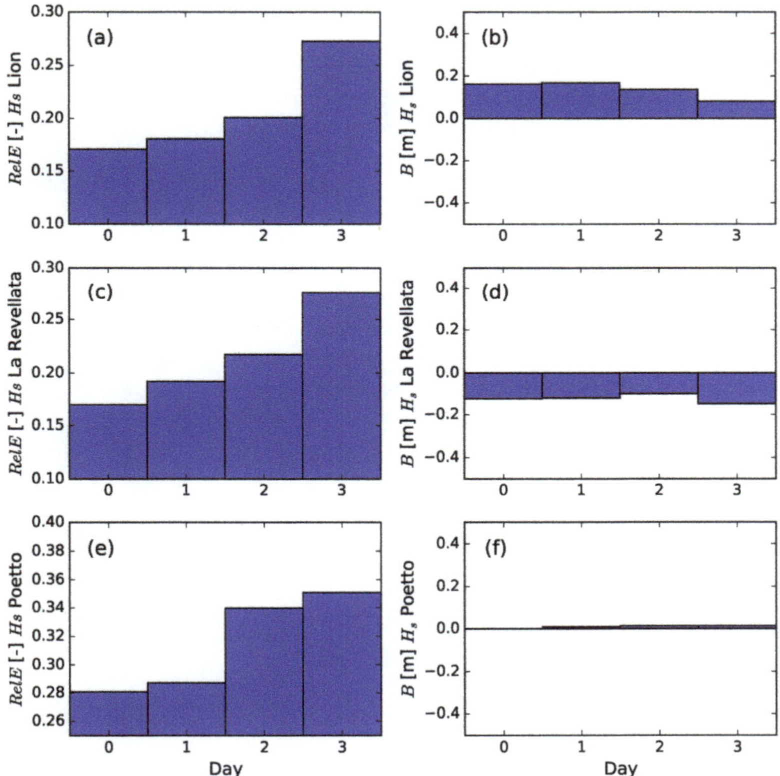

Figure 9. Performance of the operational wave system over the time window of the prediction. Relative error of significant wave height H_s versus day from the start of the simulation at Lion buoy (**a**), La Revellata buoy (**c**) and Poetto (**e**). Bias of H_s versus day from the start of the simulation at Lion buoy (**b**), La Revellata buoy (**d**) and Poetto (**f**).

4. Discussion

The results presented in Section 3 show that the operational wave system performs better in deep waters and intermediate waters than in the nearshore. This outcome is in agreement with Ravdas et al. [10] who found relatively low correlations at coastal locations. Whereas correlation coefficients for wind speed and significant wave height are 0.91 and 0.93 in deep waters at the Lion buoy site, they drop to 0.62 and 0.85 when computed data are compared with measurements in the nearshore at Poetto beach. The fact that in the nearshore the correlation coefficient of significant wave height is larger than that of the wind speed suggests that local meteorological forcing, playing an important role in wave dynamics, are not completely addressed by the GFS model. The GFS is a global meteorological model that, due to the adopted horizontal resolution (0.25°), may not be the most appropriate tool for forecasting coastal wind dynamics. Besides the fact that complex land topographies play a significant role in the evolution of coastal winds, it is worth mentioning that it is in the spring season that the difference between the sea water temperature and the temperature of the lower layers of the atmosphere is strongest. This leads to intense coastal sea breezes blowing onshore at the central part of the day, resulting in marked daily oscillations in the wind speed and consequently in the significant wave height signals (see Figures 5 and 6). These factors together make the prediction of waves in coastal waters particularly challenging, especially in the spring season.

To investigate where the mentioned discrepancies in spectral bulk wave parameters come from, Figure 10 displays the observed and computed 2D wave spectra in the nearshore at the time instant of the maximum significant wave height recorded during the period of the experiment (see panel d). 1D frequency spectra are shown in the panel (c) of the Figure. The operational system predicts wave conditions with a mean propagation direction more southern oriented with respect to measurements (142° predicted versus 115° measured). This discrepancy is likely to be related to complex coastal bottom features leading to significant refraction processes that cannot be completely addressed even with the adoption of the small scale numerical grid. In fact, the spatial resolution of the fine grid (0.017°) seems to be able to address the main features of the wave propagation pattern in the nearshore but is likely to be unable to catch the detailed wave transformation in shallow waters. The use of a finer grid, proceeding from a detailed bathymetric survey, with a spatial resolution of the order of 100 m is expected to improve the characterization of nearshore and shallow water waves. However, one should bear in mind that this different set-up would bring a drawback represented by an increase of the computational cost. Due to the large number of different strategies and configuration settings that can be adopted to increase the characterization of coastal wave processes, we avoid providing here a quantitative estimation of the required increase of the computational effort.

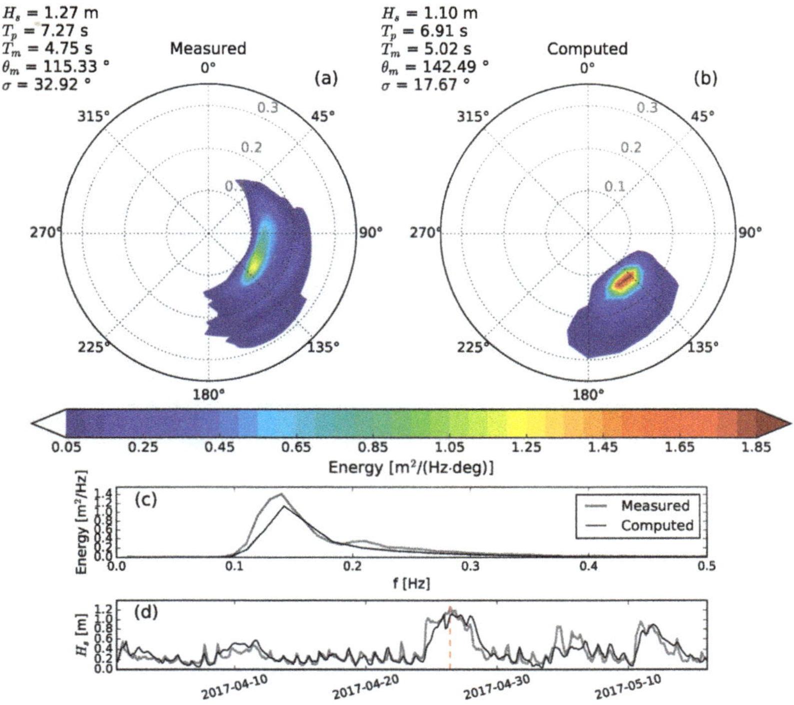

Figure 10. Measured (**a**) and predicted (**b**) two-dimensional wave spectra at the time instant of the measured maximum H_S. One-dimensional frequency spectra are plotted in panel (**c**). Panel (**d**) displays the time series of H_S. The time instant of the measured maximum H_S is highlighted by a dashed red line in panel (**d**).

From the observation of Figure 10, it is evident how wave energy is spread across a wide range of directions in the observed spectrum whereas computed energy tends to be concentrated around the main peak (simulated wave directional spread is 17.67° whereas the AWAC registered a direction spread of 32.92°). Negative biases in directional spread predictions have been reported by Crosby et al. [38] who ascribed this discrepancy to a weak computed coastal reflection. In our case, the spectral shape suggests that the directional spread underestimation is likely to be a consequence of the wind underestimation by the GFS model. In fact, Figure 6 reveals a significant underestimation of wind speed at Poetto when the maximum wave height occurs (morning of 26 April). Note also that the computed 1D spectrum lacks the secondary peak at 0.21 Hz that is clearly observable in the measurements (Figure 10, panel c). The underestimation of locally generated sea waves is likely to be responsible for the overestimation of the mean period. Since the wave results rely upon wind predictions, the use of a more accurate atmospheric model with a resolution able to address the effects induced by complex coastal topographies is expected to improve the characterization of sea wave components mainly driven by local meteorological forcing.

Relevant insight into the capability of the model to simulate nearshore wave processes can be achieved by the observation of Figure 11 that plots the time evolution of the measured and computed 1D frequency spectra at Poetto beach. Although discrepancies can be detected, the model satisfactorily reproduces the main evolution patterns of observed frequency spectra. In particular, moderate wave energy at high frequencies that appears in the period with absence of ground swells (see the period from 15 April 2017 to 23 April 2017) is likely to be generated by local wind breezes that blow over a limited fetch. In addition, the distribution of wave energy during the energetic events seems to be well addressed by the model, although the observation of panel (d) reveals that energy tends to be overpredicted at frequencies below the peak frequency. An analogous discrepancy was reported by Rogers and Vledder [39] who suggested that such overprediction is a common feature of models that implement the Discrete Interaction Approximation (DIA). Finally, the underprediction of high frequency energy is likely to be the result of the poor characterization of wind speed in coastal areas already mentioned.

The magnitude of the statistical error indicators of significant wave height found in the present study in intermediate and deep waters is on the same order of magnitude of that obtained by state-of-the-art wave systems recently developed for the Mediterranean basin [2,10]. In particular, Ravdas et al. [10] reported an averaged scatter index of 0.25 similar to the scatter index we obtained at La Revellata buoy (0.23). In that study, they compared the results of the operation wave system, based on the community Wave Model (WAM) forced by forecast winds from the European Centre for Medium-Range Weather Forecasts (ECMWF), with measured wave data proceeding from regional and national wave buoy nets deployed in water depths in the order of, or larger than, 100 m. Due to extensive measured dataset, they were able to provide assessment metrics for different Mediterranean Sea subregions. In contrast with the methodology followed by Ravdas et al. [10] who applied a computational grid with constant spatial resolution of $1/24°$ covering the whole Mediterranean basin, in this study we have nested a coastal fine grid (resolution $1/60°$) into a deep-water coarse grid (resolution $1/6°$). Our multi-grid approach is intended to improve the characterization of the coastal wave processes relevant for the beach monitoring program.

While most of the wave operational systems present in the literature are tested against data (from wave riders or satellite) retrieved in deep or intermediate water [2,6,10], we include an evaluation of our operational wave system against wave measurements collected at 18 m depth. Among the few examples of operational wave systems tested in shallow waters, we can compare our results with Ponce de Leon and Orfila [11]. Ponce de Leon and Orfila [11] used three different wind datasets and two phase-averaged spectral wave models with a grid resolution of 1500 m to simulate waves around Mallorca Island (Spain, Mediterranean Sea) under the effect of fast-changing sea breezes. Thus, their wave conditions, characterized by a combination of moderate Mediterranean swells and locally-generated wind waves, were similar to those we have observed in our experiment at Poetto beach. Testing their system with wave data at a water depth of 135 m, they obtained a scatter index ($SI = 0.35$) higher than that we have obtained at the same depth ($SI = 0.23$ at La Revellata). However, when the evaluation was extended to shallower depths (28 m depth), the performance of their system decreased ($SI = 0.4$). A decrease in precision in shallower depths is also found in our work, but the SI for the system presented in this paper remains below 0.35. Although the quality of the predictions in coastal waters is not as excellent as that in intermediate and deep waters, the multi-grid approach adopted in this study allows the specification of the main wave propagation features in the presence of complex shoreline geometries and under the effect of coastal breezes.

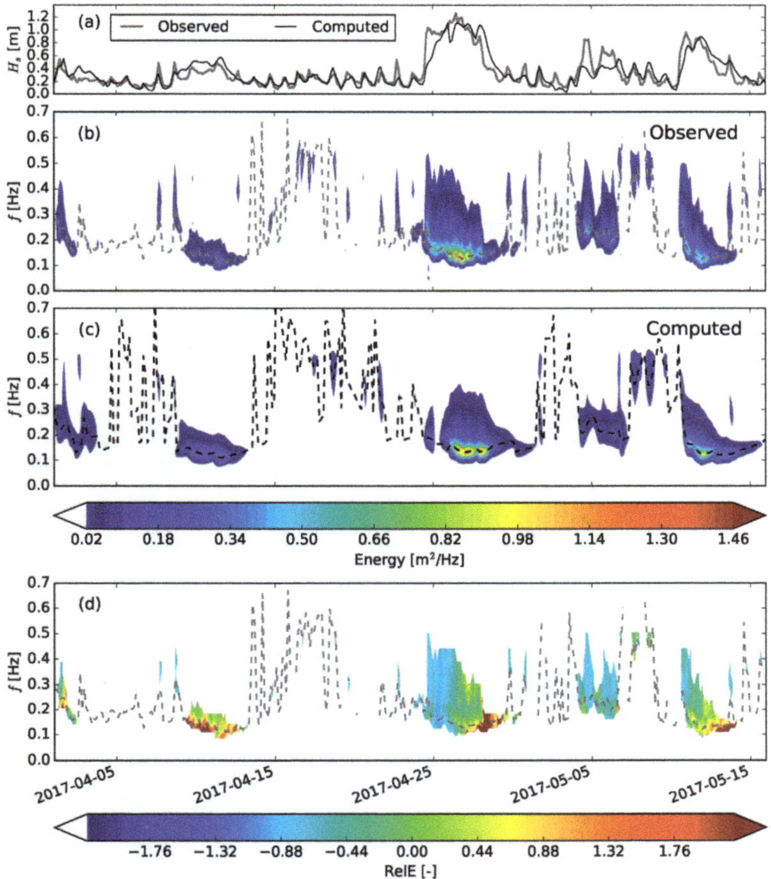

Figure 11. Temporal variation of 1D frequency spectra. (**a**) H_s time series. (**b**) Observed frequency spectra. (**c**) Computed frequency spectra. (**d**) Relative error of spectral energy density. Grey and black dashed lines are, respectively, the observed and computed peak frequency.

5. Conclusions

This work assesses the performance of an operational wave system included in a beach monitoring program for the Poetto beach (Southern Sardinia). The operational wave system aims to extend and complement in situ observations and the information collected by a video camera installation. The wind field from the atmospheric model GFS represents the forcing of the wave model WWIII. The wave model runs once a day to predict the wave evolution for five days ahead in the entire Mediterranean Sea using a multi-grid approach in which high spatial resolution is achieved in the coastal seas surrounding the islands of Sardinia and Corsica. The reliability of the operational system is tested by comparing the computed results with measurements collected at different water depths.

In general terms, the operational wave system presented in this work is able to predict the wave evolution in deep and intermediate waters with excellent accuracy (relative error of significant wave height is equal to 17%). However, the relative error of significant wave height rises to 28% in front of Poetto beach at 18 m depth. Moreover, the model underestimates the directional spread, especially in the nearshore (the bias is $-8°$ in 130 m depth and $-25°$ at 18 m depth). The analysis of spectral results suggest that the implementation of few improvements is advisable to refine the characterization of wave transformation processes in coastal and shallow waters. The main improvements can be identified as: an increased resolution of the computational grid (built from a detailed bathymetric survey) in coastal areas and the coupling with an atmospheric model able to catch local effects. These changes in the system configuration would have the associated drawback represented by an increase of the computational cost. Reducing the error in coastal waters would contribute to the reduction of the uncertainty of the operational wave system applications in coastal hazard and management.

Even with the mentioned limits, the information provided by the operational wave system can integrate measurements collected by in situ and remote systems inside a multi-platform monitoring program. By making this environmental dataset available to coastal scientists and managers, the monitoring program is of crucial importance to plan economical activities in coastal areas with applications including flood and erosion risk assessment, beach management and support to stakeholder activities. The dataset produced through the monitoring tasks is especially important in areas subject to considerable anthropological pressure, such as the beach systems of the metropolitan area of Cagliari, where the knowledge of wave forcing is a key factor to understand and predict short and long-term geomorphological processes.

Author Contributions: A.R. was responsible for the conceptualization, investigation and analysis; A.R. and C.B. wrote the original draft; M.P. and D.T. were responsible for the experimental methodology; A.I. and S.D.M. acquired funding and supervised the investigation; all authors reviewed and approved the manuscript; A.R. presented part of this work at the Coastlab 2018 conference.

Funding: This work was supported by: (1) Regione Autonoma Sardegna under L.R. 7/2007, "Promozione della ricerca scientifica e dell'innovazione tecnologica in Sardegna" for NEPTUNE and NEPTUNE 2 projects; (2) Consorzio di Bonifica Sardegna Meridionale (CBSM) under project number COM-CONV-2016DEMURO-CBSM-ARU.A.00.21. Marinella Passarella gratefully acknowledges Sardinia Regional Government for the financial support of her PhD scholarship (P.O.R. Sardegna F.S.E. Operational Programme of the Autonomous Region of Sardinia, European Social Fund 2007-2013 - Axis IV Human Resources, Objective 1.3, Line of Activity 1.3.1.). Daniele Trogu acknowledges the support by the University of Cagliari ("Borse di studio di Ateneo" - Dottorati di Ricerca XXXIV ciclo, voce CO.AN.A.06.01.01.01.01.02).

Acknowledgments: The authors would like to thank the Battellieri di Cagliari and Marco Piroddi for their assistance during the field campaign. Gilbert Emzivat of Meteo France and Alain Le Berre of CEREMA are acknowledged for providing the buoy data. The authors acknowledge the organizers of the Coastlab 2018 conference (22–26 May, Santander, Spain) where part of this research was first presented. The suggestions and comments made by the reviewers significantly improved the manuscript.

Conflicts of Interest: The authors declare no conflict of interest.

References

1. O'Reilly, W.; Olfe, C.B.; Thomas, J.; Seymour, R.; Guza, R. The California coastal wave monitoring and prediction system. *Coast. Eng.* **2016**, *116*, 118–132. [CrossRef]
2. Mentaschi, L.; Besio, G.; Cassola, F.; Mazzino, A. Performance evaluation of Wavewatch III in the Mediterranean Sea. *Ocean Model.* **2015**, *90*, 82–94. [CrossRef]
3. Sandhya, K.; Murty, P.; Deshmukh, A.N.; Nair, T.B.; Shenoi, S. An operational wave forecasting system for the east coast of India. *Estuar. Coast. Shelf Sci.* **2018**, *202*, 114–124. [CrossRef]
4. Sembiring, L.; van Ormondt, M.; van Dongeren, A.; Roelvink, D. A validation of an operational wave and surge prediction system for the Dutch coast. *Nat. Hazards Earth Syst. Sci.* **2015**, *15*, 1231–1242. [CrossRef]
5. Liberti, L.; Carillo, A.; Sannino, G. Wave energy resource assessment in the Mediterranean, the Italian perspective. *Renew. Energy* **2013**, *50*, 938–949. [CrossRef]
6. Gudes Soares, C.; Rusu, L.; Bernardino, M.; Pilar, P. An operational wave forecasting system for the Portuguese continental coastal area. *J. Oper. Oceanogr.* **2011**, *4*, 17–27. [CrossRef]
7. Brambilla, W.; van Rooijen, A.; Simeone, S.; Ibba, A.; De Muro, S. Field Observations, Video Monitoring and Numerical Modeling at Poetto Beach, Italy. *J. Coast. Res.* **2016**, *2*, 825–829. [CrossRef]
8. Atkinson, A.L.; Power, H.E.; Moura, T.; Hammond, T.; Callaghan, D.P.; Baldock, T.E. Assessment of runup predictions by empirical models on non-truncated beaches on the south-east Australian coast. *Coast. Eng.* **2017**, *119*, 15–31. [CrossRef]

9. Passarella, M.; Goldstein, E.B.; De Muro, S.; Coco, G. The use of genetic programming to develop a predictor of swash excursion on sandy beaches. *Nat. Hazards Earth Syst. Sci* **2018**, *18*, 599–611. [CrossRef]
10. Ravdas, M.; Zacharioudaki, A.; Korres, G. Implementation and validation of a new operational wave forecasting system of the Mediterranean Monitoring and Forecasting Centre in the framework of the Copernicus Marine Environment Monitoring Service. *Nat. Hazards Earth Syst. Sci.* **2018**, *18*, 2675–2695. [CrossRef]
11. Ponce de Leon, S.; Orfila, A. Numerical study of the marine breeze around Mallorca Island. *Appl. Ocean Res.* **2013**, *40*, 26–34. [CrossRef]
12. De Muro, S.; Porta, M.; Passarella, M.; Ibba, A. Geomorphology of four wave-dominated microtidal Mediterranean beach systems with Posidonia oceanica meadow: A case study of the Northern Sardinia coast. *J. Maps* **2017**, *13*, 74–85. [CrossRef]
13. De Muro, S.; Porta, M.; Pusceddu, N.; Frongia, P.; Passarella, M.; Ruju, A.; Buosi, C.; Ibba, A. Geomorphological processes of a Mediterranean urbanized beach (Sardinia, Gulf of Cagliari). *J. Maps* **2018**, *14*, 114–122. [CrossRef]
14. De Muro, S.; Ibba, A.; Simeone, S.; Buosi, C.; Brambilla, W. An integrated sea-land approach for mapping geomorphological and sedimentological features in an urban microtidal wave-dominated beach: a case study from S Sardinia, western Mediterranean. *J. Maps* **2017**, *13*, 822–835. [CrossRef]
15. Strazzera, E.; Cherchi, E.; Ferrini, S. *A* Choice Modelling Approach for Assessment of Use and Quasi-Option Values in Urban Planning for Areas of Environmental Interest; Sustainability Indicators and Environmental Valuation Working Papers 42903; Fondazione Eni Enrico Mattei (FEEM): Milano, Italy, 2008.
16. Buosi, C.; Cherchi, A.; Ibba, A.; Marras, B.; Marrucci, A.; Schintu, M. Benthic foraminiferal assemblages and sedimentological characterisation of the coastal system of the Cagliari area (southern Sardinia, Italy). *Boll. Soc. Paleontol. Ital.* **2013**, *52*, 1–9. [CrossRef]
17. Schintu, M.; Marrucci, A.; Marras, B.; Galgani, F.; Buosi, C.; Ibba, A.; Cherchi, A. Heavy metal accumulation in surface sediments at the port of Cagliari (Sardinia, western Mediterranean): Environmental assessment using sequential extractions and benthic foraminifera. *Mar. Pollut. Bull.* **2016**, *111*, 45–56. [CrossRef] [PubMed]
18. Simeone, S.; De falco, G.; Como, S.; Olita, A.; De Muro, S. Deposition dynamics of banquettes of Posidonia oceanica in beaches. *Rend. Online Soc. Geol. Ital.* **2008**, *3*, 726–727.
19. Vacchi, M.; De Falco, G.; Simeone, S.; Montefalcone, M.; Morri, C.; Ferrari, M.; Bianchi, C.N. Biogeomorphology of the Mediterranean Posidonia oceanica seagrass meadows. *Earth Surf. Process. Landf.* **2017**, *42*, 42–54. [CrossRef]
20. De Muro, S.; Pusceddu, N.; Buosi, C.; Ibba, A. Morphodynamics of a Mediterranean microtidal wave-dominated beach: Forms, processes and insights for coastal management. *J. Maps* **2017**, *13*, 26–36. [CrossRef]
21. Buosi, C.; Tecchiato, S.; Pusceddu, N.; Frongia, P.; Ibba, A.; De Muro, S. Geomorphology and sedimentology of Porto Pino, SW Sardinia, western Mediterranean. *J. Maps* **2017**, *13*, 470–485. [CrossRef]
22. Ruju, A.; Ibba, A.; Porta, M.; Buosi, C.; Passarella, M.; De Muro, S. The role of hydrodynamic forcing, sediment transport processes and bottom substratum in the shoreward development of Posidonia oceanica meadow. *Estuar. Coast. Shelf Sci.* **2018**, *212*, 63–72. [CrossRef]
23. Passarella, M. On the Prediction of Swash Excursion and the Role of Seagrass Beach-Cast Litter: Modelling and Observations. Ph.D. Thesis, University of Cagliari, Cagliari, Italy, 2019.
24. Holman, R.; Stanley, J. The history and technical capabilities of Argus. *Coast. Eng.* **2007**, *54*, 477–491. [CrossRef]
25. Kroon, A.; Davidson, M.; Aarninkhof, S.; Archetti, R.; Armaroli, C.; Gonzalez, M.; Medri, S.; Osorio, A.; Aagaard, T.; Holman, R.; Spanhoff, R. Application of remote sensing video systems to coastline management problems. *Coast. Eng.* **2007**, *54*, 493–505. [CrossRef]
26. Archetti, R.; Paci, A.; Carniel, S.; Bonaldo, D. Optimal index related to the shoreline dynamics during a storm: the case of Jesolo beach. *Nat. Hazards Earth Syst. Sci.* **2016**, *16*, 1107–1122. [CrossRef]
27. Tolman, H.L. A Third-Generation Model for Wind Waves on Slowly Varying, Unsteady, and Inhomogeneous Depths and Currents. *J. Phys. Oceanogr.* **1991**, *21*, 782–797. [CrossRef]
28. Besio, G.; Mentaschi, L.; Mazzino, A. Wave energy resource assessment in the Mediterranean Sea on the basis of a 35-year hindcast. *Energy* **2016**, *94*, 50–63. [CrossRef]
29. Perez, J.; Menendez, M.; Losada, I.J. GOW2: A global wave hindcast for coastal applications. *Coast. Eng.* **2017**, *124*, 1–11. [CrossRef]

30. Buosi, C.; Ibba, A.; Passarella, M.; Porta, M.; Ruju, A.; Trogu, D.; De Muro, S. Geomorphology, beach classification and seasonal morphodynamic transition of a Mediterranean gravel beach (Sardinia, Gulf of Cagliari). *J. Maps* **2019**. [CrossRef]
31. Ilic, S.; van der Westhuysen, A.; Roelvink, J.; Chadwick, A. Multidirectional wave transformation around detached breakwaters. *Coast. Eng.* **2007**, *54*, 775–789. [CrossRef]
32. Sartini, L.; Mentaschi, L.; Besio, G. Evaluating third generation wave spectral models performances in coastal areas. An application to Eastern Liguria. In Proceedings of the OCEANS 2015-Genova, Genoa, Italy, 18–21 May 2015; pp. 1–10.
33. Amante, C.; Eakins, B. *ETOPO1 1 Arc-Minute Global Relief Model: Procedures, Data Sources and Analysis*; Technical Report; NOAA Technical Memorandum NESDIS NGDC-24; NOAA, National Geophysical Data Center, Marine Geology and Geophysics Division: Boulder, CO, USA, 2009.
34. Hasselmann, S.; Hasselmann, K. Computations and Parameterizations of the Nonlinear Energy Transfer in a Gravity-Wave Spectrum. Part I: A New Method for Efficient Computations of the Exact Nonlinear Transfer Integral. *J. Phys. Oceanogr.* **1985**, *15*, 1369–1377. [CrossRef]
35. Tolman, H.L. *User Manual and System Documentation of WAVEWATCH III Version 4.18*; Technical Report; NOAA/NWS/NCEP: 5830 University Research Court, College Park, MD, USA, 2014.
36. Garratt, J.R. Review of Drag Coefficients over Oceans and Continents. *Mon. Weather Rev.* **1977**, *105*, 915–929. [CrossRef]
37. Bernier, N.B.; Alves, J.H.G.M.; Tolman, H.; Chawla, A.; Peel, S.; Pouliot, B.; Bélanger, J.M.; Pellerin, P.; Lépine, M.; Roch, M. Operational Wave Prediction System at Environment Canada: Going Global to Improve Regional Forecast Skill. *Weather Forecast.* **2016**, *31*, 353–370. [CrossRef]
38. Crosby, S.C.; O'Reilly, W.C.; Guza, R.T. Modeling Long-Period Swell in Southern California: Practical Boundary Conditions from Buoy Observations and Global Wave Model Predictions. *J. Atmos. Ocean. Technol.* **2016**, *33*, 1673–1690. [CrossRef]
39. Rogers, W.E.; Vledder, G.P.V. Frequency width in predictions of windsea spectra and the role of the nonlinear solver. *Ocean Model.* **2013**, *70*, 52–61. [CrossRef]

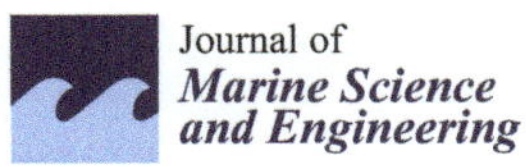
Journal of
Marine Science and Engineering

Article

Experimental Study on Toe Scouring at Sloping Walls with Gravel Foreshores

M. Salauddin * and J. M. Pearson

School of Engineering, University of Warwick, Coventry CV47AL, UK
* Correspondence: m.salauddin@warwick.ac.uk

Received: 18 May 2019; Accepted: 25 June 2019; Published: 27 June 2019

Abstract: Sea defences, such as urban seawalls can fail due to the development of a scour hole at the toe of the structure. The scour depth or the information on ground levels at the structure toe is required for the sustainable management of coastal defences, due to its influence on the structural performance. This research reports and summarises the main findings of a new laboratory study on toe scouring at a smooth sloping wall with permeable gravel foreshore. A set of small-scale laboratory experiments of wave-induced scouring at sloping seawalls were conducted. Two gravel sediments of prototype d_{50} values of 13 mm and 24 mm were used to simulate the permeable 1:20 (V:H) gravel beach configurations in the front of a smooth 1 in 2 sloping wall. Each experiment comprised of a sequence of around 1000 random waves of a JONSWAP energy spectrum with a peak enhancement factor of 3.3. The relationship of the scour depth with toe water depth, Iribarren number, and wall slope were investigated from the test results of this work and through a comparison with available datasets in the literature. The results of this study showed that the relative toe water depth and Iribarren number influence the relative toe scour depth at a sloping structure on a shingle beach. Within the experimental limitations, the maximum toe scour depths were observed for the experiments under spilling and plunging wave attack.

Keywords: physical model experiments; scouring; shingle foreshore; sloping wall

1. Introduction

Toe scouring at a coastal structure is usually defined as the development of a scour hole at the toe of the structure. This may eventually lead to the deterioration and damage of the structure through the redistribution of the sediment near the toe of the structure under the wave action over time [1,2]. Sea defences, such as urban seawalls in the United Kingdom (UK), can fail because of this toe scouring phenomenon, see [3–5]. In addition, climate change is believed as one of the most crucial challenges encountering mankind [6] in today's world. The combined influence of climate change and global sea level rise, which is expected as 4 mm/year for the 21st century [7] by the IPCC (Intergovernmental Panel on Climate Change) set a long-standing threat of coastal hazards on the sea defences in nearshore regions. The scour depth or the information on ground levels at the structure toe is required for the sustainable management of coastal defences in response to climate change, due to its influence on the structural performance [8].

Many researchers have performed laboratory and field research over the years, to improve the understanding of this phenomenon and, to provide design guidance on toe scouring at coastal structures with sandy foreshores, for example [4,9–12]. In 2008, [1] provided an overview of experimental and field studies on toe scouring at vertical breakwaters on sandy foreshore configurations. Some numerical researches have been performed in recent years to understand the scouring patterns at coastal structures in addition to physical studies, see [13–16].

Investigations on vertical walls showed that the toe scour depth and relative toe water depth is strongly correlated for sandy beach configurations, see [1,17]. For instance, for the prediction of scour depth at vertical seawalls with sandy beach, [17] and [8] established an empirical relationship (Equation (1)) between the dimensionless scour depth (S_t/H_s) and relative toe water depth (h_t/L_{0m}). For a known beach slope, these authors suggested another empirical equation to predict the toe scour depth at vertical walls with sandy beaches, see Equation (2).

$$\frac{S_{tmax}}{H_s} = 4.5e^{-8\pi(h_t/L_m+0.01)}\left(1 - e^{-6\pi(h_t/L_{0m}+0.01)}\right) \quad [-0.013 \le h_t/L_{0m} \le 0.18] \quad (1)$$

$$\frac{S_t}{H_s} = 6.8(0.207\ln(\alpha) + 1.51)e^{-5.85k_m h_t}\left(1 - e^{-3k_m h_t}\right) - 0.137 \quad [-0.04 \le h_t/L_{0m} \le 0.12] \quad (2)$$

in which S_{tmax} is the maximum toe scour depth, S_t is the toe scour depth, H_s significant wave height defined as highest one-third of wave heights = $H_{1/3}$, α is the beach slope, h_t is the toe water depth, and L_{0m} is the deep-water wave length based on mean wave period T_m.

Powell et al. [18] conducted a laboratory study with prototype sediment diameters of $5 < d_{50} < 30$ mm in a model scale of 1:17 and derived a dimensionless scour plot for the approximation of scour depth at the vertical walls on shingle beds. Based on a laboratory study on scour depths at vertical structures with a sandy slope and two gravel slopes, [19] proposed an empirical relationship between the maximum wave height at the toe of the slope, submergence of the berm, and local wave length to estimate the scour depth at this type of sea defence. Nevertheless, these empirical predictions have not been validated with other datasets.

Recently, [20] performed an extensive laboratory investigation on the toe scour depths at vertical breakwaters with two permeable shingle beaches. The authors showed that there is a strong relationship between scour depths with relative toe water depth and Iribarren number for shingle foreshores, as observed by [12] on sandy beaches. As expected, the maximum scour depths were reported for spilling and plunging waves ($0.005 \le h_t/L_{0m} \le 0.04$).

In comparison to the vertical walls, there have been surprisingly few studies devoted to the toe scouring at sloping walls with both sandy and permeable shingle beach configurations. In 2006, [12] investigated scouring at a 1 in 2 sloping wall, as well as at a plain vertical wall on a sandy beach. The authors postulated that toe scouring are independent of the slope of the structure and claimed that the scour depths at sloping walls do not differ from those reported at vertical walls. To date, little knowledge is available regarding the scouring at sloping walls with permeable shingle beach configurations.

This study investigated the toe scouring at sloping walls on shingle foreshores. In addition to scour depths, the profile of foreshore slope after a wave attack was also investigated. This research reports and summarizes the main findings of a new laboratory study on toe scouring at a sloping seawall with the permeable shingle foreshore slope. A set of small scale laboratory experiments of wave-induced scouring at sloping seawalls has been conducted in a wave flume at the University of Warwick. Two gravel sediments of prototype d_{50} values of 13 mm and 24 mm were used to simulate the permeable 1:20 (V:H) gravel beach configurations at smooth 1 in 2 sloping walls. The findings of this laboratory study provide the data and comprehensive knowledge that are required to understand the toe scouring phenomenon at a sloping structure with a gravel foreshore.

2. Laboratory Set-Up

The small-scale physical experiments were performed in the wave channel within the school of engineering at the University of Warwick. The two-dimensional (2D) wave flume is approximately 22 m long, 0.60 m wide, and has an operating depth of 0.40 m–0.70 m. The wave channel is equipped with an active absorbing-piston type wave paddle, enables simulating both uniform and random sea state within the flume. The physical experiments were carried out on a plain (prior to start the test) 1 in 20 permeable shingle beach in front of a smooth impermeable 1:2 (V:H) sloping seawall. In Figure 1, a layout of laboratory set-up that was used in this work to carry out physical experiments on scouring at 1 in 2 smooth sloping walls is presented.

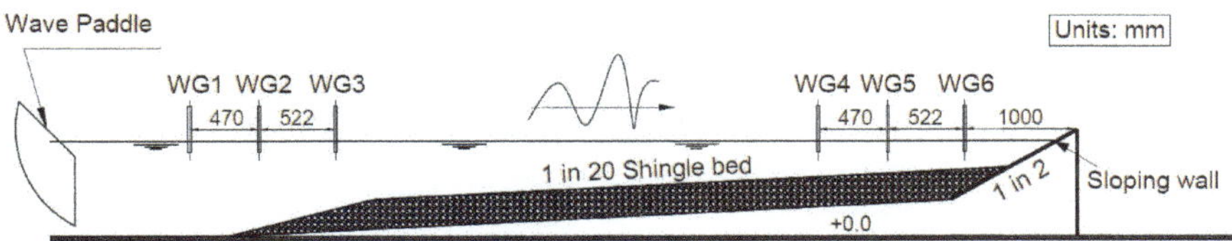

Figure 1. Laboratory set-up used for physical experiments on scouring at a 1 in 2 sloping wall.

In generally, the correct representation of beach permeability, the threshold of sediment mobility, and the relative magnitudes of onshore/offshore movement are the three main requirements to reproduce a gravel beach within the laboratory, as suggested by Powel [21]. The model sediments have only two parameters (size and specific gravity); therefore, the proper representation of all three criteria is hardly achievable within the laboratory. Powel [21] reported that the crushed anthracite with a specific gravity of 1.39 is able to satisfy most of the requirements by modelling the appropriate threshold of motion and onshore/offshore movement. As following the approach that was recommended by Powel [21], the filtered anthracite coal of a specific gravity of 1.40 (T/m^3) was used to represent the gravel beach materials. Two different sediment sizes of gravel were used in this work, scaled adapting the methodology, as suggested by [21]. The tested model bed materials d_{50} of 2.10 mm and 4.20 mm represent the prototype grain size d_{50} of 13 mm and 24 mm, respectively, at a 1 in 50 scale with a prototype specific gravity of 2.65 (T/m^3), see details in [20]. It is important to note that any reference to the size of sediment within this paper has only been referred as prototype values, unless otherwise stated.

A 1:50 length scale was applied in this study to generate random wave conditions within the wave flume. Two constant nominal wave steepnesses (s_{op} = 0.02 and 0.05) were tested to cover both wind sea state as well as swell sea conditions. A matrix of 120 test conditions (significant wave heights, wave steepnesses, crest freeboards, water depths, and gravel sizes) were covered to investigate scouring at the toe of the sloping structure, see Table 1. The test matrix that was followed within this study was developed by following the available guidelines for typical small-scale investigations, as reported in [22,23]. For example, the inshore significant wave heights that were tested in this study varied from 50 mm to 150 mm, which is comparable to every good small-scale investigation. Furthermore, each experiment comprised of a sequence of around 1000 random waves of a JONSWAP (Joint North Sea Wave Project) energy spectrum, with a peak enhancement factor γ = 3.3 to represent a typical storm of 3 h.

The incident wave characteristics (wave heights and periods) were measured by determining the free water surface elevations at the six different locations along the wave flume during the experiments. The well-established three-point technique developed by Mansurd and Funke [24] was adapted to separate the incident waves from reflected waves. One set of three wave gauges were placed in relatively deep water (close to wave generator) to determine the deep-water wave characteristics. The next set of three gauges were fixed near the toe of the sloping wall to determine the incident wave conditions at the structure. The wave gauge (WG6) near the toe of the sloping wall was set by adapting the technique of Klopman and Van der Meer [25], which helps to reduce the effect of a reflective wall on the measurements. Further, the measurements of wave conditions were also repeated without the structure in place, in order to decrease any probable uncertainty in the measurement of the incident wave characteristics induced by the reflection.

Table 1. Overview of test conditions.

Structural and Bed Configuration	Toe Water Depth, h_t [mm]	Crest Freeboard, R_c [mm]	Wave Height, H_{m0} [mm]	Wave Steepness, s_{op} [-]	Wave Period, T_p [s]
Sloping wall on a shingle bed d_{50} = 13 mm	60	190	50–160	0.02	1.27–2.26
				0.05	0.80–1.43
	75	245	50–160	0.02	1.27–2.26
				0.05	0.80–1.43
	100	150	50–160	0.02	1.27–2.26
				0.05	0.80–1.43
	150	100	50–160	0.02	1.27–2.26
				0.05	0.80–1.43
	180	140	50–160	0.02	1.27–2.26
				0.05	0.80–1.43
	200	50	50–160	0.02	1.27–2.26
				0.05	0.80–1.43
Sloping wall on a shingle bed d_{50} = 24 mm	60	190	50–160	0.02	1.27–2.26
				0.05	0.80–1.43
	75	245	50–160	0.02	1.27–2.26
				0.05	0.80–1.43
	100	150	50–160	0.02	1.27–2.26
				0.05	0.80–1.43
	150	100	50–160	0.02	1.27–2.26
				0.05	0.80–1.43
	180	140	50–160	0.02	1.27–2.26
				0.05	0.80–1.43
	200	50	50–160	0.02	1.27–2.26
				0.05	0.80–1.43

The scour depths were calculated with the use of a depth point gauge before and at the end of the experiment. The foreshore bed was reshaped to the initial uniform bed profile of 1 in 20 permeable gravel slope in the front of the structure, prior to the start run of each test. The scour depth at the toe of sloping wall (S_t) and maximum scour depth (S_{max}) were measured for each experiment. In addition, scour depths were also measured at several locations along the foreshore slope to determine the bed profile of gravel beach after a random wave attack. This was executed with the use of the depth point gauges at fixed locations, and consequent analysis was performed.

3. Results and Discussion

3.1. Inshore Wave Conditions

For this study, the distributions of measured incident wave heights at deep water are plotted for each experiment and are compared with the expected Rayleigh distribution. Figure 2 represents the distribution of wave heights for deep water conditions (close to wave paddle) for two test configurations, and also compares the results with the estimated Rayleigh distribution. It is clearly seen from the graphs that the resulting data points show a good agreement with the Rayleigh distribution for both low and high wave steepness However, the measured wave heights were found to deviate from the expected Rayleigh distribution for very high waves, which may be a result of the limitations of paddle generation, such as the wave paddle stroke or wave breaking on the paddle. For the tested conditions, the reflection co-efficient varied from 0.25 to 0.45 at a relatively deep water, near the structure the observed values were in the range from 0.37 up to 0.79 near the structure. This is the reason why the experiments were also carried out without the structure in place to remove the structure induced reflection in the measurements.

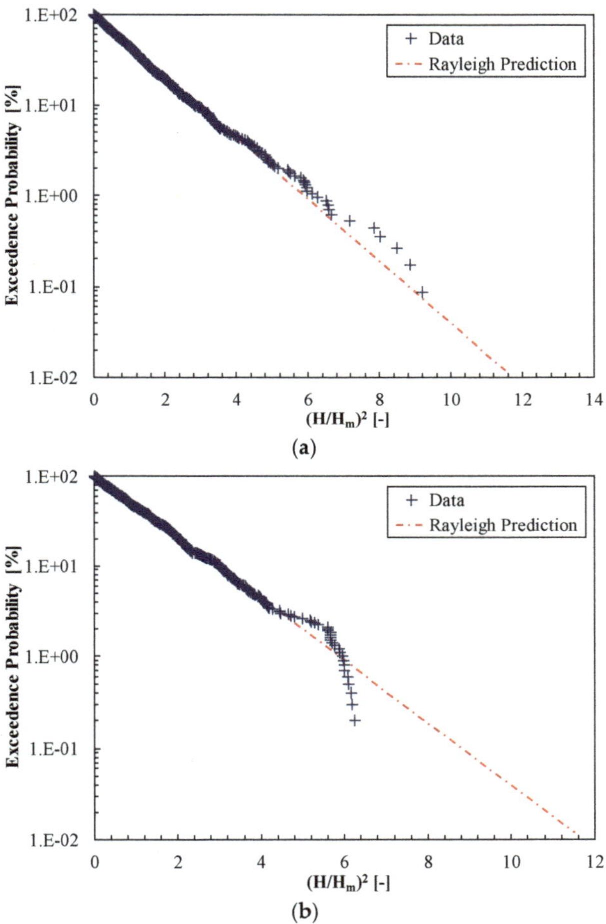

Figure 2. Comparison of the distribution of measured wave heights with the Rayleigh predictions- (**a**) $s_{m\text{-}1,0} = 0.02$, $H_{m0} = 0.06$ m (**b**) $s_{m\text{-}1,0} = 0.06$, $H_{m0} = 0.078$ m.

3.2. Relationship between Development of Scour Depth and Storm Duration

The development of a scour depth has strong correlation with the storm duration or the number of waves [5]. In 2000, Sumer and Fredsøe [26] reported that the peak scour depth occurs between 1000–2000 irregular waves at a 1 in 1.2 seawall on a sandy bed. Recent investigations by [20] on the vertical breakwaters with shingle foreshores showed that the greatest scour depth occurs at around 1000 random waves.

In Figure 3, the time development of maximum scour depth is presented by plotting the measured scour depths against the number of waves. The experiments were performed with approximately 3000 random waves to observe the development of scour depths, where the measurements were conducted at around 1000, 2000, and 3000 waves, respectively. From the graph, it is noticeable that the maximum value of scour depth occurs at approximately 1000 waves, and it reveals similar features as reported by [20] for vertical walls.

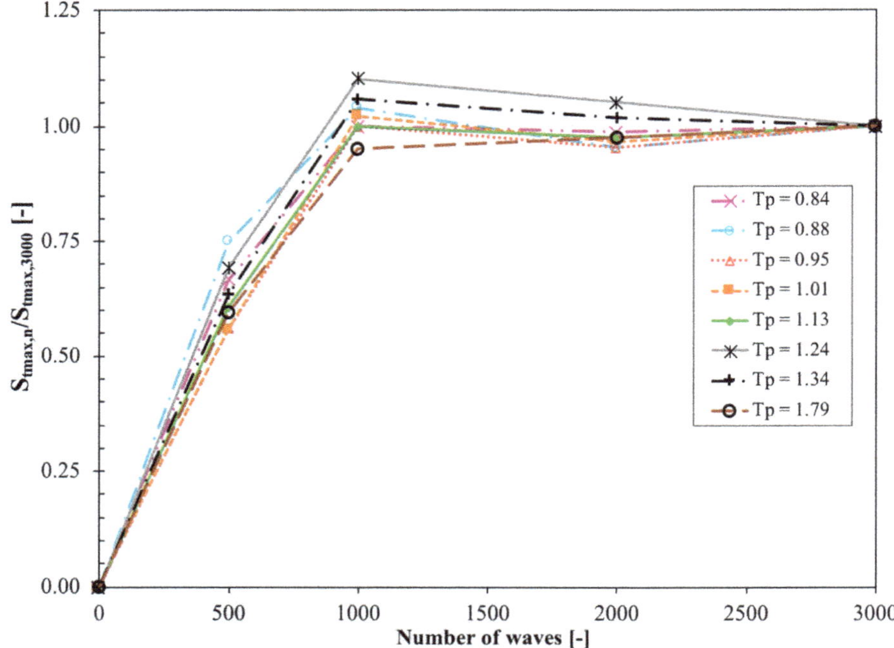

Figure 3. Relationship between development of scour depth and storm duration.

3.3. Variation of Bed Level for a Sloping Wall with a Shingle Foreshore

The variation of bed level (final elevation–initial elevation of foreshore) due to the random wave attack was inspected for the tested conditions within this study. In Figure 4, two examples of the measured bed level changes for swell (s_{op} of 0.02) and storm (s_{op} of 0.05) wave conditions are presented for six different toe water depths (h_t = 0.06 m; 0.075 m; 0.10 m; 0.15 m; 0.16 m; 0.18 m). The data points correspond to positive values of bed changes denote accretion, whereas negative values indicate the scouring at the structure. Figure 4 demonstrates that the maximum accretion or scouring at the structure occurs for the lowest toe water depth for a known wave condition. To cite an example, if we look at the measured data points corresponding to swell waves (s_{op} of 0.02) in Figure 4a, it is noticeable that the greatest accretion of 0.098 m occurs at the lowest water depth of 0.06 m.

From the graphs, it is also observed that the storm waves generate scouring, while the swell waves provide accretion at the structure for a certain water depth. This may happen due to the two different wave characteristics, which actually dominate the movement or transport of beach materials. Generally, steep storm waves generate scouring at gravel beaches through offshore sediment transport, whereas swell waves (long waves) provide accretion of onshore sediment transport of bed materials [27]. Similar characteristics of accretion and erosion with respect to long and short waves were also observed by [20] for plain vertical walls on shingle foreshores.

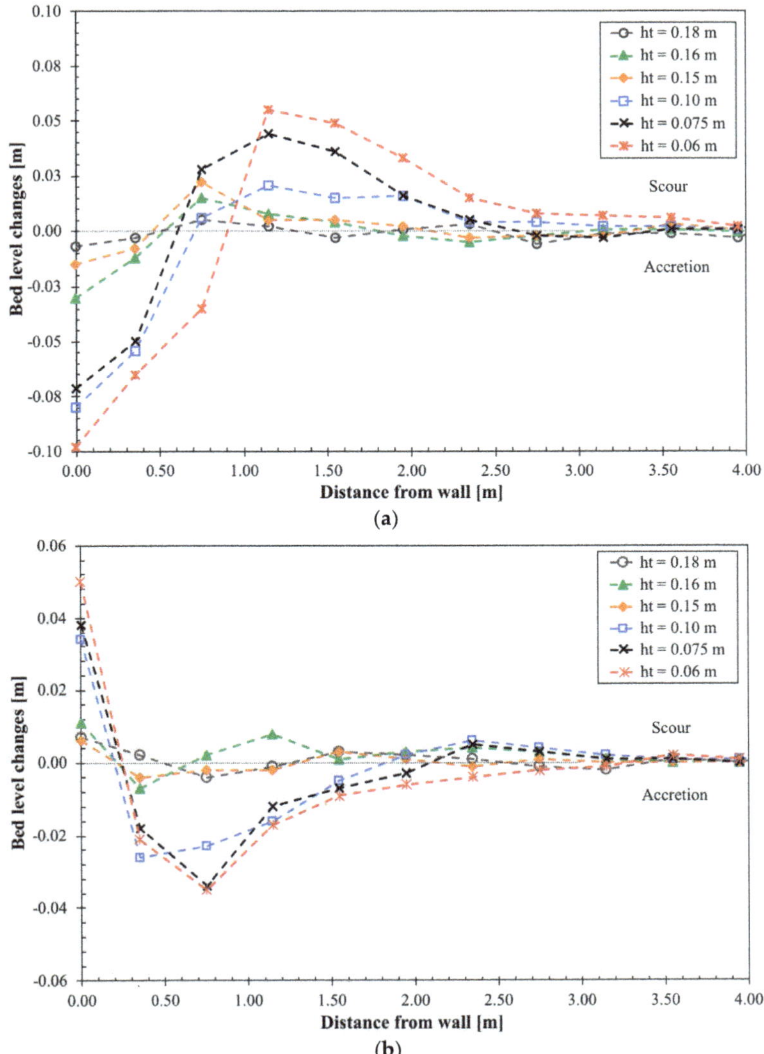

Figure 4. Bed level changes after 1000 waves—(**a**) $s_{op} = 0.02$, $H_{m0} = 0.07$ m (**b**) $s_{op} = 0.05$, $H_{m0} = 0.10$ m.

3.4. Variation of Toe Scour Depth with Relative Water Depth

Figure 5 shows the variation of the measured relative scour depth with relative water depth at the toe of the structure. The dashed line represents the trend of scour depths at seawalls with sandy foreshores, as reported by [8]. The negative values of relative toe water depth indicate the presence of an extended beach above the still water level prior to the start of the experiment, whereas negative non-dimensional toe scour depths denote the accretion at the structure.

The resulting data points show that the maximum scouring at the toe of the structure occurs under the spilling and plunging wave conditions ($0.005 \leq h_t/L_{0m} \leq 0.04$). Similar trends of scour depths under spilling and plunging impacts are also noticeable from the scouring predictions of [8], see Figure 5. For the tested conditions, the maximum erosion at the toe of the structure is observed $S_t/H_{1/3} = 0.93$ at a relative toe water depth (h_t/L_{0m}) of around 0.01 and the maximum accretion is noted $S_t/H_{1/3}$ = 1.51 at a relative toe water depth (h_t/L_{0m}) of about 0.03.

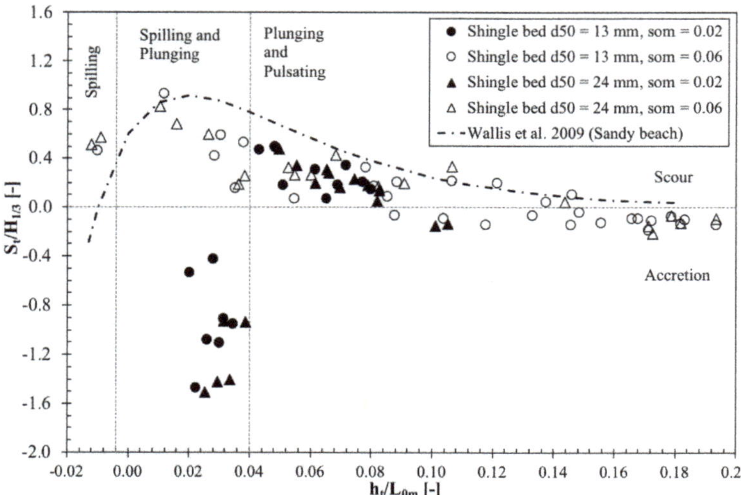

Figure 5. Variation of non-dimensional toe scour depth at a sloping wall with relative toe water depth.

In Figure 5, another clear aspect is that, under plunging and pulsating (surging) breakers (h_t/L_{0m} > 0.04), the scour depths continued to decrease with the increase of the relative toe water depth. For the data points corresponding to relatively higher toe water depths (h_t/L_{0m} > 0.10), accretion at the toe of the structure is noticeable from the graph. These characteristics were also reported by [1,12] for a plain vertical wall with a sandy foreshore slope.

For the tested spilling and plunging conditions, accretion at the slope mainly occurred for long waves with relatively low wave steepness, whereas scouring was mostly observed for short waves with relatively high wave steepness. In Figure 5, if we consider the data points that correspond to an average wave steepness of 0.02 under spilling and plunging conditions, the accretion at the structure is observed, while the experiments with relatively high wave steepness of 0.06 give scouring at the structure for a certain relative toe water depth. This can be also related with reality, where accretion at the structure is mostly observed for long waves through the onshore sediment transport of beach materials and toe scouring phenomenon under storm sea conditions due to the offshore sediment transport.

3.5. Variation of Toe Scour Depth with Iribarren Number

The Iribarren number or breaker parameter is generally expressed as a relationship between the structure slope and wave steepness (Equation (3)) enables distinguishing the non-breaking and breaking waves.

$$I_r = \frac{\tan \propto}{\sqrt{\frac{H_{1/3}}{L_{0m}}}} \tag{3}$$

To observe the influence of Iribarren number (I_r) on toe scouring, the measured relative scour depths are plotted ($S_t/H_{1/3}$) against the Iribarren number in Figure 6. One of the clear conclusions from the graph is that there is considerable variation in scour depths for similar values of the Iribarren number. This may be a result of the variation of toe water depths that has not been fully considered in Figure 6.

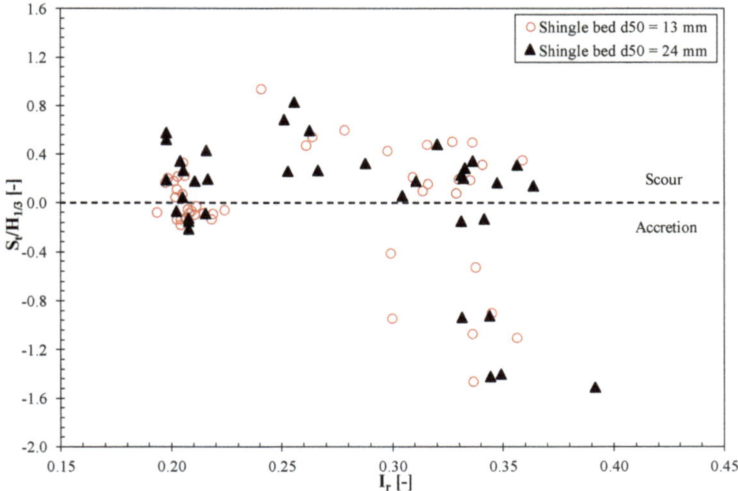

Figure 6. Variation of non-dimensional toe scour depth with Iribarren number.

3.6. Variation of Toe Scour Depth with Relative Toe Water Depth and Iribarren Number

The measured values of non-dimensional scour depths are plotted against dimensionless toe water depth with the data categorized into two ranges of Iribarren number to investigate the combined influence of relative water depth and Iribarren number on the scour depths, as following:

- $0.20 < I_r < 0.30$ and
- $I_r > 0.30$

The graph demonstrates that, for any known value of relative toe water depth (h_t/L_{0m}), the greatest scour depths occur for the larger Iribarren numbers, see Figure 7. Similar trends of scour depths were observed by [12] for a sloping seawall on a sandy beach foreshore.

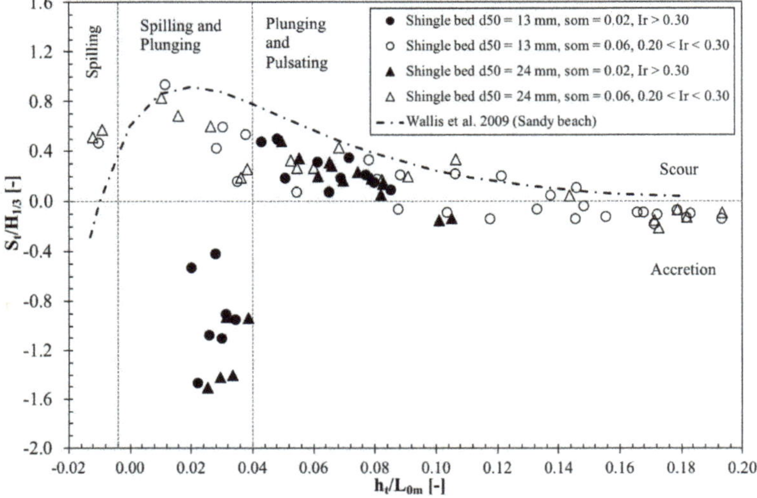

Figure 7. Variation of non-dimensional toe scour depth with Iribarren number.

3.7. Variation of Toe Scour Depth with Wall Slope

To investigate the influence of wall slope on the scour depth at coastal structures, the resulting foreshore profiles of this study at 1:2 smooth sloping walls have been compared with the dataset of the plain vertical walls, as reported by [20], see Figure 8. It is important to note that, for both sloping and vertical structures, the experiments were performed with the same permeable shingle materials and inshore wave conditions, but with different structural configurations. As observed in Figure 8, the measured scour depths at sloping walls within this study do not remarkably differ from those that were reported at vertical walls by [20]. Similar characteristics of scour depths with respect to wall slope were also observed by [12] on sandy slope based on a laboratory study at 1 in 2 sloping wall and a plain vertical wall with sandy foreshore.

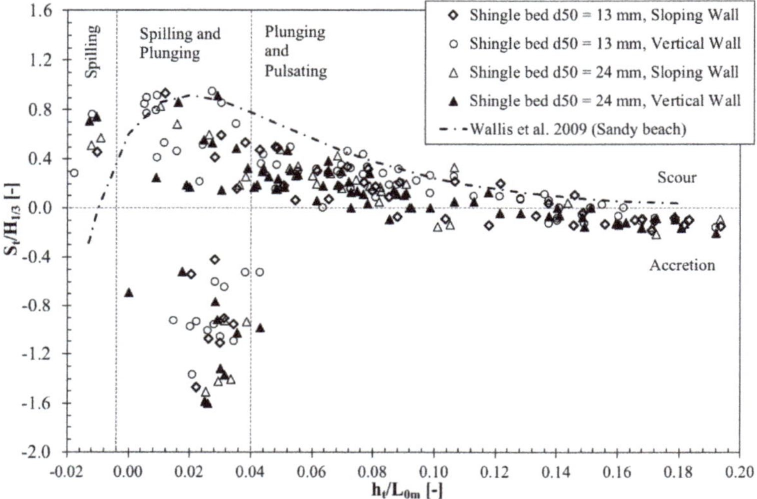

Figure 8. Variation of non-dimensional toe scour depth with wall slope.

4. Conclusions

This laboratory research investigated the mechanics of scour at a smooth impermeable 1 in 2 sloping wall with a uniform 1:20 permeable shingle foreshore slope. A set of small scale physical tests were carried out to measure the scour depths and the bed level changes for two different gravel foreshores. The relationship of scour depth with toe water depth, Iribarren number, and wall slope were investigated from the test results of this work and through a comparison with the available dataset in the literature. The following conclusions can be drawn based on the described study:

- The results of this study showed that the relative toe scour depth at a sloping structure on a shingle beach, is influenced by the relative toe water depth and Iribarren number.
- Within the experimental limitations, the maximum toe scour depths were observed for the experiments under the spilling and plunging wave attack ($0.005 \leq h_t/L_{0m} \leq 0.04$).
- It was found that the scour depths at the sloping walls within this study do not remarkably differ from those reported at vertical walls with similar permeable foreshore slopes, which indicates that the scour depths are independent of the slope of the structure.

It is important to note the test conditions, such as water depths, wave conditions, etc. that were followed within this study were developed by adapting the available guidelines for two-dimensional small-scale investigations, as reported in [22,23] to make it comparable with typical physical model study. It is expected that the results of these two-dimensional hydraulic tests can be related to the prototype conditions; however, the further large-scale investigations would always be recommended.

Author Contributions: Conceptualization, M.S. and J.P.; Laboratory set-up, M.S.; Formal analysis, M.S.; Supervision—J.P.; Writing—original draft preparation, M.S.; Writing—review and editing, M.S and J.P.

Funding: This research received no external funding.

Acknowledgments: M. Salauddin would like to thank the University of Warwick Graduate School for sponsoring his PhD study through the Chancellor's International Scholarship scheme. The financial support of the Natural Environmental Research Council (Grant Ref.: NE/RE003645/1) through providing the advanced sophisticated wave analysis tools, is gratefully acknowledged. The laboratory work was financially supported through the Leverhulme Trust Senior Research Fellowships scheme (2016/2017) of the Royal Academy of Engineering The authors would also like to acknowledge the technical support of Mr. Ian Baylis of Warwick Water Laboratory in the preparation of laboratory set-up. We would also like to thank the three anonymous reviewers for their insight full comments and suggestions on the earlier draft of the manuscript.

Conflicts of Interest: This research is a part of a PhD study. The authors declare no conflict of interest. The funders had no role in the design of the study; in the collection, analyses, or interpretation of data; in the writing of the manuscript, or in the decision to publish the results.

References

1. Müller, G.; Allsop, W.; Bruce, T.; Kortenhaus, A.; Pearce, A.; Sutherland, J. The occurrence and effects of wave impacts. *Proc. ICE Marit. Eng.* **2008**, *160*, 167–173. [CrossRef]
2. Pearson, J.M. Overtopping and Toe Scour at Vertical Seawalls. In Proceedings of the 9th International Conference on Coasts, Marine structures and Breakwaters: Adapting to Change, Scotland, UK, 16–18 September 2009; Institution of Civil Engineers, Thomas Telford Ltd.: Edinburgh, UK, 2010; pp. 598–608.
3. Powell, K.A. *Toe Scour at Sea Walls Subject to Wave Action: A Literature Review*; Report SR 119; HR Wallingford: Wallingford, UK, 1987.
4. Fowler, J.E. *Scour Problems and Methods for Prediction of Maximum Scour at Vertical Seawalls*; Technical Report CERC-92-16; US Army Corps of Engineers Waterways Experiment Station, Coastal Engineering Research Center: Vicksburg, MS, USA, 1992.
5. Sutherland, J.; Brampton, A.H.; Motyka, G.; Blanco, B.; Whitehouse, R.J.W. *Beach Lowering in Front of Coastal Structures—Research Scoping Study*; Report FD1916/TR; Defra—Flood Management Division: London, UK, 2003. Available online: http://sciencesearch.defra.gov.uk/ (accessed on 14 February 2017).
6. Cohen-Shacham, E.; Walters, G.; Janzen, C.; Maginnis, S. *Nature-Based Solutions to Address Global Societal Challenges*; IUCN: Gland, Switzerland, 2016.
7. Church, J.A.; Clark, P.U.; Cazenave, A.; Gregory, J.M.; Jevrejeva, S.; Levermann, A.; Merrifield, M.A.; Milne, G.A.; Nerem, R.S.; Nunn, P.D. Sea level change. In *Climate Change 2013: The Physical Science Basis Contribution of Working Group I to the Fifth Assessment Report of the Intergovernmental Panel on Climate Change*; Stocker, T.F., Qin, D., Plattner, G.-K., Tignor, M., Allen, S.K., Boschung, J., Nauels, A., Xia, Y., Bex, V., Midgley, P.M., Eds.; Cambridge University Press: Cambridge, UK; New York, NY, USA, 2013.
8. Wallis, M.; Whitehouse, R.; Lyness, N. Development of guidance for the management of the toe of coastal defence structures. In Proceedings of the Presented in the 44th Defra Flood and Coastal Management Conference, Telford, UK, 30 June –2 July 2009.
9. Xie, S.L. Scouring Patterns in Front of Vertical Breakwaters and Their Influences on the stability of the Foundation of the Breakwaters. Master's Thesis, Department of Civil Engineering, Delft University of Technology, Delft, The Netherlands, 1981.
10. Kraus, N.C.; Smith, J.M. *SUPERTANK Laboratory Data Collection Project. Volume 1: Main Text*; Technical Report CERC-94-3; US Army Corps of Engineers Waterways Experiment Station, Coastal Engineering Research Center: Vicksburg, MS, USA, 1994; pp. 2191–2204.
11. Pearce, A.; Sutherland, J.; Müller, G.; Rycroft, D.; Whitehouse, R. Scour at a seawall-field measurements and physical modelling. In Proceedings of the 30th International Conference on Coastal Engineering, ASCE, San Diego, CA, USA, 2–8 September 2006.
12. Sutherland, J.; Obhrai, C.; Whitehouse, R.; Pearce, A. Laboratory tests of scour at a seawall. In Proceedings of the 3rd International Conference on Scour and Erosion, CURNET, Technical University of Denmark, Gouda, The Netherlands, 3–5 July 2007.

13. Jayaratne, R.; Premaratne, B.; Adewale, A.; Mikami, T.; Matsuba, S.; Shibayama, T.; Esteban, M.; Nistor, I. Failure Mechanisms and Local Scour at Coastal Structures Induced by Tsunami. *Coast. Eng. J.* **2016**, *58*, 1640017. [CrossRef]
14. Pourzangbar, A.; Saber, A.; Yeganeh-Bakhtiary, A.; Ahari, L.R. Predicting scour depth at seawalls using GP and ANNs. *J. Hydroinform.* **2017**, *19*, 349–363. [CrossRef]
15. Tahersima, M.; Yeganeh-Bakhtiary, A.; Hajivalie, F. Scour pattern in front of vertical breakwater with wave overtopping. *J. Coast. Res.* **2011**, *64*, 598.
16. Tofany, N.; Ahmad, M.; Kartono, A.; Mamat, M.; Mohd-Lokman, H. Numerical modeling of the hydrodynamics of standing wave and scouring in front of impermeable breakwaters with different steepnesses. *Ocean Eng.* **2014**, *88*, 255–270. [CrossRef]
17. Sutherland, J.; Brampton, A.H.; Obrai, C.; Dunn, S.; Whitehouse, R.J.W. *Understanding the Lowering of Beaches in Front of Coastal Defence Structures, Stage 2—Research Scoping Study*; Report FD1927/TR; Defra—Flood Management Division: London, UK, 2008. Available online: http://sciencesearch.defra.gov.uk/ (accessed on 20 February 2017).
18. Powell, K.A.; Lowe, J.P. The scouring of sediments at the toe of seawalls. In Proceedings of the Hornafjordor International Coastal Symposium, Hofn, Iceland, 20–24 June 1994; pp. 749–755.
19. Jayaratne, R.; Mendoza, E.; Silva, R.; Gutiérrez, F. Laboratory Modelling of Scour on Seawalls. In Proceedings of the Presented in the International Conference on Coastal Structures, Boston, MA, USA, 9–11 September 2015.
20. Salauddin, M.; Pearson, J.M. Wave overtopping and toe scouring at a plain vertical seawall with shingle foreshore: A Physical model study. *Ocean Eng.* **2019**, *171*, 286–299. [CrossRef]
21. Powell, K.A. *Predicting Short Term Profile Response for Shingle Beaches*; Report SR 219; HR Wallingford: Wallingford, UK, 1990.
22. Eurotop. EurOtop II—Manual on Wave Overtopping of Sea Defences and Related Structures: An Overtopping Manual Largely Based on European Research, But for Worldwide Application. 2016. in press. Available online: www.overtopping-manual.com (accessed on 21 January 2017).
23. Wolters, G.; Van Gent, M.; Allsop, N.W.H.; Hamm, L.; Mühlestein, D. HYDRALAB III: Guidelines for physical model testing of rubble mound breakwaters. In Proceedings of the 9th International Conference on Coasts, Marine Structures and Breakwaters: Adapting to Change, Edinburgh, UK, 16–18 September 2009; pp. 659–670.
24. Mansard, E.P.D.; Funke, E.R. The measurement of incident and reflected spectra using a least squares method. *Coast. Eng.* **1980**, *1*, 154–172.
25. Klopman, G.; Van der Meer, J.W. Random wave measurements in front of reflective structures. *J. Waterw. Port Coast Ocean Eng.* **1990**, *125*, 39–45. [CrossRef]
26. Sumer, B.M.; Fredsøe, J. Experimental study of 2D scour and its protection at a rubble-mound breakwater. *Coast. Eng.* **2000**, *40*, 59–87. [CrossRef]
27. Sherman, D.J. Gravel beaches. *Natl. Geogr. Res. Explor.* **1991**, *7*, 442–452.

Journal of
Marine Science and Engineering

Article

Statistical Analysis of the Stability of Rock Slopes

Marcel R.A. van Gent [1,*], Ermano de Almeida [1,2] and Bas Hofland [1,2]

1 Department of Coastal Structures & Waves, Deltares, 2629HV Delft, The Netherlands
2 Department of Hydraulic Engineering, Delft University of Technology, 2628CN Delft, The Netherlands; E.deAlmeida@tudelft.nl (E.d.A.); B.Hofland@tudelft.nl (B.H.)
* Correspondence: Marcel.vanGent@deltares.nl; Tel.: +318-8335-8246

Received: 26 November 2018; Accepted: 1 March 2019; Published: 6 March 2019

Abstract: Physical model tests were performed in a wave flume at Deltares with rock armoured slopes. A shallow foreshore was present. At deep water, the same wave conditions were used, but by applying different water levels, the wave loading on the rock armoured slopes increased considerably with increasing water levels. This allowed an assessment of the effects of sea level rise. Damage was measured by using digital stereo photography (DSP), which provides information on each individual stone that is displaced. Two test series were performed five times. This allowed for a statistical analysis of the damage to rock armoured slopes, which is uncommon due to the absence of statistical information based on a systematic repetition of test series. The statistical analysis demonstrates the need for taking the mean damage into account in the design of rock armoured slopes. This is important in addition to characterising the damage itself by erosion areas and erosion depths. The relation between damage parameters, such as the erosion area and erosion depth, was obtained from the tests. Besides tests with a straight slope, tests with a berm in the seaward slopes were also performed. A new method to take the so-called length effect into account is proposed to extrapolate results from physical model tests to real structures. This length effect is important, but is normally overlooked in the design of rubble mound structures. Standard deviations based on the presented model tests were used.

Keywords: stability; erosion; rock slopes; sea level rise; repetition tests; berm; wave flume; length effect

---

1. Introduction

Due to sea level rise, the wave loading on sea defences can become more severe. This is especially the case for sea defences in shallow water. Depth-limited waves at the base of structures become larger when water levels increase under equal wave conditions in deep water. Thus, not only is there a direct influence of increasing water levels, which cause sea defences to become more vulnerable to wave overtopping, but wave loading is also affected by rising sea levels. Increased wave loading on structures affects wave overtopping and also the stability of the armour. Here, the effects of increasing water levels on the stability of rock armoured slopes have been studied under conditions that are equal in deep water, but more severe at the base of the structure.

Early stability formulae for armour layers were derived by [1] and [2]. For most significant port structures in deep water, concrete armour units are used in the armour layer for which the stability formula based on [2] is the state-of-the-art stability formula. The stability of rock armour layers in deep water is less relevant for practical applications since rock armour layers are often applied in shallower water. Shallow water or very shallow water can be defined as conditions where, during design, the offshore wave height reduces to 90% or less at the base of the structure. For such structures with rock armoured slopes, stability formulae based on [3] can be applied, as also recommended in [4]. In [5], it was shown that for shallow water, the spectral significant wave height ($H_s = H_{m0}$) is a better parameter to describe the influence of the wave height than the significant wave height from the time domain ($H_s = H_{1/3}$). In the present study, rock armour layers are also applied in shallow water, and both measures to characterize the significant wave height are presented in the present research. The stability of rock armoured slopes was studied by performing physical model tests in a wave flume. Tests with rock armoured slopes were performed with straight slopes and with a berm in the seaward slope. Damage was measured using digital stereo photography (DSP) and characterised by using various damage parameters. The tests on the straight slopes were performed five times, which allows for a statistical analysis of the results by taking into account the spreading in the test results (using the mean values, μ, the standard deviations, σ, and the variations, σ/μ). Besides the spreading of the test results in the wave flume (see also [6]), in reality, spreading of the damage also occurs. For longer structures, the risk that, at a particular location along this structure, the damage exceeds a specified maximum allowable damage increases. In the present paper, guidance for how to take this length effect into account in the design of rock armoured slopes is provided. The importance of the length effect is illustrated in an example based on the proposed method and the statistical information that was obtained from the physical model tests by repeating test series with the same conditions.

2. Physical Model Tests

Physical model tests were performed in the Scheldt Flume of Deltares, Delft (width 1 m, height 1.2 m, length 110 m of which 55 m was used in this study). The wave board is equipped with active reflection compensation (ARC). This means that the motion of the wave board compensates for the waves reflected by the structure, preventing them from re-reflecting back into the model. The wave board is equipped with second order wave steering. This means that second order effects of the first higher and lower harmonics are considered in the wave board motion. This wave generation system ensures that the generated waves resemble waves that occur in nature.

Figure 1 shows the cross-section of the test set-up in the flume. A horizontal shallow foreshore was present in front of the structure. The length of the horizontal foreshore (4 m) is long enough to ensure that the depth-limited waves that reach the structure are adapted to the (shallow) water at the base. Therefore, the foreshore is representative of other shallow foreshores, where the incident wave conditions at the base (as applied in most existing design guidelines for rock slopes) are adapted to the local water depth. The transition slope from deep water to this shallow horizontal foreshore was close to 1:15. Five structure configurations were tested. One with a straight slope (see Figure 2) and four with a berm (see Figure 3): A narrow (0.08 m, i.e., a width of 5 stone diameters: 5 D_{n50}) and a wide berm (0.16 m, i.e., 10 D_{n50}), each at two different levels. The tested structures consisted of an armour layer of rock with a D_{n50} of 0.0163 m, a grading of $D_{n85}/D_{n15} = 1.25$, density of $\rho_s = 2710$ kg/m^3 and a layer thickness of two diameters, a filter layer of rock with D_{n50} of 0.0094 m, and a layer thickness of two diameters, on a 1:3 slope with an impermeable core. The crest level was fixed at 0.296 m above the base of the structure (non-overtopped). The statistical analysis in the present research focuses on the structure configuration with a straight 1:3 rock slope. The tests investigated the range of damage levels that is also relevant for other structure configurations with rock armour layers (i.e., $S < 12$). The actual damage levels depend, for instance, on the slope and permeability of the structure, but the variation in the damage levels is assumed not to depend on the slope and permeability of the structure; the statistics of the 1:3 slope are considered representative for rock slopes with damage levels of $S < 12$.

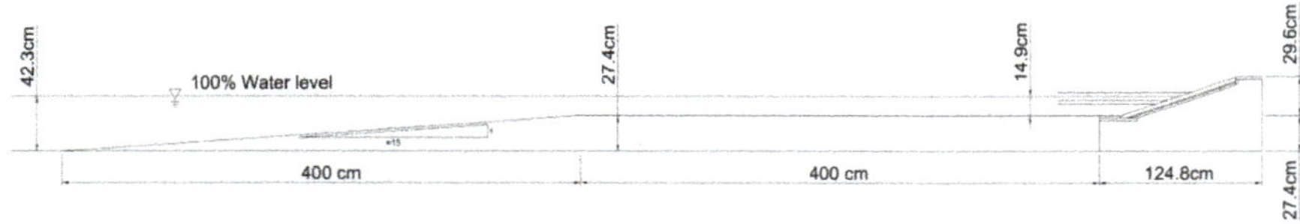

Figure 1. Foreshore and structure geometry in the wave flume.

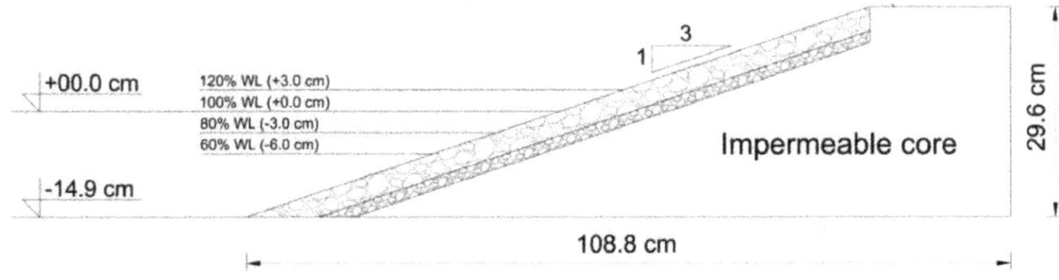

Figure 2. Structure geometry with a straight slope.

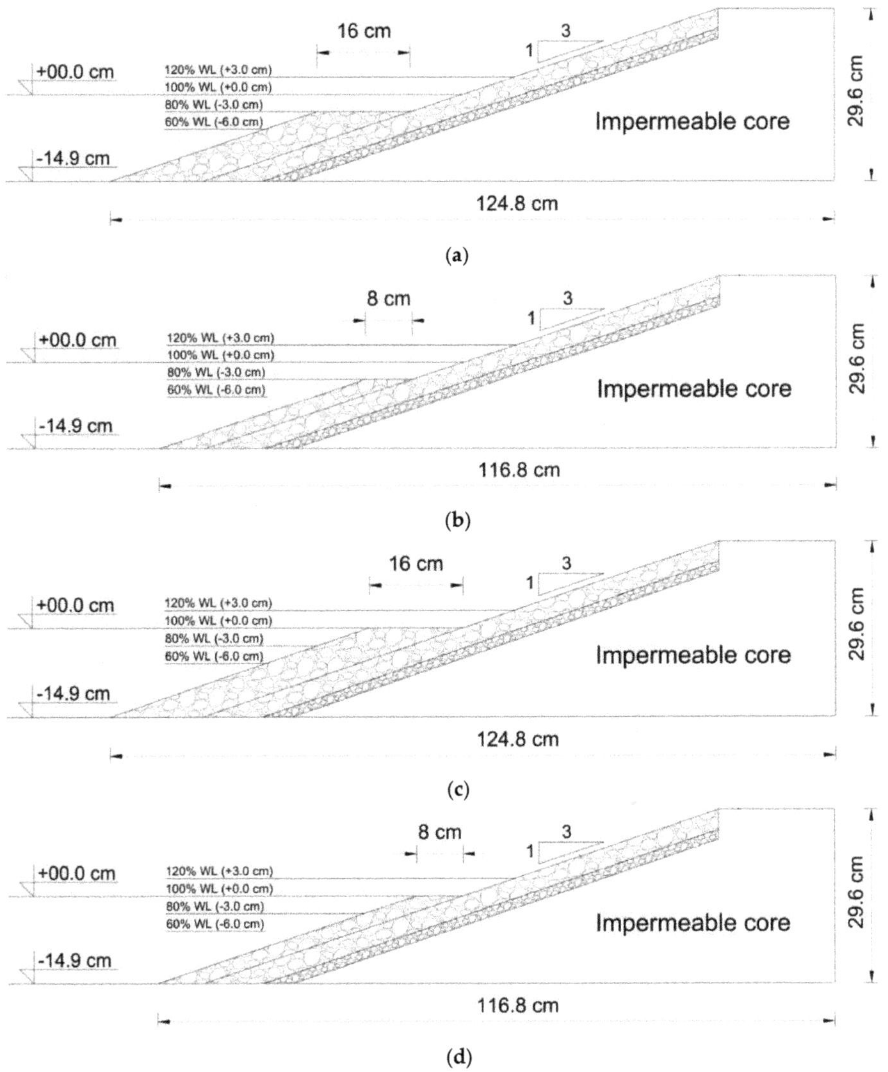

(a)

(b)

(c)

(d)

Figure 3. Structure geometries with berms (**a**): Wide berm at a level of "80% of the 100% Water level (WL)"; (**b**): Narrow berm at "80% of the 100% WL"; (**c**): Wide berm at a level of the "100% WL"; (**d**): Narrow berm at the "100% WL").

The waves were measured in deep water and in front of the base of the structure. At each location, the incident waves were obtained from an array of three wave gauges. The incident deep water conditions were always the same, namely a spectral significant wave height of H_{m0} = 0.088 m and a peak wave period of T_p = 1.27 s (wave steepness s_{op} = 0.035). Standard JONSWAP spectra were used. The number of waves was N = 1000. Each structure was tested with four different water levels (WL), resulting in water depths at still water at the base of the structure of d = 0.089 m, 0.119 m, 0.149 m, and 0.179 m (i.e., 0.03 m increase in each case). Table 1 shows the mean values of the measured wave conditions in deep water and at the base ($H_{1/3}$ denotes the significant wave height from the time-domain analysis and $T_{m-1,0}$ is the spectral wave period characterising the effects of the wave energy spectra on the armour stability, wave run-up, wave overtopping, etc.).

Table 1. Measured wave conditions at deep water and at the base.

Water Level		**Deep**		**Toe**				
Notation	D (m)	H_{m0} (m)	T_p (s)	H_{m0} (m)	$H_{1/3}$ (s)	$H_{2\%}$ (s)	T_m (s)	$T_{m-1,0}$ (s)
60%	0.089	0.088	1.27	0.039	0.030	0.035	1.21	1.59
80%	0.119	0.089	1.27	0.052	0.041	0.048	1.25	1.51
100%	0.149	0.089	1.27	0.065	0.052	0.061	1.23	1.41
120%	0.179	0.088	1.27	0.073	0.061	0.074	1.23	1.38

Six test series were performed, Series S1 and S2 with the structure in Figure 2, and Series S3 to S6 with one of the berm configurations in Figure 3. Except for Series S2, all series consisted of four increasing water level conditions (see Table 1) without any repair of the slope after each test, thus the resulting damage values are cumulative damage values. Series S1 with a uniform slope was performed five times. In Series S2 only the 100% condition was tested, which enabled a comparison of the results with (Series S1) and without (Series S2) milder conditions prior to the design condition. This test was also performed five times. Damage to the rock armoured slope was measured using the digital stereo photography technique (DSP) as described in [7,8]. With this measurement technique, all displaced stones can be identified by comparing the images (i.e., a 3D digital representation of the slope) after a test run with the images of the initial slope, see Figure 4. Based on this technique, all required parameters characterising the damage can be determined.

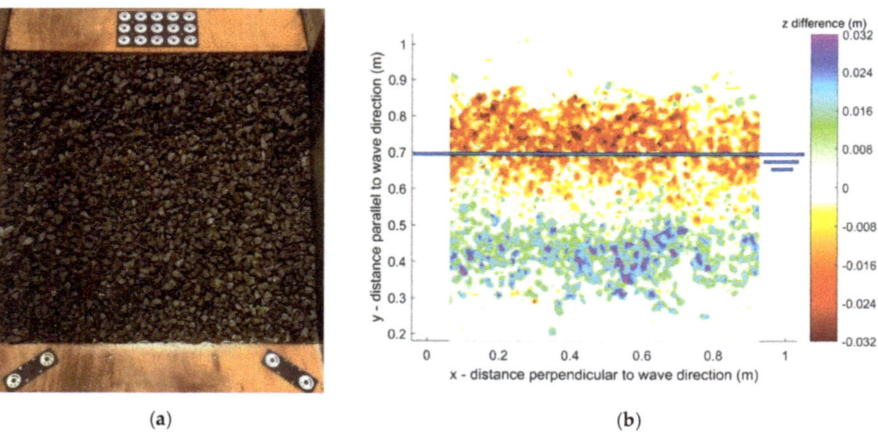

Figure 4. Image of the rock armoured slope (**a**) and an example of the measurement (compared to the initial situation) of the distribution of damage using digital stereo photography (DSP), where the warm colours denote erosion and the cold colours denote accretion (**b**).

Figure 5 shows the non-dimensional erosion and deposition profiles for Series S1 and S2, averaged over five tests, including the 90% confidence intervals. On the horizontal axis, y/D_{n50} denotes the intersection of the original profile of the armour and the still water level for the 100% condition. For all conditions, erosion occurs around the still water level, with the maximum erosion depth slightly above the still water level.

The following parameters were used to characterise the damage to the rock armoured slopes:

$S =$ $A_e/D_{n50}{}^2$ where A_e is the eroded area comparing the profiles before the tests and after a test [9] in the tests based on the average profile over the test section.

$E_{2D} =$ d_e/D_{n50} where d_e is the maximum depth of erosion perpendicular to the slope [6] in the tests based on the average profile over the test section.

$E_{3D,m} =$ d_e/D_{n50} where d_e is the maximum depth of erosion perpendicular to the slope, based on a moving average over a circular area of m D_{n50} [7,8].

Here, for the parameters, S and E_{2D}, the values averaged over a width of 0.88 m (i.e., 54 stones with D_{n50} of 0.0163 m) were used. For the parameter, $E_{3D,m}$, two values for m were used, m = 1 and m = 5.

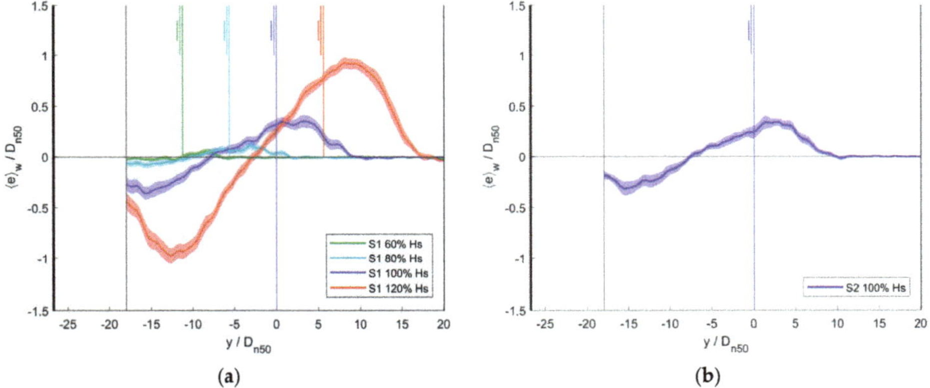

Figure 5. Non-dimensional erosion (positive values) and deposition (negative values) profiles averaged over five tests (**a**) Series S1, for four different water levels; (**b**) Series S2 for the 100% water level), including the 90%-confidence interval.

Table 2 shows the test results based on the width of the measurement section (i.e., 54 stones). Since the tests in Series S1 and Series S2 were performed five times, the mean values (μ) and the standard deviations are presented (σ). The tests with berm configurations (Series S3 to S6) were not repeated (thus the numbers denote the actual test results; h_b denotes the water depth above the berm for the 100% water level condition); the damage values are provided for the lower part of the slope, including the berm (i.e., from the base of the structure to start of the upper slope), and for the upper slope. The S-values for the entire slope are the summation of the two values. For the other damage parameters, the values for the entire slope are the maximum values of the two values (since maximum damage occurs either at the lower slope or at the upper slope).

Table 2. Test results (μ denotes the mean value of the five tests and σ denotes the corresponding standard deviation; 'lower' denotes the part of the slope below the berm plus the berm, and 'upper' denotes the slope above the berm).

Series	Condition	Slope	S		E_{2D}		$E_{3D,1}$		$E_{3D,5}$	
			μ	σ	μ	σ	μ	σ	μ	σ
S1	60%	Uniform 1:3	0.3	0.044	0.07	0.027	1.07	0.050	0.17	0.016
	80%		0.9	0.278	0.16	0.024	1.15	0.096	0.35	0.044
	100%		3.7	0.413	0.41	0.044	1.45	0.077	0.73	0.093
	120%		11.4	0.445	0.96	0.086	2.13	0.220	1.31	0.120
S2	100%	Uniform 1:3	3.4	0.888	0.39	0.064	1.51	0.148	0.74	0.060
			S		E_{2D}		$E_{3D,1}$		$E_{3D,5}$	
			lower	upper	lower	upper	lower	upper	lower	upper
S3	60%	Wide berm $B = 0.16$ m, $h_b = 0.03$ m	0.1	0.0	0.03	0.00	1.13	0.12	0.18	0.01
	80%		1.1	0.0	0.15	0.01	1.11	0.25	0.28	0.02
	100%		1.7	0.3	0.25	0.08	1.35	1.15	0.44	0.20
	120%		1.5	1.0	0.25	0.16	1.35	1.15	0.38	0.38
S4	60%	Narrow berm $B = 0.08$ m, $h_b = 0.03$ m	0.4	0.0	0.07	0.01	0.93	0.15	0.15	0.03
	80%		0.8	0.0	0.13	0.01	1.07	0.72	0.32	0.08
	100%		2.8	0.1	0.51	0.03	1.40	0.95	0.81	0.18
	120%		2.9	1.9	0.58	0.31	1.40	1.32	0.86	0.53
S5	60%	Wide berm $B = 0.16$ m, $h_b = 0$ m	0.6	0.1	0.11	0.01	1.09	0.23	0.22	0.04
	80%		0.6	0.1	0.14	0.03	1.09	0.91	0.28	0.11
	100%		0.6	0.3	0.16	0.05	1.24	1.07	0.24	0.18
	120%		1.2	3.4	0.19	0.45	1.27	1.30	0.31	0.74
S6	60%	Narrow berm $B = 0.08$ m, $h_b = 0$ m	0.2	0.1	0.04	0.01	0.97	0.25	0.17	0.03
	80%		0.7	0.3	0.14	0.12	1.10	1.11	0.28	0.20
	100%		1.2	2.0	0.23	0.31	1.15	1.35	0.37	0.68
	120%		0.1	4.6	0.04	0.53	1.11	1.51	0.24	0.74

3. Analysis of Results

The following observations can be made based on the tests results summarized in Table 2:

- Increasing water levels increase the wave loading and consequently the damage to the rock armour slopes. This increase in damage depends on the parameter that is used to characterise the damage. In the present tests, the damage increased by a factor of 44 for S, 15 for E_{2D}, 2 for $E_{3D,1}$, and 8 for $E_{3D,5}$.
- Comparing the damage values after the 100% condition in Series S1 (with the two preceding conditions) with the 100% condition in Series S2 (without the preceding conditions) indicates that the influence of the preceding conditions on the damage values (μ) is not statistically significant. The spreading (σ) in the damage is clearly larger for the one without preceding conditions, except for the last damage parameter ($E_{3D,5}$).
- The variations (σ/μ) in the test results for the damage parameters, S and E_{2D}, are relatively large for low damage values. For low damage values (60% and 80% conditions), the variations for these damage parameters, S and E_{2D}, are, on average, comparable, but are larger than for the damage parameters, $E_{3D,1}$ and $E_{3D,5}$. Within the range of damage that is normally considered relevant, $2 < S < 12$ (where $S = 2$ is characterised as the start of the damage and $S = 12$ as the failure for the tested 1:3 slope), the variations in the damage values are comparable for the various damage parameters (S, E_{2D}, $E_{3D,1}$, and $E_{3D,5}$). However, for the tests without milder conditions prior to the design condition (i.e., 100% water level in Series S2), the variations in the damage parameters, $E_{3D,1}$ and $E_{3D,5}$, are clearly less than for the damage parameters, S and E_{2D}. As will be discussed

in the following analysis, these observations are affected by the width of the test section over which the damage is determined (the above observations are valid for a width of 54 stones).

- For damage values larger than $S = 1$, adding a berm leads to lower damage values than in the tests with a straight slope. For wider berms, the effect of the berm increases (i.e., less damage). The structures with a submerged berm ($h_b = 0.03$ m) show less damage to the upper slope and more damage to the lower slope than the structures with the water level at the berm ($h_b = 0$ m). This is similar to results shown in [10].
- If damage is characterised by E_{2D} or $E_{3D,5}$, the berm does not always reduce the damage values, which is also the case for higher amounts of damage (e.g., the 100% condition). The wide berms with a width of 10 stone diameters reduce the damage most effectively, irrespective of the damage parameter that is used.

As also discussed in [11], berms can be added to existing straight slopes to reduce overtopping and damage from increasing wave loading or water levels without increasing the crest level. Since the level of berms can be increased relatively easily once sea water levels increase, adding and/or modifying a berm can be a useful measure for climate adaptation.

Figure 6a shows the measured damage values, S (based on the eroded area), versus E_{2D} (based on the erosion depth). This figure shows a clear trend between these two parameters. The relevant range of damage is often between the start of the damage ($S = 2$) and failure ($S = 12$); within this range, the trend can be simplified to $S = 10\ E_{2D}$ for the tested configurations (1:3 slopes, with and without a berm). Figure 6b shows the measured damage values, S (based on the eroded area), versus $E_{3D,5}$. This figure also shows a clear trend between these two parameters. Within the relevant range of damage, this trend can be simplified to $S = 8\ (E_{3D,5})^2$ for the tested configurations (1:3 slopes, with and without a berm).

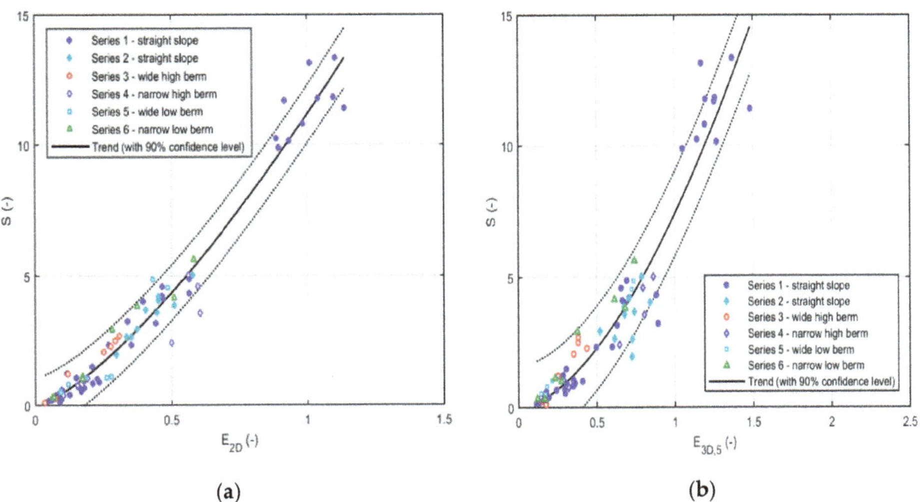

Figure 6. (**a**) Measured damage values, S (based on the eroded area) versus E_{2D} (based on the erosion depth); (**b**) damage values, S (based on the eroded area) versus $E_{3D,5}$ (characterisation width 27 stone diameters).

The test results match reasonably well with existing empirical expressions for rock slope stability, with or without a berm. In [12], these comparisons were shown: For the straight slopes, the measured data is within the 90%-error bands of predictions using the stability formulae for plunging and surging waves by [3]. For the configurations with berms, the measured damage is on average somewhat lower for the part below the berm and somewhat higher for the part above the berm than predicted with the stability formulae by [10].

In the present tests, the damage was determined over a width that is equivalent to 54 stones of 0.0163 m (i.e., somewhat less than the width of the flume due to the use of markers close to the glass walls of the flume for the digital stereo photography technique). Series S1 was performed five times. Therefore, the mean damage values, the standard deviations, and the variations (i.e., the standard deviation, σ, divided by the mean values, μ) could be determined over the width of 54 stones. The statistical analysis can also be performed over a smaller width of the test section, for instance, over 10 × 27 stones instead of 5 × 54 stones (*S* (54) means that the mean and standard deviations of the damage are determined over a width of 54 stones, and *S* (27) means that the mean and standard deviations of the damage are determined over a width of 27 stones, etc.). The influence of the width of the test section (characterisation width) on the statistical values was analysed. Figure 7, Figure 8, and Figure 9 show, respectively, the mean damage values, the standard deviations, and the variations as a function of the width over which the damage was determined. These figures illustrate that the damage values were affected by the width over which the damage was determined. This is not only valid for the present tests, but for all physical model tests with rock slopes.

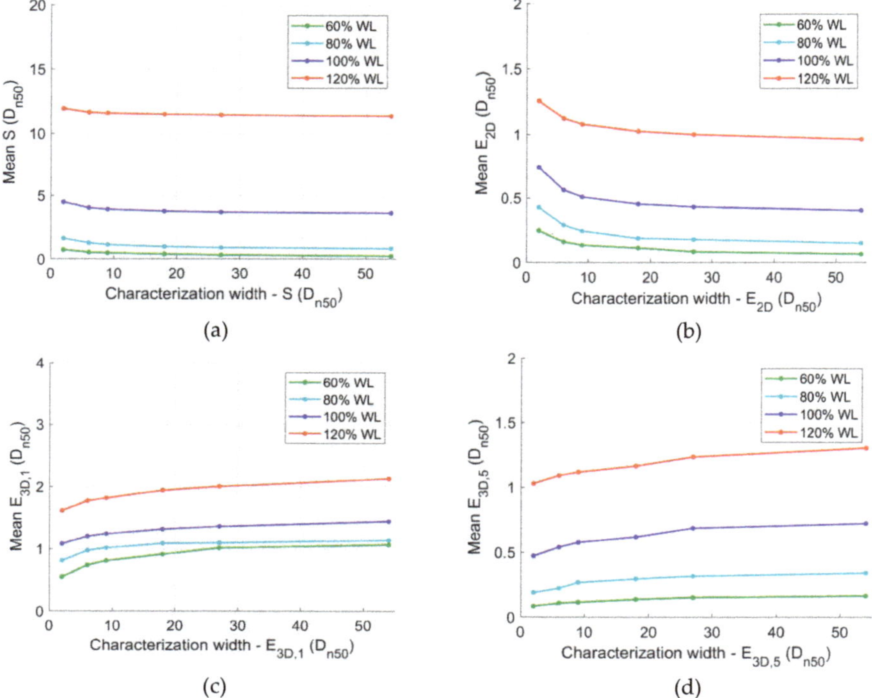

Figure 7. Mean damage values versus the characterisation width, i.e., the width over which the damage is determined (for Series S1); in each panel, another damage parameter was used. (**a**) *S* parameter; (**b**) E2D parameter; (**c**) $E_{3D,1}$ parameter; (**d**)$E_{3D,5}$ parameter.

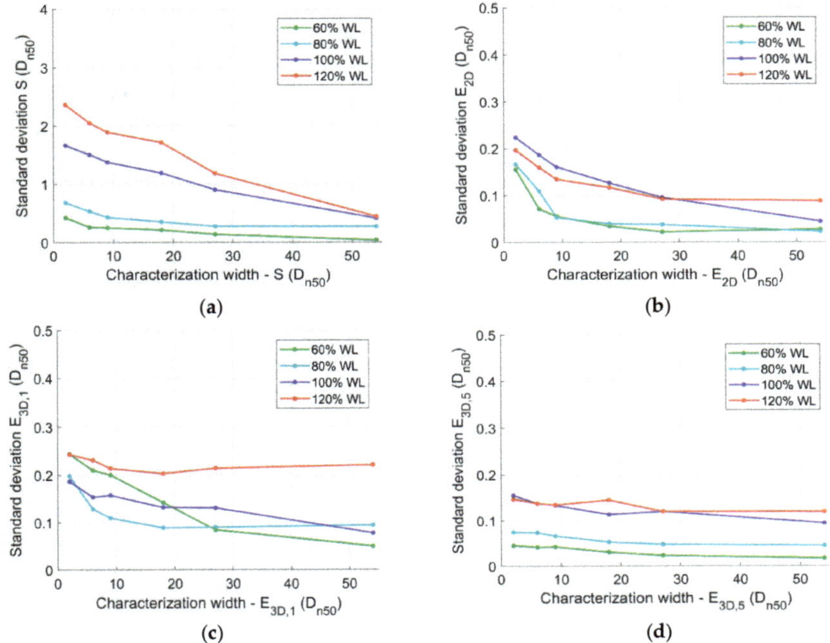

Figure 8. Standard deviations of the damage values versus characterisation width, i.e., the width over which the damage is determined (for Series S1). (**a**) S; (**b**) E_{2D}; (**c**) $E_{3D,1}$; (**d**) $E_{3D,5}$.

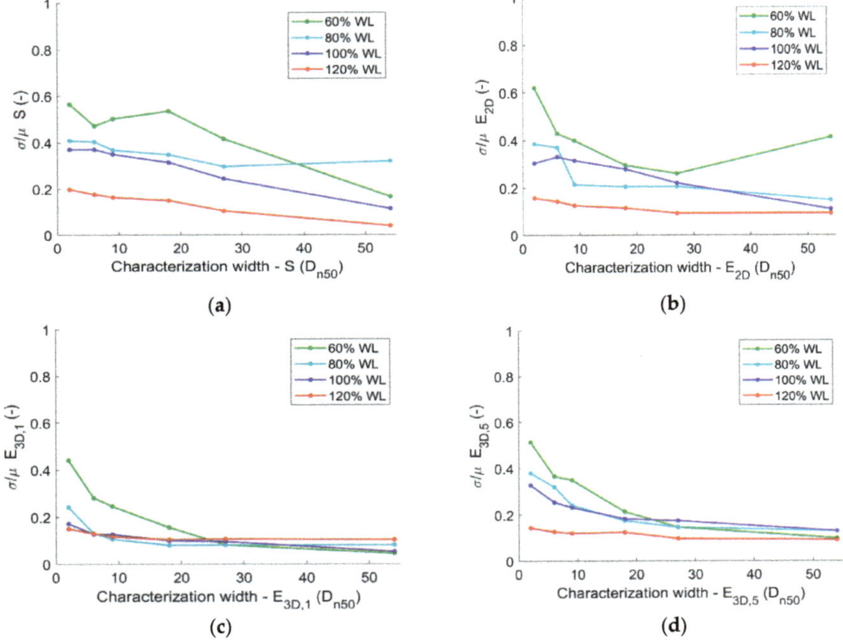

Figure 9. Variation (i.e., standard deviation divided by mean) in the damage versus characterisation width, i.e., width over which the damage is determined (for Series S1). (**a**) S; (**b**) E_{2D}; (**c**) $E_{3D,1}$; (**d**) $E_{3D,5}$.

Figure 8 shows that the width over which the damage is obtained (i.e., the characterisation width) also affects the standard deviations of the damage parameters. Figure 8a shows that for the damage parameter, S, the standard deviations were highly affected by the width of the section over which the damage was determined. Figure 8b,d show that the standard deviations of the parameters, E_{2D} and $E_{3D,5}$, are more or less constant for a section with a width of 25 stones or more. Figure 8c shows that for the parameter, $E_{3D,1}$, there is not a clear trend; for some conditions, the values decrease for wider sections while for other conditions, the values are constant for wider sections.

Figure 9 shows the variation as a function of the width over which the damage was obtained. For the lower damage values (i.e., at the 60% and 80% water levels), the variability of the damage parameters, $E_{3D,1}$ and $E_{3D,5}$, was less than for the damage parameters, S and E_{2D}. For the higher damage values (i.e., at the 100% and 120% water levels) and the largest width of the test section (e.g., a width of 54 stones), there was no clear difference between the variation values of the various damage parameters. Figure 9 shows that the variation of the damage parameters, $E_{3D,1}$ and $E_{3D,5}$, does not vary significantly for various widths over which the damage was obtained, as long as the width was about 25 stones or more. Note that for $E_{3D,1}$ and $E_{3D,5}$, the mean values increase for a larger width, such that for equal values of the spreading, the variations decrease for a larger width. As also discussed in [13,14], it is recommended that damage is determined in a wave flume over a width of about 25 stones.

Besides the width over which the damage was determined in the tests affecting the damage values, there is also a difference between the width of the structure in the tests (scaled to prototype) and the width of the structure in reality. The present tests can be used to take this effect into account. This is the so-called length effect and this will be discussed in the following section.

4. Length Effect

Damage is normally measured over a specific width in a wave flume (here, 54 stones of 0.0163 m). In reality, the width of the structure is likely to be longer than the width (scaled to prototype dimensions) applied in the tests. Therefore, the mean and maximum damage in the real structure can deviate from the mean and maximum damage measured in the tests. This is called the length effect of the structure. In this section, the subscript, m, refers to model test results and the subscript, p, refers to real (prototype) structures. Given that, in this study, tests have been repeated, the mean values and standard deviations of the damage can be determined. These can be used to extrapolate the results in a narrow section (i.e., the test section) to a wider section, the width of the real structure. Several assumptions are made in the following method of which the most important are that damage is normally distributed and that the spreading in the present tests is representative of other structures with rock slopes.

Due to the assumption that the damage is normally distributed, the probability that at any location within the test section the damage, S, is larger than the damage, S_m, obtained over the entire width of the measurement section is 50%, thus $P(S > S_m) = 0.5$ (S_m characterises the mean damage in the model test section since it is based on the averaged profile over the width of the test section). For practical situations in physical model tests, this is considered as a reasonable assumption for damage numbers of $S_m \geq 2$. The probability that within this test section there is at any location damage that exceeds a pre-described critical damage, S_{crit} (i.e., the maximum allowable damage) is denoted by $P(S > S_{crit})$.

$$P(S > S_{crit}) = 1 - F(S_m) \quad (1)$$

where $F(S_m)$ is the cumulative normal distribution, with the (mean) value, S_m, and standard deviation, σ, from the model tests (In Excel: $F(S_m)$ = NORM.DIST (S_{crit}; S_m; σ; TRUE)).

To calculate the length effect due to the difference between the width of the test section, the scaled to prototype scale, L_m, and the width of the real structure, L_p, the ratio of these two, $r = L_p/L_m$, is used. The probability that in the real (prototype) structure a critical damage is exceeded at any location is:

$$P(S_p > S_{crit}) = 1 - [F(S_m)]^r \quad (2)$$

where $F(S_m)$ is the same cumulative normal distribution, with the (mean) value, S_m, and the standard deviation, σ, from the model tests.

If a model test is performed without any repetitions, the standard deviation, σ, can be estimated based on the present tests. The standard deviation and variation (σ/μ, where μ = S_m) appeared to be dependent on the amount of damage and on the width of the test section (i.e., the number of stones, N_{st}, within the width of the test section). For the four different damage parameters' standard deviations obtained from the tests (see Table 2 and Figure 8), a simple fit can be used for the damage levels between the start of the damage and failure (i.e., for S between $2 < S < 12$). These fits are given in Equations (3a)–(3d) and Figure 10. Note that these expressions for the standard deviations include not only the damage level, but also the number of stone, N_{st}, within the width of the test section. The latter influence was not included in [10].

$$S : \sigma = 3900\, S^{0.2}\, (N_{st} + 50)^{-2} \text{ for } 2 \leq S \leq 12 \quad (3a)$$

$$E_{2D} : \sigma = 600\, E_{2D}{}^{0.2}\, (N_{st} + 50)^{-2} \text{ for } 0.2 \leq E_{2D} \leq 1.2 \quad (3b)$$

$$E_{3D,1} : \sigma = 3600\, E_{3D,1}{}^{1.5}\, (N_{st} + 200)^{-2} \text{ for } 1.2 \leq E_{3D,1} \leq 2.2 \quad (3c)$$

$$E_{3D,5} : \sigma = 13000\, E_{3D,5}{}^{0.2}\, (N_{st} + 300)^{-2} \text{ for } 0.3 \leq E_{3D,5} \leq 1.4 \quad (3d)$$

Assuming that the standard deviations in the present tests are representative for all rock slopes, the above mentioned expressions for the probability that a critical value is exceeded can be used for other physical models representing real structures using only the parameters, r (for that particular structure), and the measured damage, S_m (for that particular test) and S_{crit} (for that particular structure in reality). The expressions relating the standard deviation to the amount of damage and the number of stones applied in the tests provide a robust estimation of the standard deviation. As can be seen in Table 2, the associated uncertainties in the estimates of the damage are large.

The width of the real (prototype) structure, L_p, can be divided into r sections with a (scaled) width of the test section, L_m. The average damage profile in each of these r sections provides r values of S_p. Analogous to what is being assumed in the test section, the probability that at any location the damage, S, is larger than the damage value of that section, S_p, is 50%, thus $P(S > S_p) = 0.5$. The maximum damage of the r values of S_p will be higher than the damage in the model, S_m. This maximum value (which is still a value based on an averaged profile over a width of L_m) can be calculated as follows:

$$S_p = F^{-1}\left(0.5^{\frac{1}{r}}\right) \quad (4)$$

where $F^{-1}(S)$ is the inverse of the cumulative normal distribution, with the value, S_m, and standard deviation, σ, both from the model tests (In Excel: S_p = NORM.S.INV(0.5^(1/r)) σ + S_m).

The above expressions (Equations (1), (2), and (4)) can also be applied for the other mentioned damage parameters, but the expression to relate the standard deviation to the amount of damage varies per damage parameter (Equations (3a)–(3d)).

These expressions (Equations (3a)–(3d)) lead to variations (σ/μ) that decrease for larger damage values using S, E_{2D}, or $E_{3D,5}$ or increase for larger values using $E_{3D,1}$. Note that [6] also derived estimates of the spread depending on the amount of damage, but those were based on individual damage profiles and not on average damage profiles.

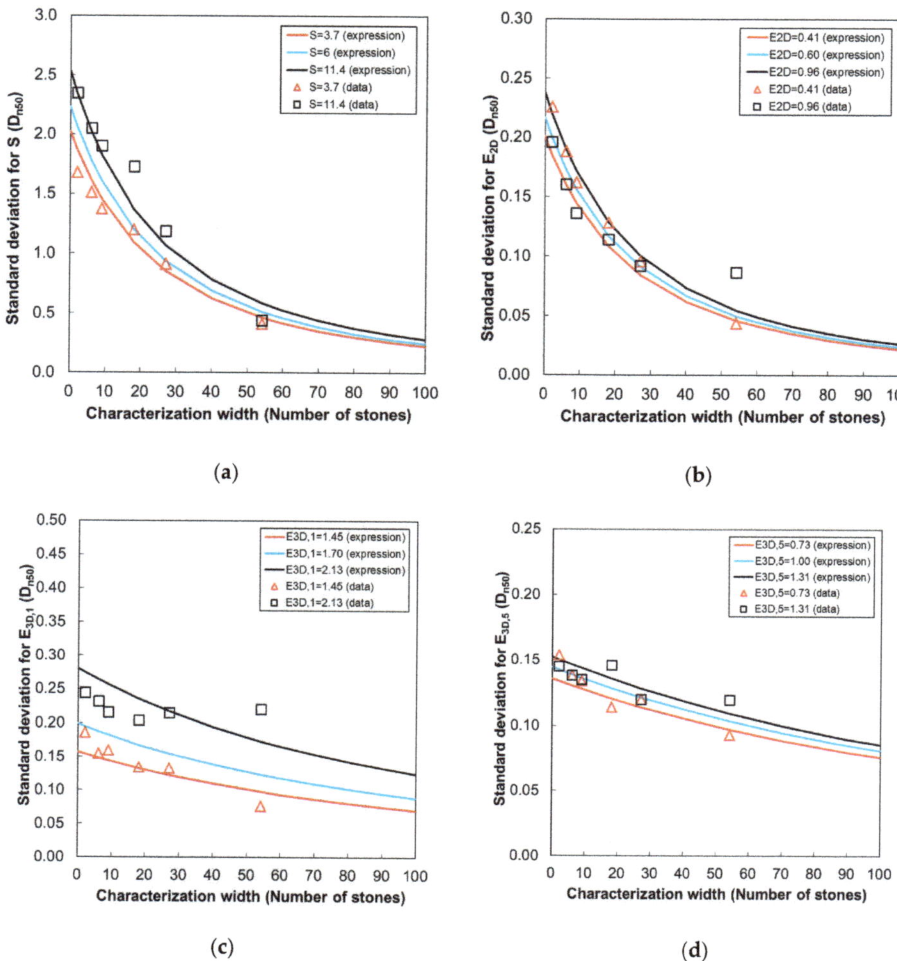

Figure 10. Fit (Equations (3a)–(3d)) to the obtained standard deviations from the measurements (standard deviations versus the number of stones over the width of the test section). (**a**) S parameter; (**b**) E_{2D} parameter; (**c**) $E_{3D,1}$ parameter; (**d**) $E_{3D,5}$ parameter.

In the above described expressions, the assumption is made that the damage to the rock slope is normally distributed and consequently the maxima are Gumbel distributed. For all mentioned damage parameters, (S, E_{2D}, $E_{3D,1}$, and $E_{3D,5}$) Equations (1), (2), and (4) can be used. In contrast to the damage parameters, S and E_{2D}, that are based on *average* damage profiles, the damage parameters, $E_{3D,1}$ and $E_{3D,5}$, are based on *maximal* values that occurred over the width of the measurement section. Since the latter two are based on maxima, extreme value analysis can be used. To account for the difference between the width of the test section, scaled to the prototype scale, L_m, and the width of the real structure, L_p, the ratio of these two, $r = L_p/L_m$, can be used to determine the damage value in the real structure, using the Gumbel distribution:

$$E_{3D} = \mu_G - \sigma_G \ \ln\left(-\ln\left(1 - \frac{1}{r}\right)\right) \tag{5}$$

where μ_G and σ_G are the location and scale parameters of the Gumbel distribution, which can be computed directly from the mean ($E_{3D,m}$) and standard deviation (σ) obtained from the model tests using $\sigma_G = \sigma\, 6^{0.5}/\pi$ and $\mu_G = E_{3D,m} - 0.577\, \sigma_G$.

To illustrate the importance of the length effect and taking into account the spreading of the results, an example is given here based on Equation 1 to 5 (**bold** are assumed input values; in italics, *red* is the outcome of this example).

L = Width of the test section in the model:	**0.8 m**
D_{n50} = Stone diameter in the model ($N_{st} = L/D_{n50}$):	**0.0125 m**
M = Model scale:	**50**
L_m = Width of the test section scaled to prototype scale (m):	40 m
L_p = Width of the real (prototype) structure (m):	**1000 m**
$R = L_p/L_m$	25
S_m = Damage observed in model test (based on the average profile):	**5**
σ = Standard deviation of the damage, using Equation (3a) with $S = S_m = 5$:	0.414
S_p = Damage (maximum of r sections) in the real (prototype) structure, using Equation (4):	*5.80*
S_{crit} = Critical damage: Maximum allowable damage in the real (prototype) structure:	**6**
$P(S_m > S_{crit})$ = Probability that S at any cross-section in a model is larger than S_{crit} (Equation (1)):	*0.008*
$P(S_p > S_{crit})$ = Probability that S at any cross-section in the real structure is larger than S_{crit} (Equation (2)):	*0.179*

This example shows that for a model test with, for instance, a measured damage of $S = 5$, there is a probability of 17.9% that in the real structure, $S = 6$ will be exceeded while the probability that $S = 6$ will be exceeded in the tests is less than 1%. The damage value for the real structure (i.e., the maximum damage based on a damage profile averaged over a width of 40 m in the real structure: The maximum value of 1000 m/40 m = 25 damage values) will be higher than the damage in the model (i.e., the damage obtained based on the average profile in the measurement section): 5.80 versus 5.

In above described example, the damage parameter, S, was used, but the same can be done for the other three damage parameters, using the earlier mentioned expressions for the standard deviation, σ, for each damage parameter.

The following example provides an illustration of the length effect using another damage parameter, namely $E_{3D,5}$.

$E_{3D,5\text{-}m}$ = Damage observed in model test (based on the average profile):	**0.60**
σ = Standard deviation of the damage, using Equation (3d) with $E_{3D,5} = 0.60$:	0.09
$E_{3D,5\text{-}p}$ = Damage in the real (prototype) structure, using Equation (4) (Normal distribution):	*0.770*
$E_{3D,5\text{-}p}$ = Damage in the real (prototype) structure, using Equation (5) (Gumbel distribution):	*0.781*
$E_{3D\text{-}crit}$ = Critical damage: Maximum allowable damage in the real (prototype) structure:	**0.8**
$P(E_{3D,5\text{-}m} > E_{3D\text{-}crit})$ = Probability that $E_{3D,5}$ in a model is larger than $E_{3D\text{-}crit}$ (Equation (1)):	*0.012*
$P(E_{3D,5\text{-}p} > E_{3D\text{-}crit})$ = Probability that $E_{3D,5}$ in the real (prototype) model is larger than $E_{3D\text{-}crit}$ (Equation (2)):	*0.260*

This example shows that for a model test in which a damage level of $E_{3D,5} = 0.60$ is measured, there is a probability of 26% that in the real structure, that is 25 times wider ($R = L_p/L_m = 25$), the critical value of $E_{3D,5} = 0.80$ will be exceeded. The damage in the real structure will be higher than the measured damage in the model; in the case when a normal distribution is used, $E_{3D,5} = 0.77$ versus $E_{3D,5} = 0.60$, while in the case when the Gumbel distribution is used, $E_{3D,5} = 0.78$ versus $E_{3D,5} = 0.60$. Thus, in this example, the expected damage values for the real (prototype) structure are not sensitive to the applied approach; for both distributions, the expected damage values, $E_{3D,5}$, in the real structure are similar and are clearly larger than that observed in the measurements.

In the approach described here, the measured spreading in Series S1 was considered as representative, rather than Series S2. In Series S1, milder wave conditions were generated prior to the design condition (100% WL). This is considered as being more representative of real structures since before reaching the peak of the storm (that is modelled in the model tests), milder wave conditions also occur earlier in a storm. In addition, it is likely that structures experience storms prior to the design storm. In Series S2, no milder conditions were present prior to the design condition. Therefore, the standard deviations in the above-mentioned approach are based on Series S1 and not on Series S2. Note that, unlike the spreading, the mean damage values obtained in Series S1 and Series S2 are rather similar.

5. Conclusions and Recommendations

Based on the physical model tests, the damage to rock armoured slopes was analysed for conditions with different water levels in still water. The wave conditions in deep water were the same, but due to the shallow foreshore, the wave conditions at the base of the structure varied considerably. Tests with a straight slope and tests with berms were performed. Those with a straight slope were performed five times and this allowed for an analysis of the spreading in the tests results. The following conclusions can be drawn based on the described study:

- Increasing water levels can increase the damage to rock armour slopes considerably for structures with a shallow foreshore. In the performed tests, the increasing water levels caused increased wave loading and consequently an increase of the damage with a factor 2 (for $E_{3D,1}$) to 44 (for S) while the wave conditions in deep water were equal. Thus, this factor depends on which parameter is used to characterise the damage.
- Wide berms (with a width of 10 stone diameters) are effective to reduce the total damage to the armour layer, irrespective of the parameter that is used to characterise the damage. Berms can be added to existing straight slopes to reduce overtopping and damage from increasing water levels and increasing wave loading without increasing the crest level. Since the level of berms can be increased relatively easily once sea water levels increase, adding and/or modifying a berm is a useful measure for climate adaptation.
- The present tests can be used to estimate the spreading of the damage to rock armoured slopes for tests that have not been repeated as many times as in the present tests. Simple expressions were derived to estimate the standard deviations and variations in the damage to rock armoured slopes.
- With the range of damage that is normally considered relevant, $2 < S < 12$ (where $S = 2$ is characterised as the start of the damage and $S = 12$ as failure for the tested 1:3 slope), the variations in the damage values are comparable for the various damage parameters (S, E_{2D}, $E_{3D,1}$, and $E_{3D,5}$). However, for the tests without milder conditions prior to the design condition (i.e., 100% water level in Series S2), the variations in the damage parameters, $E_{3D,1}$ and $E_{3D,5}$, are clearly less than for the damage parameters, S and E_{2D}.
- If conditions with lower waves precede the design conditions, the magnitude of the damage is similar to the situation where no conditions with milder waves precede the design condition. However, the variability in the damage increases if no conditions with milder waves precede the design condition.
- Since in 2D physical model tests, structures are generally less wide than the real (prototype) structures that they resemble, the difference between the scaled width of the structure in the model and the width of the structure in reality needs to be taken into account. Expressions to take this length effect into account were proposed.

Since the spreading of the damage to rock armoured slopes is rather large, future research should aim to study the spreading of the damage to rock armoured slopes in more detail with even more repetitions than performed here. It is recommended to take the spreading of the damage to rock armoured slopes into account in physical model tests and in reality; in investigations where only one test is performed, the presented tests can provide an estimate of the spreading. It is recommended that the length effect is considered if the structure, in reality, is wider than the scaled width of the model in a wave flume.

Author Contributions: Conceptualization, Methodology, Validation and Investigation: M.R.A.v.G., E.d.A. and B.H.; Formal Analysis, Data Curation: M.R.A.v.G. and E.d.A.; Writing-Original Draft Preparation: M.R.A.v.G.; Resources, Funding Acquisition: M.R.A.v.G. and B.H.; Writing-Review & Editing: M.R.A.v.G., E.d.A. and B.H.

Funding: This project received co-funding from the European Union's Horizon 2020 research and innovation programme under grant agreement No 654110, HYDRALAB+.

Acknowledgments: We would like to thank Sofia Caires (Deltares) for her valuable contribution to the present work. Most of the described research described here has first been presented and published [15] at the Coastlab 2018 conference (May 22–26, Santander, Spain); the conference is acknowledged for providing the platform to present the described research.

Conflicts of Interest: The authors declare no conflict of interest.

References

1. Iribarren, R. *Una Formula Para el Calcula de los Diques de Escollera (A Formula for the Calculation of Rock-Fill Dikes)*; Pasajes: Madrid, Spain, 1938.
2. Hudson, R.Y. Laboratory investigation of rubble-mound breakwaters. *J. Waterw. Harb. Div.* **1959**, *85*, 93–121.
3. Van Gent, M.R.A.; Smale, A.J.; Kuiper, C. Stability of rock slopes with shallow foreshores. In Proceedings of the 4th International Coastal Structures Conference, Portland, OR, USA, 26–30 August 2003.
4. CIRIA; CUR; CETMEF. *The Rock Manual. The Use of Rock in Hydraulic Engineering*, 2nd ed.; C683; CIRIA: London, UK, 2007.
5. Herrera, M.P.; Gómez-Martín, M.E.; Medina, J.R. Hydraulic stability of rock armors in breaking wave conditions. *Coast. Eng.* **2017**, *127*, 55–67. [CrossRef]
6. Melby, J.A.; Kobayashi, N. Progression and variability of damage on rubble mound breakwaters. *J. Waterw. Port Coast. Ocean Eng.* **1998**, *124*, 286–294. [CrossRef]
7. Hofland, B.; van Gent, M.; Raaijmakers, T.; Liefhebber, F. Damage evolution using the damage depth. In Proceedings of the Coastal Structures 2011, Yokohama, Japan, 5–9 September 2011.
8. Hofland, B.; Disco, M.; van Gent, M.R.A. Damage characterization of rubble mound roundheads. In Proceedings of the Coastal 2014, Varna, Bulgaria, 29 September–2 October 2014.
9. Broderick, L.L. Riprap Stability versus Monochromatic and Irregular Waves. Ph.D. Thesis, Oregon State University, Corvallis, OR, USA, 1984.
10. Van Gent, M.R.A. Rock stability of rubble mound breakwaters with a berm. *Coast. Eng.* **2013**, *78*, 35–45. [CrossRef]
11. Van Gent, M.R.A.; Lim, L. Incorporating effects of oblique waves in the design of coastal protection structures under sea level rise. In Proceedings of the Sustainable Built Environment Conference (SBE16), Singapore, 5–6 September 2016.
12. De Almeida, E. Damage Assessment of Coastal Structures in Climate Change Adaptation. Master's Thesis, Delft University of Technology, Delft, The Netherlands, 2017.
13. De Almeida, E.; van Gent, M.R.A.; Hofland, B. Damage characterization of rock armoured slopes. In Proceedings of the Coastal 2018, Santander, Spain, 3–7 May 2018.
14. De Almeida, E.; Van Gent, M.R.A.; Hofland, B. Damage characterization of rock slopes. *J. Mar. Sci. Eng.* **2019**, *7*, 10. [CrossRef]
15. Van Gent, M.R.A.; de Almeida, E.; Hofland, B. Statistical analysis of the stability to rock armoured slopes. In Proceedings of the Coastal 2018, Santander, Spain, 3–7 May 2018.

Journal of
Marine Science and Engineering

Article

Wave Overtopping over Coastal Structures with Oblique Wind and Swell Waves

Ivo M. van der Werf and Marcel R.A. van Gent *

Deltares, PO Box 177, 2600 MH Delft, The Netherlands; Ivo.vanderWerf@deltares.nl
* Correspondence: Marcel.vanGent@deltares.nl; Tel.: +31-88-335-8246

Received: 14 November 2018; Accepted: 3 December 2018; Published: 6 December 2018

Abstract: Most guidelines on wave overtopping over coastal structures are based on conditions with waves from one direction only. Here, wave basin tests with oblique wave attack are presented where waves from one direction are combined with waves from another direction. This is especially important for locations where wind waves approach a coastal structure under a specific direction while swell waves approach the coastal structure under another direction. The tested structure was a dike with a smooth and impermeable 1:4 slope. The test programme consisted of four types of wave loading: (1) Wind waves only: "sea" (approaching the structure with an angle of 45°), (2) Wind waves and swell waves from the same direction (45°), (3) Wind waves and swell waves, simultaneously from two different directions (45° and −45°, thus perpendicular to each other), and (4) Wind waves, simultaneously from two different directions (45° and −45°, thus perpendicular to each other). Existing guidelines on wave overtopping have been extended to predict wave overtopping discharges under the mentioned types of wave loading (oblique sea and swell conditions).

Keywords: overtopping; dikes; sea defenses; bimodal seas; swell; oblique waves; crossing seas; wave basin

---

1. Introduction

The prediction of the amount of wave overtopping during storms is important for the design and evaluation of flood protection structures, for instance to determine the crest height, the protection of the crest and inner slope of dikes, and the volumes of water that need to be drained from the landward side of the structure.

Guidelines on wave overtopping over coastal structures are based on conditions with waves from one direction only, either waves approaching structures perpendicular or structures with an oblique wave attack. No guidelines exist for coastal structures where waves from one direction are combined with waves from another direction. A simultaneous wave attack by wave fields with different wave directions is of special importance for locations where wind waves approach a coastal structure under a specific direction while swell waves approach the coastal structure under another direction. Such bimodal seas with sea states from different directions (crossing seas) have, for instance, been discussed in [1–3]. Although in practice directional spreading (short-crested waves) of each of the sea states may also play a role, the present study focuses on bimodal seas with long-crested waves from two directions.

Most methods to characterize wave overtopping focus primarily on mean overtopping discharges (e.g., the empirical expressions by [4,5] and the artificial neural network by [6]. Individual overtopping events can be characterized by using estimates of water layer thicknesses and velocities during overtopping events as described by [7–10]. The present work focusses on the mean overtopping discharge as parameter to characterize wave overtopping. Two existing empirical expressions to estimate the mean overtopping discharge for waves from one direction are mentioned here. A widely used method to estimate the mean wave overtopping discharge for structures in relatively deep water is based on [4]:

$$\frac{q}{\sqrt{gH_{m0}^3}} = \frac{0.067}{\sqrt{\tan\alpha}}\gamma_b\xi_{m-1,0}\exp\left(-\frac{4.75}{\gamma_b\gamma_f\gamma_\beta\gamma_v}\frac{R_c}{H_{m0}\xi_{m-1,0}}\right) \tag{1}$$

with a maximum of

$$\frac{q}{\sqrt{gH_{m0}^3}} = 0.2\exp\left(-\frac{2.6}{\gamma_f\gamma_\beta\gamma_v}\frac{R_c}{H_{m0}}\right) \tag{2}$$

where q is the mean wave overtopping discharge, g is the acceleration due to gravity, $\xi_{m-1,0}$ is the breaker parameter based on the spectral significant wave height at the toe of the structure H_{m0} and the spectral wave period $T_{m-1,0}$ (see [11,12]), R_c is the crest height relative to the water level, α is the slope angle, and γ_b, γ_f, γ_β and γ_v are influence factors for berms, roughness of the slope, the angle of wave attack, and crest elements respectively. Since the combination of these influence factors has not been validated systematically, the combination of influence factors has been set at $\gamma_b\ \gamma_f\ \gamma_\beta\ \gamma_v \geq 0.4$, see [4], to avoid too large reductions that have not been based on data. The reliability of Equations (1) and (2) can be expressed by assuming the coefficients 4.75 and 2.6 as normally distributed stochastic parameters with standard deviations $\sigma_{4.75} = 0.5$, $\sigma_{2.6} = 0.35$.

Another method to estimate the mean wave overtopping discharge for structures in relatively deep water is described in [5] and mentioned in [13]:

$$\frac{q}{\sqrt{gH_{m0}^3}} = \frac{0.023}{\sqrt{\tan\alpha}}\gamma_b\xi_{m-1,0}\exp\left[-\left(\frac{2.7}{\gamma_b\gamma_f\gamma_\beta\gamma_v}\frac{R_c}{H_{m0}\xi_{m-1,0}}\right)^{1.3}\right] \tag{3}$$

with a maximum of

$$\frac{q}{\sqrt{gH_{m0}^3}} = 0.09\exp\left[-\left(\frac{1.5}{\gamma_f\gamma_\beta\gamma_v}\frac{R_c}{H_{m0}}\right)^{1.3}\right] \tag{4}$$

The reliability of Equations (3) and (4) can be expressed by assuming the coefficients 0.023, 2.7, 0.09 and 1.5 as normally distributed stochastic parameters with standard deviations $\sigma_{0.023} = 0.5$, $\sigma_{2.7} = 0.2$, $\sigma_{0.09} = 0.0135$ and $\sigma_{1.5} = 0.15$. Note that the influence factors γ_b, γ_f, γ_β and γ_v are based on expressions like Equations (1) and (2) with these reduction factors to the power -1 while in Equations (3) and (4) the same reduction factors are applied to a different power (-1.3) resulting in a much larger impact of the reduction factors than can be justified based on the data on which these factors have been based. In order to apply the same influence factors as derived based on Equations (1) and (2), Equations (5) and (6) need to be used instead of Equations (3) and (4):

$$\frac{q}{\sqrt{gH_{m0}^3}} = \frac{0.023}{\sqrt{\tan\alpha}}\gamma_b\xi_{m-1,0}\exp\left[-\frac{3.6}{\gamma_b\gamma_f\gamma_\beta\gamma_v}\left(\frac{R_c}{H_{m0}\xi_{m-1,0}}\right)^{1.3}\right] \tag{5}$$

with a maximum of

$$\frac{q}{\sqrt{gH_{m0}^3}} = 0.09\exp\left[-\frac{1.7}{\gamma_f\gamma_\beta\gamma_v}\left(\frac{R_c}{H_{m0}}\right)^{1.3}\right] \tag{6}$$

Note that overtopping data do hardly justify the change from the power 1 to 1.3; based on overtopping data [14] concluded that the power 1.0 is better than the power 1.3. It is proposed to apply a maximum reduction due to combinations of various influence factors, similar to [4]: $\gamma_b\ \gamma_f\ \gamma_\beta\ \gamma_v \geq 0.4$ since the combination of reduction factors has not been validated systematically. Note that effects of wind on wave overtopping have not been incorporated in these empirical expressions. Some guidance on effects of wind on wave overtopping over sloping structures is provided in [15].

Most of the data on mean overtopping discharges and individual overtopping events are based on physical model tests with a perpendicular wave attack. Oblique waves lead to lower overtopping discharges and less severe overtopping events than perpendicular wave attack. In [16] prediction formulae for the influence of oblique wave attack (influence factor γ_β) on the mean overtopping discharge over impermeable structures (e.g., dikes) were derived. In [16] a reduction factor is included for wave obliquity γ_b, where β represents the angle of wave incidence (with $\beta = 0°$ for perpendicular wave attack and $\beta = 90°$ for waves propagating in the longitudinal direction of the structure). In the prediction formulae distinction is made between long-crested waves (unidirectional waves) and short-crested waves (multi-directional waves). For long-crested waves approaching impermeable structures, [16] proposed:

$$\begin{array}{ll} \gamma_\beta = 1 & \text{for } 0° \leq \beta \leq 10° \\ \gamma_\beta = \cos^2(\beta - 10°) & \text{for } 10° \leq \beta \leq 50° \\ \gamma_\beta = 0.6 & \text{for } \beta \geq 50° \end{array} \quad (7)$$

For short-crested waves approaching impermeable structures, [16] proposed:

$$\gamma_\beta = 1 - 0.0033 \times \beta \qquad \text{for } \beta \leq 80° \quad (8)$$

These reduction factors for oblique wave attack on impermeable structures can be applied in combination with various formulae for predicting the mean wave overtopping discharge over impermeable structures. To account for the effects of oblique wave attack over permeable structures expressions were derived by [17–19]. In [17], tests with long-crested waves on various types of armour slopes were described. For long-crested waves approaching permeable structures with rock armour layers, [17] proposed:

$$\gamma_\beta = \cos^{1/3} \beta \qquad \text{for } 0° \leq \beta \leq 75° \quad (9)$$

In [19] different expressions were proposed based on tests within the range of $0° \leq \beta \leq 60°$. For long-crested waves approaching permeable structures, [19] proposed:

$$\gamma_\beta = 1 - 0.0077 \times \beta \qquad \text{for } \beta \leq 60° \quad (10)$$

For short-crested waves approaching permeable structures, [19] proposed:

$$\gamma_\beta = 1 - 0.0058 \times \beta \qquad \text{for } \beta \leq 60° \quad (11)$$

The effect of oblique wave attack on the overtopping discharge can also be predicted by using a data-driven neural network technique [6] for various types of structures (e.g., dikes and breakwaters).

Figure 1 shows a comparison of methods that predict the reduction in wave overtopping due to oblique wave attack. Note that these methods were developed for different wave loading (short-crested or long-crested) or different types of structures (impermeable or permeable), only [17,19] proposed expressions for mean overtopping discharges over permeable structures with long-crested waves. In the expression by [17] the influence of wave obliquity is much smaller than in the expression by [19]. The comparison shows that in all expressions the influence of wave obliquity on mean wave overtopping discharges is rather large; note that the reduction factor is present in the exponential part of the equations (Equations (1)–(6)). The comparison also shows that the mutual differences between expressions for the same situation (i.e., the blue and red dashed lines for permeable structures and long-crested waves) are even more significant, indicating that there can be a rather large uncertainty around proposed expressions, and/or that the influence factor for oblique waves depends on other aspects than those present in the expressions.

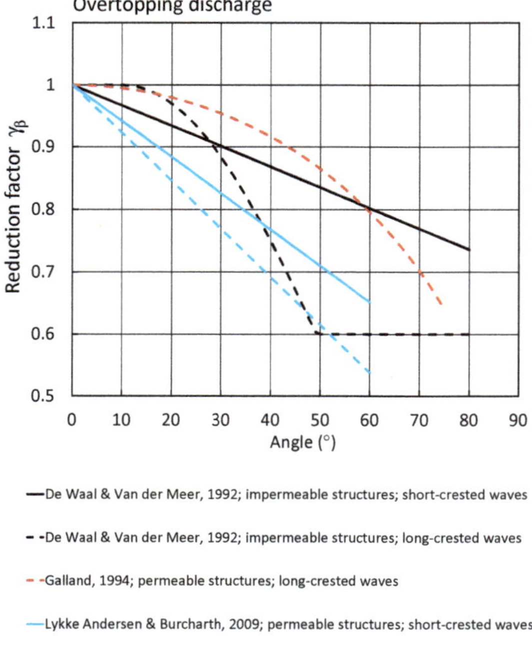

Figure 1. Comparison of methods [16,17,19] describing the influence of oblique waves on wave overtopping (figure based on [20]).

As mentioned before, existing guidelines on wave overtopping over coastal structures are based on conditions with waves from one direction only, either waves approaching structures perpendicular or structures with oblique wave attack. For coastal structures where waves from one direction are combined with waves from another direction no guidelines exist. The present research aims at providing a guideline to predict the mean overtopping discharge over dikes for one wave field and also for two wave fields at the same time. The combination of a sea state with swell is one of those combinations of two wave fields.

2. Physical Model Tests

Physical model tests have been performed in the Delta Basin of Deltares, Delft. This size of the basin is 50 m × 50 m. Two multi-directional wave generators have been used, one with length of 26 m and another of 40 m. Both wave generators are equipped with an active reflection compensation system (see for details of the system [21]). The tested structure had a very smooth impermeable 1:4 slope made of wood with a crest level of 1.2 m above the horizontal foreshore. The water depth was 0.9 m in all tests (R_c = 0.3 m). Long-crested waves were used in all tests. Although the wave generators can generate waves under an angle, in the present tests all waves were generated perpendicular to the wave board leading to all waves approaching the structure under an angle of 45°. All wave energy spectra were JONSWAP spectra (γ = 3.3), for the wind waves and for the swell waves. Although for swell waves a Pierson-Moskowitz spectrum (thus a peak enhancement factor of (γ = 1.0 instead of 3.3) is more appropriate for swell conditions since the spectrum of swell generally is a fully developed and a Pierson–Moskowitz spectrum is assumed to represent fully developed conditions in deep water [22]. However, for consistency reasons the same shape was used for all wave energy spectra. Besides that, in [11,12] it was shown that the spectral shape does not affect wave run-up and wave overtopping as long as the spectral wave period of the spectrum $T_{m-1,0}$ is the same. Waves were measured in front of the toe of the structure with a directional wave gauge (GRSM). The directional wave gauge consisted of a standard wave gauge combined with a two-component velocity sensor. The signals of these instruments were jointly analysed to produce an incident and reflected wave spectrum using the maximum entropy method based on [23]. The wave steepness of the wind waves measured at the toe was in the range of s_p = 0.018 to 0.052. The wave steepness of the swell waves was in the range of s_p = 0.0005 to 0.040. Note that the lower range of the wave steepness of the swell component is very low. The selected wave conditions cover a fairly wide range of relevant values of the wave steepness. Mean wave overtopping discharges were measured over a width of 1 m at the middle of the structure (with a width of 21.7 m), see Figure 2. Visually it was observed that variations of wave overtopping occurred along the crest of the structure but no reasons for this variation along the structure could be determined.

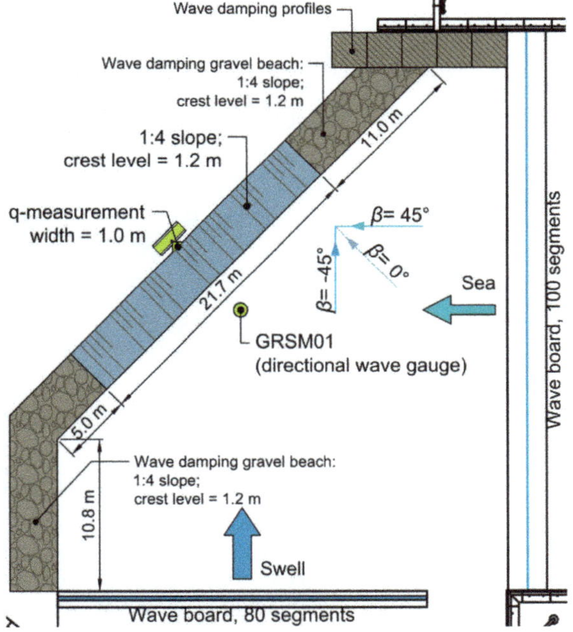

Figure 2. Model set-up in the wave basin.

The test programme consisted of four types of wave loading:

- Wind waves only: "sea" (45°).
- Wind waves and swell waves from the same direction: "Sea + swell" (45°).
- Wind waves and swell waves, simultaneously from two different directions: "Sea + swell perpendicular" (45° and −45°; crossing seas).
- Wind waves, simultaneously from two different directions: "Sea + sea perpendicular" (45° and −45°; crossing seas).

The two types of wave loading with crossing sea states (see for instance Figure 3) can be seen as a combination of wind waves and swell waves, but in the third type of wave loading, the wave steepness of the swell waves is considerably lower than in the fourth type of wave loading. Since in the fourth type of wave loading the generated wave heights and wave steepness of both sea states are similar, in the following we refer to them as two sea states of wind waves ("sea + sea"), to distinguish those from the third type of wave loading (referred to as "sea + swell"). Although the conditions of the two sea states ("sea + sea") could have been varied mutually, such a variation is assumed not to be essential.

(a)

(b)

Figure 3. Wave basin tests with waves from two directions; (**a**) two wave generators with structure at the right, (**b**) two sea states approaching the structure under 45°.

Table 1 shows the measured wave conditions and measured wave overtopping discharges. Each condition was tested once. For the second and third series with swell waves (with peak wave periods of the swell waves that are a factor 3—or more—larger than the peak wave periods of the sea waves), the measured wave conditions at the toe of the structure have been decomposed into a sea component and a swell component where the swell component is based on the energy in the low-frequencies and the sea component on the energy in the higher frequencies. The frequency separating the two components is the position with minimum energy between the energy peaks of the individual sea states, which corresponds closely with the frequency in the middle of the two peaks (see Figure 4 for an illustration of the double peaked spectrum and the separation frequency).

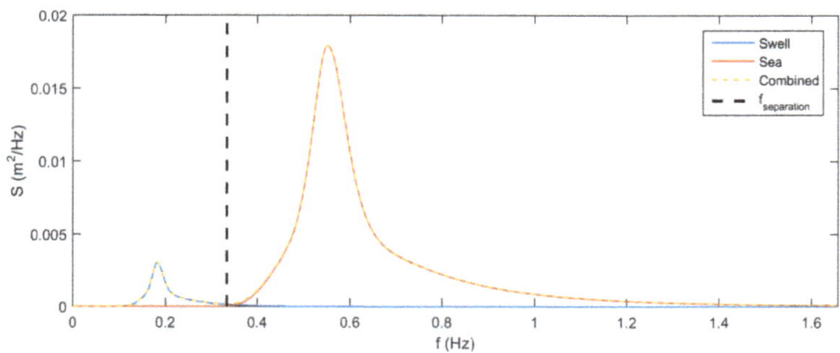

Figure 4. Double peaked spectrum and separation frequency (for Test S3-02).

Table 1. Measured wave conditions and measured wave overtopping discharges.

	Test	Sea			Swell			Overtopping
		H_{m0}	T_p	$T_{m-1,0}$	H_{m0}	T_p	$T_{m-1,0}$	q
		(m)	(s)	(s)	(m)	(s)	(s)	(l/s/m)
Sea	S1-01	0.168	1.728	1.562	0	0	0	0.009
	S1-02	0.171	1.593	1.539	0	0	0	0.015
	S1-03	0.210	1.748	1.700	0	0	0	0.086
	S1-04	0.225	1.762	1.766	0	0	0	0.111
Sea + Swell	S2-01	0.182	1.586	1.570	0.080	4.783	4.855	0.082
	S2-02	0.234	1.762	1.745	0.094	4.833	5.278	0.387
Sea + swell perpendicular	S3-01	0.167	1.706	1.535	0.075	4.801	4.765	0.061
	S3-02	0.166	1.590	1.516	0.075	4.802	4.785	0.065
	S3-03	0.205	1.744	1.659	0.091	4.844	5.160	0.221
	S3-04	0.229	1.769	1.741	0.055	4.966	5.606	0.195
	S3-05	0.226	1.781	1.739	0.056	5.021	5.659	0.200
	S3-06	0.169	1.604	1.534	0.032	5.063	5.471	0.020
	S3-07	0.166	1.619	1.527	0.024	4.889	5.720	0.023
	S3-08	0.167	1.610	1.523	0.061	4.804	4.700	0.029
	S3-09	0.168	1.584	1.512	0.076	4.800	4.737	0.070
	S3-10	0.208	2.408	2.207	0.031	4.982	6.714	0.386
	S3-11	0.159	2.386	1.995	0.029	5.052	6.014	0.122
	S3-12	0.137	1.892	1.782	0.026	4.865	5.369	0.029
Sea + sea perpendicular	S4-01	0.172	1.575	1.522	0	0	0	0.017
	S4-02	0.210	1.745	1.670	0	0	0	0.107
	S4-03	0.231	1.791	1.773	0	0	0	0.238
	S4-04	0.096	1.147	1.109	0	0	0	0.000
	S4-05	0.118	1.278	1.224	0	0	0	0.000
	S4-06	0.164	2.364	2.030	0	0	0	0.159
	S4-07	0.136	1.291	1.219	0	0	0	0.000
	S4-08	0.143	1.898	1.733	0	0	0	0.035

3. Analysis of Test Results

Table 1 and Figure 5 show the test results. In Figure 5 the test results are presented where on the horizontal axis the relative free board ($R_c/(H_{m0}\ \xi_{m-1,0}\ \gamma_\beta)$ is shown (H_{m0} and $\xi_{m-1,0}$ are based on the conditions of the wind waves: "sea") and on the vertical axis the relative overtopping rate $q/(gH_{m0})^{0.5}\times(\tan\alpha)^{0.5}/\xi_{m-1,0}$.

Adding swell from the same direction to these sea conditions leads to larger overtopping discharges (compare Test S1-02 to Test S2-01 or Test S1-04 to Test S2-02 in Table 1). If swell is added from the perpendicular direction (i.e., sea under an angle of 45° and swell from an angle of −45°) the overtopping discharge is also larger than for sea waves only (compare for instance Test S1-01 to Test S3-01 or Test S1-02 to Test S3-02, or Test S1-04 to Test S3-04 in Table 1). In the upper graph of Figure 5 the "sea" conditions and the "sea + swell" conditions are shown. Adding the swell conditions to the wind waves ("sea") leads to a higher amount of wave energy approaching the structure. Therefore, it is not surprising that the overtopping is larger for the "sea + swell" conditions.

Comparing sea and swell conditions from the different directions to sea and swell conditions from the same direction does not lead to significantly different wave overtopping discharges (compare Test S2-01 to Test S3-09 in Table 1). The middle graph of Figure 5 shows the "sea + swell" conditions where they approach the structure in the same direction and the "sea + swell" conditions where the "sea" and "swell" are perpendicular to each other. The direction of the swell (same direction or perpendicular to the wind waves) does not seem to be very important; the (limited) conditions of "sea + swell" conditions where they approach the structure in the same direction (green triangles) correspond to conditions that are in the higher range of overtopping discharges of "sea" and "swell" perpendicular to each other (upper red dots).

In the lower graph of Figure 5 the "sea" conditions and the "sea + sea" conditions (crossing seas) are shown, where the total amount of incident waves energy (thus irrespective of the exact direction) is accounted for in the incident wave height H_{m0}. Two sea states with waves from two different directions (i.e., 45° and −45°) lead to somewhat larger overtopping discharges compared to waves from one direction only (compare also Test S1-02 to Test S4-01 or Test S1-03 to Test S4-02 in Table 1).

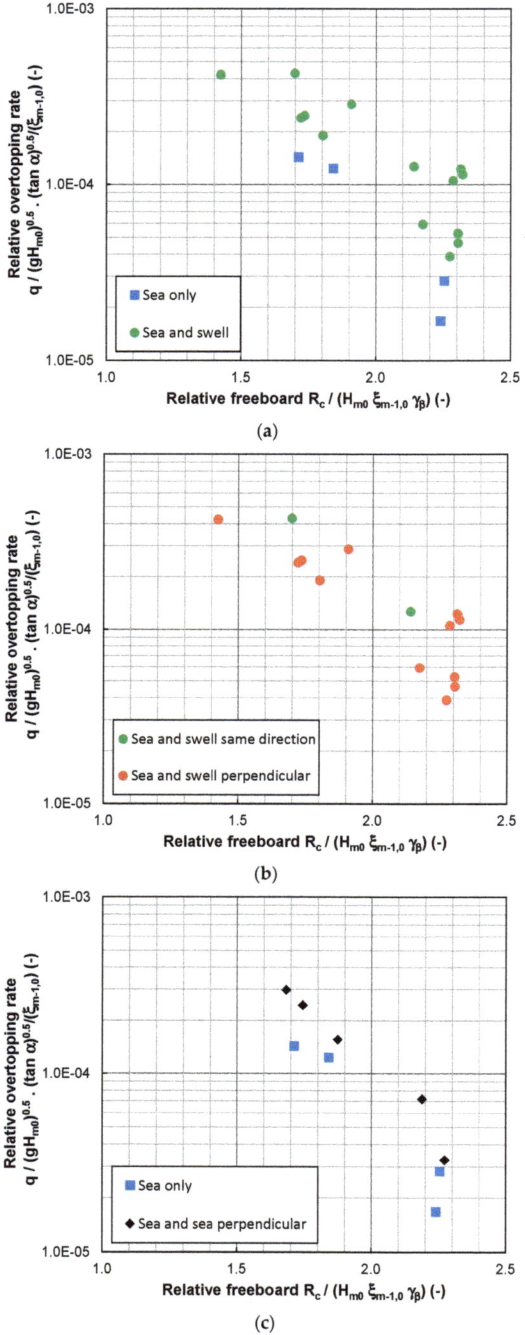

Figure 5. Comparison of overtopping discharges between various types of wave loading; (**a**) "sea" conditions versus "sea + swell" conditions; (**b**) "sea + swell" where they approach the structure in the same direction and "sea + swell" conditions where the "sea" and "swell" are perpendicular to each other; (**c**) "sea" conditions versus "sea + sea" conditions.

This analysis indicates that:

(1) The presence of swell leads to more overtopping.

(2) The direction of the swell (same direction or perpendicular to wind waves) seems to be less relevant for wave overtopping discharges.

(3) Two sea states with waves from two different directions (i.e., 45° and −45°) seem to lead to somewhat larger overtopping discharges compared to waves from one direction (i.e., 45°) with the same total incident wave height at the toe.

The test results are compared to existing empirical expressions to estimate wave overtopping discharges. All tested wave conditions were with long-crested waves under an angle of 45° and with a smooth, impermeable slope. Therefore, Equation (7) has been used to account for an oblique wave attack. This leads to a reduction factor of γ_β = 0.67 for wave obliquity. First, Equations (1) and (2) have been used. Since there is a systematic difference between the measured overtopping discharges and the predictions for wind waves only, the coefficient 4.75 needs to be adapted to 3.4 to account for the bias. The bias may be caused by variations of wave overtopping along the crest of the structure with locations that have an overtopping discharge that is on average larger and locations that are on average lower than the mean discharge along the structure. This would mean that the overtopping box is located at a position with an overtopping discharge that is larger than the average overtopping discharge along the structure. Equations (1) and (2) have been adapted to account for the second wave field which is swell with a low wave steepness. For the wave height H_{m0} and the surf similarity parameter $\xi_{m-1,0}$ the parameters of the wind waves ("sea") have been used. The swell (with a very low wave steepness in the range of s_p = 0.0005 to 0.0025) has been incorporated by treating the swell as a water level increase leading to a fictitious lower freeboard. The resulting expression is as follows:

$$\frac{q}{\sqrt{gH_{m0}^3}} = \frac{0.067}{\sqrt{\tan\alpha}}\gamma_b\xi_{m-1,0}\exp\left(-\frac{c_1}{\gamma_b\gamma_f\gamma_\beta\gamma_v}\frac{(R_c - c_2 H_{m0-Swell})}{H_{m0}\xi_{m-1,0}}\right) \tag{12}$$

where H_{m0} is the spectral significant wave height of the wind waves ("sea"), $H_{m0-\text{Swell}}$ is the spectral significant wave height of the swell, and $\xi_{m-1,0}$ is the breaker parameter based on the wind waves ("sea"). This expression is considered valid only for swell conditions with a wave steepness of $s_{m-1,0} \leq 0.005$ or $s_p \leq 0.005$. As mentioned previously, the bias of the measured discharge was incorporated by changing the coefficient c_1 from 4.75 to 3.45. For practical applications, it is advised to use the coefficient 4.75 since this coefficient appeared to be optimal for all available data for sloping structures. The coefficient c_2 was calibrated based on the present data-set: c_2 = 0.5.

Figure 6 shows the calculated versus the measured overtopping discharges on a linear scale, for each of the four types of wave loading. This figure confirms that the presence of swell leads to more overtopping and that Equation (12) accounts for this increased overtopping due to swell (with a low wave steepness). Two sea states with waves from two different directions (i.e., 45° and −45°, denoted by black diamonds in Figure 6) lead to somewhat larger overtopping discharges compared to waves from one direction (i.e., 45°, denoted by blue squares in Figure 6). The differences are in the order of magnitude of 35% (i.e., 35% more overtopping for these crossing sea states than for wind waves from one direction only). This additional overtopping can be incorporated in Equation (12) by adding a fifth influence factor with a value of $\gamma_\#$ =1.05 for crossing sea states to the four reduction factors γ. However, it remains unknown to what extent the additional overtopping discharge for crossing sea states depends on the wave angle of both sea states.

Equation (12) is an extension of Equation (1) to account for swell conditions. Compared to Equation (1), Equation (5) is a slightly different expression based on [5]. Within the range of the tested single sea states, Equation (5) leads to very similar results compared to Equation (1) for single sea states. Nevertheless, Equation (5) can also be extended to account for swell waves in the same way as Equation (1) was extended to Equation (12). Equation (13) shows the extension of Equation (5) to account for swell waves:

$$\frac{q}{\sqrt{gH_{m0}^3}} = \frac{0.023}{\sqrt{\tan\alpha}}\gamma_b\xi_{m-1,0}\exp\left[-\frac{c_1}{\gamma_b\gamma_f\gamma_\beta\gamma_v}\left(\frac{(R_c - c_2H_{m0-Swell})}{H_{m0}\xi_{m-1,0}}\right)^{1.3}\right] \quad (13)$$

As mentioned before the bias of the measured discharge can be incorporated by changing the coefficient c_1 (here from 3.6 to 2.7). For practical applications it is advised to use the coefficient 3.6 since this coefficient appeared to be optimal for all available data for sloping structures. The coefficient c_2 was calibrated based on the present data-set: $c_2 = 0.5$. Figure 7 shows the calculated versus the measured overtopping discharges on a linear scale, for each of the four types of wave loading. Since the performance of the original Equations (1) and (5) for single sea states is very similar in the tested range, also the performance of the extended Equations (12) and (13) to account for swell waves, is similar (i.e., the differences between Figures 6 and 7 are small).

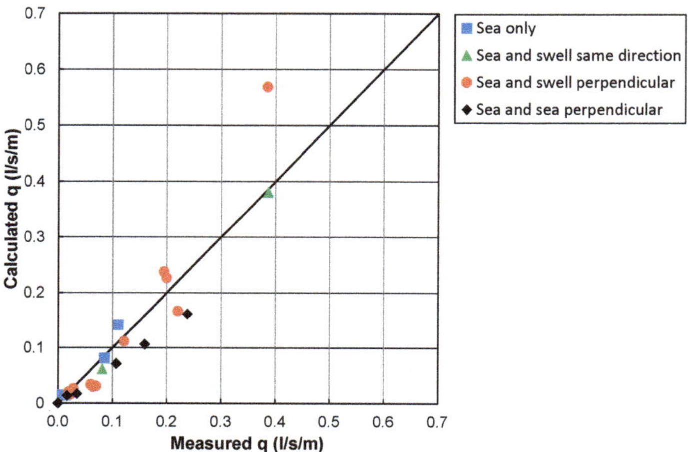

Figure 6. Measured versus calculated wave overtopping, using Equation (12).

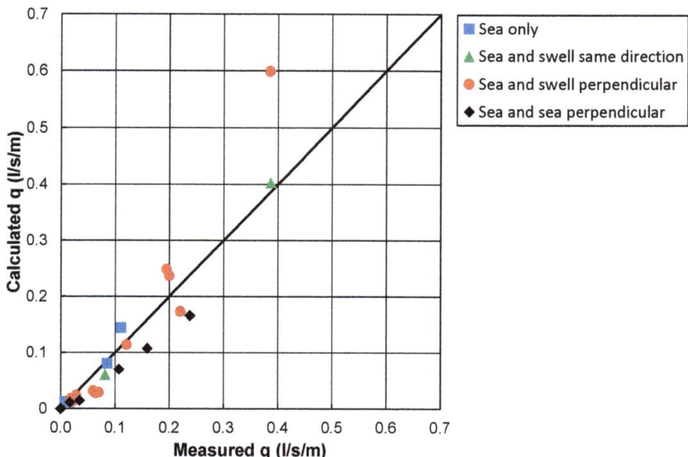

Figure 7. Measured versus calculated wave overtopping, using Equation (13).

4. Conclusions, Recommendations, and Future Research

Wave overtopping over dikes with oblique wind and swell waves has been studied by performing physical model tests with various types of wave loading. The analysis of the test results shows that the presence of swell (with a low steepness; $s_{m-1,0} \leq 0.005$ or $s_p \leq 0.005$) leads to more wave overtopping. This effect has been accounted for in new empirical formulae for the mean wave overtopping discharge. The direction of these swell waves with a low steepness (same direction as wind waves or perpendicular to wind waves) is less relevant for wave overtopping discharges. Two sea states with wind waves from two different directions (i.e., $\beta = 45°$ and $\beta = -45°$ in the present tests) seem to lead to somewhat larger overtopping discharges compared to wind waves from one direction only (i.e., $\beta = 45°$ in the present tests). In our tests, this additional wave overtopping was about 35%, which can be represented by an influence factor for crossing seas of $\gamma_{\#} = 1.05$ in an empirical formulae. However, it remains unknown to what extent the additional overtopping for crossing seas depends on the wave angle of both sea states. If besides a certain sea state also a second wave field is present, a swell condition (with a low wave steepness) leads to more additional wave overtopping than a second wave field with the same wave height and a higher wave steepness).

Using the generally applied wave overtopping formula in [4] as a basis, the extended wave overtopping formulae are as follows:

$$\frac{q}{\sqrt{gH_{m0}^3}} = \frac{0.067}{\sqrt{\tan\alpha}}\gamma_b\xi_{m-1,0}\exp\left(-\frac{4.75}{\gamma_b\gamma_f\gamma_\beta\gamma_{\#}\gamma_v}\frac{(R_c - 0.5H_{m0-Swell})}{H_{m0}\xi_{m-1,0}}\right) \tag{14}$$

with a maximum of

$$\frac{q}{\sqrt{gH_{m0}^3}} = 0.2\exp\left(-\frac{2.6}{\gamma_f\gamma_\beta\gamma_{\#}\gamma_v}\frac{(R_c - 0.5H_{m0-Swell})}{H_{m0}}\right) \tag{15}$$

where $\xi_{m-1,0}$ is the breaker parameter based on the wind waves at the toe of the structure, $H_{m0-Swell}$ is based on the energy in the swell component (with peak wave periods that are a factor 3 or more larger than the peak wave period of the wind waves, and a wave steepness that is smaller than 0.005), and $\gamma_{\#} = 1.05$ is to account for additional wave overtopping for situations in which two sea states approach the structure simultaneously under different angles ("crossing seas"). Since Equations (14) and (15) are extensions of a presently applied wave-overtopping formula for single sea states, the field of application of Equations (14) and (15) includes single sea states, crossing sea states, and sea states combined with swell.

Although Equations (14) and (15) are preferred, alternative expressions to estimate wave overtopping discharges are as follows:

$$\frac{q}{\sqrt{gH_{m0}^3}} = \frac{0.023}{\sqrt{\tan\alpha}}\gamma_b\xi_{m-1,0}\exp\left[-\frac{3.6}{\gamma_b\gamma_f\gamma_\beta\gamma_{\#}\gamma_v}\left(\frac{(R_c - 0.5H_{m0-Swell})}{H_{m0}\xi_{m-1,0}}\right)^{1.3}\right] \tag{16}$$

with a maximum of

$$\frac{q}{\sqrt{gH_{m0}^3}} = 0.09\exp\left[-\frac{1.7}{\gamma_f\gamma_\beta\gamma_{\#}\gamma_v}\left(\frac{(R_c - 0.5H_{m0-Swell})}{H_{m0}}\right)^{1.3}\right] \tag{17}$$

In the present study the bimodal waves with sea states that approach the structure under different angles indicate that such crossing seas lead to somewhat larger wave overtopping discharges than if waves approach the structure under one angle. Since only perpendicular sea states (i.e., $\beta = 45°$ and $\beta = -45°$) have been tested it is advised to analyse the influence of crossing seas also for other wave angles. Since only long-crested waves have been studied it is advised to analyse also the influence of short-crested waves for such crossing seas.

Several influence factors are present in empirical overtopping formulae (e.g., friction, oblique waves, berms, crest elements, and crossing waves). These influence factors are based on physical model tests to quantify these influence factors. However, the combination of these influence factors has not been validated systematically and therefore the accuracy of the term $\gamma_b\ \gamma_f\ \gamma_\beta\ \gamma_\#\ \gamma_v$ is unknown. It is advised to validate the combination of the various influence factors. As long as such a validation has not been performed, it is advised to apply a maximum influence of the reduction factors to wave overtopping $\gamma_b\ \gamma_f\ \gamma_\beta\ \gamma_v \geq 0.4$ (as proposed in [4]) to avoid too large reductions that have not been based on data.

In the present study the mean overtopping discharge has been studied. The interaction of waves from two different directions may also be of importance for individual overtopping events. It is advised to analyze such effects on volumes of individual overtopping waves and other parameters describing individual overtopping events like water layer thicknesses and velocities during overtopping events.

Author Contributions: I.M.v.d.W. and M.v.G. were responsible for the conceptualization, investigation and analysis. M.v.G. wrote the original draft, was responsible for the experimental methodology, and supervised the investigation. I.M.v.d.W. reviewed the original draft and presented the work at the Coastlab 2018 conference.

Funding: This research received no external funding.

Acknowledgments: Most of the described research has first been presented and published [24] at the Coastlab 2018 conference (22–26 May, Santander, Spain); the conference is acknowledged for providing the platform to present the described research.

Conflicts of Interest: The authors declare no conflict of interest.

References

1. Young, I.A.; Verhagen, L.A.; Banner, M.L. A note on the bimodal directional spreading of fetch—Limited wind waves. *J. Geophys. Res.* **1995**, *100*, 773–778. [CrossRef]
2. Petrova, P.; Tayfun, M.A.; Soares, C.G. The effect of third-order nonlinearities on the statistical distributions of wave heights, crests and troughs in bimodal crossing seas. *J. Offshore Mech. Arct. Eng.* **2013**, *135*, 021801. [CrossRef]
3. Petrova, P.G.; Guedes Soares, C. Distributions of nonlinear wave amplitudes and heights from laboratory generated following and crossing bimodal seas. *Nat. Hazards Earth Syst. Sci.* **2014**, *14*, 1207–1222. [CrossRef]
4. TAW. Technical Report Wave Run-up and Wave Overtopping at Dikes. In *Technical Advisory Committee on Flood Defence (TAW)*; TAW: Delft, The Netherlands, 2002.
5. Van der Meer, J.W.; Bruce, T. New physical insights and design formulas on wave overtopping at sloping and vertical structures. *J. Waterway Port Coastal Ocean Eng.* **2014**, *140*, 04014025. [CrossRef]
6. Van Gent, M.R.A.; van den Boogaard, H.F.P.; Pozueta, B.; Medina, J.R. Neural network modelling of wave overtopping at coastal structures. *Coast. Eng.* **2007**, *54*, 586–593. [CrossRef]
7. Schüttrumpf, H. Wave Overtopping Flow at Seadikes—Experimental and Theoretical Investigations (In German: Wellenüberlaufströmung bei Seedeichen—Experimentelle und Theoretische Untersuchungen). Ph.D. Thesis, Technische Universität Braunschweig, Braunschweig, Germany, 2001.
8. Van Gent, M.R.A. Low-exceedance Wave Overtopping Events—Estimates of Wave Overtopping Parameters at the Crest and landward Side of Dikes. In *Delft Cluster Report DC030202/H3803*; Delft Cluster: Delft, The Netherlands, 2001.
9. Van Gent, M.R.A. Wave overtopping at dikes. *Proc. ICCE* **2002**, *2*, 2203–2215.
10. Schüttrumpf, H.; van Gent, M.R.A. Wave overtopping at seadikes. In Proceedings of the Coastal Structures Conference 2003, Portland, Oregon, 26–30 August 2003.
11. Van Gent, M.R.A. Wave run-up on dikes with shallow foreshores. *J. Waterways Port Coast. Ocean Eng.* **2001**, *127*, 254–262. [CrossRef]
12. Van Gent, M.R.A. Coastal flooding initiated by wave overtopping at sea defences. In Proceedings of the Coastal Disasters 2002, San Diego, CA, USA, 24–27 February 2002; pp. 223–237.

13. Van der Meer, J.W.; Allsop, N.W.H.; Bruce, T.; De Rouck, J.; Kortenhaus, A.; Pullen, T.; Schüttrumpf, H.; Troch, P.; Zanuttigh, B. (Eds.) *EurOtop: Manual on Wave Overtopping of Sea Defences and Related Structures;* 2016; Available online: http://www.overtopping-manual.com/ (accessed on 19 January 2018).
14. Gallach Sánchez, D. Experimental Study of Wave Overtopping Performance of Steep low-Crested Structures. Ph.D. Thesis, Universiteit Gent, Ghent, Belgium, 2018.
15. Wolters, G.; van Gent, M.R.A. Maximum wind effect on wave overtopping of sloped coastal structures with crest elements. *Proc. Coast. Struct.* **2007**, 1263–1274. [CrossRef]
16. De Waal, J.P.; Van der Meer, J.W. Wave run-up and overtopping on coastal structures. *Proc. ICCE* **1992**, *2*, 1758–1771.
17. Galland, J.-C. Rubble mound breakwater stability under oblique waves: And experimental study. *Proc. ICCE* **1994**, *1994*, 1061–1074.
18. Hebsgaard, M.; Sloth, P.; Juul, J. Wave overtopping of rubble mound breakwaters, *Proc. ICCE 1998*, Copenhagen. *World Sci.* **1998**, *2*, 2235–2248.
19. Lykke Andersen, T.; Burcharth, H.F. Three-dimensional investigations of wave overtopping on rubble mound structures. *Coast. Eng.* **2009**, *56*, 180–189. [CrossRef]
20. Van Gent, M.R.A. Oblique wave attack on rubble mound breakwaters. *Coast. Eng.* **2014**, *88*, 43–54. [CrossRef]
21. Wenneker, I.; Meesters, J.; Hoffmann, R.; Francissen, D. Active wave absorption system ARCH. In Proceedings of the 3rd International Conference on the Application of Physical Modelling to Port and Coastal Protection, Barcelona, Spain, 28–30 September–1 October 2010.
22. Holthuijsen, L.H. *Waves in Oceanic and Coastal Waters;* Cambridge University Press: Cambridge, UK, 2007.
23. Massel, S.R.; Brinkman, R.M. On the determination of directional wave spectra for practical applications. *Int. J. Appl. Ocean Res.* **1998**, *20*, 357–374. [CrossRef]
24. Van der Werf, I.M.; van Gent, M.R.A. Wave overtopping over dikes with oblique wind and swell waves. In Proceedings of the Coastlab 2018, Santander, Spain, 22–26 May 2018.

MDPI
St. Alban-Anlage 66
4052 Basel
Switzerland
Tel. +41 61 683 77 34
Fax +41 61 302 89 18
www.mdpi.com

Journal of Marine Science and Engineering Editorial Office
E-mail: jmse@mdpi.com
www.mdpi.com/journal/jmse

www.ingramcontent.com/pod-product-compliance
Lightning Source LLC
LaVergne TN
LVHW070107170726
843515LV00007B/1629